COMMAND
LEGACY

COMMAND LEGACY

A Tactical Primer for Junior Leaders

Second Edition, Revised

LT. COL. RAYMOND A. MILLEN, USA

Potomac Books, Inc.
Washington, D.C.

Library of Congress Cataloging-in-Publication Data
Millen, Raymond A.
  Command legacy : a tactical primer for junior leaders / Lt. Col. Raymond A. Millen. — 2nd ed., rev.
     p. cm.
  Includes bibliographical references and index.
  ISBN 978-1-59797-207-9 (pbk. : alk. paper)
  1. United States. Army. Infantry—Drill and tactics. 2. Infantry drill and tactics. I. Title.
  UD160.M55 2008
  356'.1154—dc22

                                                                    2008030132

Printed in the United States of America on acid-free paper that meets the American National Standards Institute Z39-48 Standard.

Potomac Books, Inc.
22841 Quicksilver Drive
Dulles, Virginia 20166

First Edition

10 9 8 7 6 5 4 3 2 1

In loving memory of my mother, Marjorie

Contents

List of Illustrations

FIGURES

TABLES

List of Abbreviations and Acronyms

1SG	First Sergeant
A/C	Aircraft
AAR	After Action Review
AASSLT	Air Assault
ACE	Ammunition, Casualties, and Equipment
ADA	Air Defense Artillery
AFV	Armor Fighting Vehicle
AG	Assistant Gunner
AI	Area of Interest
AO	Area of Operation
AP	Armor Piercing
APC	Armor Personnel Carrier
ARTEP	Army Training Evaluation Program
AT	Antitank
BDA	Battle Damage Assessment
BDU	Battle Dress Uniform
BMNT	Beginning Morning Nautical Twilight
BN	Battalion
BRP	Building Rally Point
C&R	Consolidation and Reorganization
C2	Command and Control
CALL	Center for Army Lessons Learned
CAS	Close Air Support
CASEVAC	Casualty Evacuation
CBT TNS	Combat Trains
CCIR	Commander's Critical Information Requirements

CCP	Communications Check Point
CDR	Commander
CFA	Covering Force Area
CH	Chapter (administrative action)
Chain	Chain of Command
CL	Class (of supply)
CO	Company
COA	Course of Action
Commo	Communications
COMSEC	Communications Security
CP	Command Post
CP	Check Point
CPOG	Chemical Protection Over-Garment
CS	Combat Support
CSM	Command Sergeant Major
CSS	Combat Service Support
DA	Department of the Army
DMP	Decision-Making Process
DS	Direct Support
DTG	Date Time Group
DZ	Drop Zone
ea	Each
EA	Engagement Area
ECC	Exercise Control Center
EEFI	Essential Elements of Friendly Information
EENT	Early Evening Nautical Twilight
EFFI	Enemy Friendly Forces Information
ENG	Engineer
EPW	Enemy Prisoner of War
E-Tool	Entrenching Tool
FA	Field Artillery
FEBA	Forward Edge of the Battle Area
FIST	Fire Support Team
FLD	TNS Field Trains
FLOT	Forward Line of Troops
FO	Forward Observer

FPF	Final Protective Fires
FPL	Final Protective Line
FRAGO	Fragmentary Order
FSE	Fire Support Element
FSO	Fire Support Officer
GPS	Global Positioning System
GR	Grenadier
GRREG	Graves Registration
GS	General Support
HE	High Explosive
HKT	Hunter-Killer Team
HQ	Headquarters
IAW	In Accordance With
IBRP	Initial Building Rally Point
IFV	Infantry Fighting Vehicle
IN	Infantry
IPB	Intelligence Preparation of the Battlefield
IR	Infra-red
JRTC	Joint Readiness Training Center
KIA	Killed in Action
Km	Kilometer
LAW	Light Antitank Weapon
LBE	Load Bearing Equipment
lbs	Pounds
LC	Line of Contact
LD	Line of Departure
LOGPAC	Logistical Package
LOGSTAT	Logistical Status Report
LP	Listening Post
LRP	Logistics Rally Point
LZ	Landing Zone
m	Meter
MCOO	Modified Combined Obstacle Overlay
ME	Main Effort
Mech	Mechanized
MEDEVAC	Medical Evacuation

METL	Mission Essential Task Listing
METT-T	Mission, Enemy, Terrain and Weather, Troops and Time Available
MG	Machine Gun
MIA	Missing in Action
MILES	Multiple Integrated Laser Engagement System
min	Minute
mm	Millimeter
MOPP	Mission-Oriented Protection Posture
Mort	Mortar
MOUT	Military Operations in Urban Terrain
MRE	Meal, Ready to Eat
MSR	Main Supply Route
MTC	Movement to Contact
MTP	Mission Training Plans
NAI	Named Area of Interest
NBC	Nuclear, Biological, and Chemical
NCO	Noncommissioned Officer
NCOIC	NCO in Charge
NCOPD	NCO Professional Development
NFA	No Fire Area
NLT	Not Later Than (time)
NOD	Night Observation Device
NTC	National Training Area
NVG	Night Vision Goggles
O/C	Observer/Controller
OBJ	Objective
OCOKA	Observation and fields of fire, COver and concealment, Key terrain, and Avenues of approach
OD	Olive Drab
OER	Officer Evaluation Report
OIC	Officer in Charge
OOTW	Operations Other Than War
OP	Observation Point
OPD	Officer Professional Development
OPFOR	Opposing Forces
OPORD	Operations Order

OPSEC	Operational Security
ORP	Objective Rally Point
PDF	Principal Direction of Fire
PERSTAT	Personal Status Report
PEWS	Platoon Early Warning System
PIR	Priority Intelligence Reporting
PL	Phase Line
PLD	Probable Line of Departure
PLT	Platoon
PLT	LDR Platoon Leader (also PL)
POC	Point of Contact
pr	Pair
PSG	Platoon Sergeant
PSN	Position
PZ	Pick-up Zone
Q-Party	Quartering Party
qt	Quart
R	Rifleman
R&S	Reconnaissance and Security
rd	Round
RFL	Restricted Fire Line
ROE	Rules of Engagement
RP	Release Point
RP	Rally Point
RTD	Return to Duty
RTO	Radio-Telephone Operator
S1	Personnel Section or Adjutant
S2	Intelligence Section or Intelligence Officer
S3	Operations Center or Operations Officer
S4	Logistics Section or Logistics Officer
SALUTE	Size, Activity, Location, Uniform, Time, and Equipment
SAN	Sanitation
SAW	Squad Automatic Weapon
SBF	Support by Fire
SCT	Scout
SE	Supporting Effort

SEAD	Suppression of Enemy Air Defense
SGT	Sergeant
SITEMP	Situation Template
SITREP	Situation Report
SOI	Signal Operating Instruction
SOSR	Suppress, Obscure, Secure Far Side, and Reduce Obstacle
SP	Start Point
SQD LDR	Squad (also SL)
Sup	Supply
T&E	Traversing and Elevation
TAC	Tactical Command Post
TAC SOP	Tactical Standing Operating Procedures
TAI	Targeted Area of Interest
TC	Tank Commander (vehicle commander)
TCP	Traffic Control Point
TLP	Troop Leading Procedures
TM LDR	Team Leader (also TL)
TOC	Tactical Operations Center
TOW	Tube-launched, Optic-sighted, Wire-guided (missile system)
TRADOC	Training and Doctrine Command
TRP	Target Reference Point
TTP	Tactics, Techniques, and Procedures
UMR	Unit Manning Roster
VT	Variable Time
WIA	Wounded in Action
wt	Weight
XO	Executive Officer

Preface

The preeminent historian John Toland once observed that history does not repeat itself—human nature does.1 This epiphany should serve as a guiding light for national security: no matter how long a period of peace lasts, no matter how optimistic military analysts are regarding future adversaries, no matter how inspiring our political leaders are regarding new world orders, one thing is clear—those nations responsible for the stability of the world will find themselves engaged in wars. Whether by design, accident, or stupidity, whether large or small, whether just or unjust, there will always be conflicts.

The terrorist attacks on 11 September 2001 underscore that reality for a global superpower. The War on Terror will likely involve the United States in counterinsurgency in some form. The surprising irony of the Iraq and Afghanistan conflicts is that maneuver units no longer bear the brunt of combat. Combat Support and Combat Service Support units can find themselves conducting operations that were once the sole province of the infantry and armor branches. In view of this new reality, this book is of substantial value to all junior leaders and not just to those assigned to maneuver units.

Forearmed with this empirical knowledge, it would seem logical for governments to maintain a modicum of military readiness in support of the national security strategy. The truth is that democracies inevitably reduce their armed forces during periods of peace to levels inadequate for national defense. Despite dire predictions and assurances to correct deficiencies, federal governments will subordinate security needs to other program needs—particularly as the peace extends over time. But the nation will still expect the military to perform in times of war, prepared or not. The exorbitant initial casualties in the Civil War, World War I, World War II, and Korea were a result of poor readiness. The Army cannot change this congressional inclination, but it can develop better methods of training.

The burden of fighting rests on the soldiers and junior leaders of small combat units, and especially on infantry units. In combat, success is rarely determined solely by whether these units have received sufficient training, equipment, or manpower, nor is it generalship, staff planning, or technology that wins battles; it is the tenacity of the ground troops. General Dwight D. Eisenhower recognized this simple truth of battlefield success in preparation for the Normandy campaign; that is why he spent so much time meeting the soldiers, showing his understanding for their sacrifices, and gaining an understanding of their mettle.2 With its own vast wealth of war fighting experience, it is amazing that the senior military leadership places such a low premium on the manning levels of combat units. Pitifully understrength combat companies are expected to perform a myriad of tasks and missions throughout the spectrum of conflict. Any arrangements intended to make up this deficiency during mobilization for a large conflict do nothing to increase the combat effectiveness of the unit because it is already too late; replacements at this stage are nothing more than cannon fodder.

I was prompted to write this book because there is no previous definitive source on small unit tactics. Existing manuals are too general, have too many doctrinal buzzwords, and are not applicable to company-level units. None addresses the details that infantrymen must consider and the tasks they must complete during any mission. Based on my own experiences in company command, I decided to write a book on tactics, techniques, and procedures that addressed the needs of junior leaders. This book represents my conclusions regarding tactical problems and missions and the link among tactical theory, doctrine, and practice. This is a single source reference for the purpose of priming junior leaders for tactical operations, instilling the concept of corporate effort, and inspiring thought. To be practical, any book on tactics must be specific, explaining in detail what needs to be done, why, when, and by whom. It must attempt to reconcile both what to think and how to think. If I have accomplished this exacting task and inspired a few to take my work to a higher level, I have been successful.

I would like to thank my two good friends at the United States Military Academy, (USMA) Preparatory School and USMA classmates and fellow officers Lieutenant Colonel John Lock and Major (Ret.) Richard Lavosky. First, to John, for giving me the encouragement and contacts to get this book published. Second, to Rich, for reading the draft and providing feedback as well as for our many discussions on tactics. I would also like to thank Colonel (Ret.) Timothy Wray and Lieutenant Colonel (Ret.) Timothy Lupfer, my military history instructors at West Point. Both taught me how to approach and make use of military history for my profession of arms. Their influence on me professionally was immense. I give great thanks to Terry Decker, who tirelessly devoted much time and effort to the promotion of my book. Last, I must recognize the patience of my wife, Janene, who not only tolerated my ruminations on the content of my book but also provided technical advice on its format and on the mysteries of word processing.

NOTES

1. John Toland, *The Rising Sun: The Decline and Fall of the Japanese Empire, 1936–1945* (New York: Random House, 1970), xxxv.

2. Stephen E. Ambrose, *Eisenhower: Soldier and President* (New York: Touchstone, 1990), 129. "One reason, more rational than emotional, that Eisenhower was concerned about his troops was his realization that while he, SHAEF, the generals, and the admirals could plan, prepare the ground, provide covering support, ensure adequate supplies, deceive the Germans, and in countless other ways try to ensure victory, in the end success rested with the footslogger carrying a rifle over the beaches of Normandy. If he was willing to drive forward in the face of German fire, Overlord would succeed. If he cowered behind the beached landing craft, it would fail. The operation all came down to that."

Chapter 1

Training and Fighting within the Army System

Achieving and maintaining a sufficient degree of combat readiness is a constant struggle for the company commander, both in peacetime and in combat. The urgency revolving around combat readiness exists because the company may find itself in a conflict within a matter of hours. All commanders realize that time is the one commodity lacking during a crisis. Either the unit is ready, or it's not. Poor readiness means a higher probability of incurring high casualties and mission failure—at least initially. Despite this realization, combat readiness continues to be a constant issue. The Army has struggled with this issue since Revolutionary war leader Friedrich Wilhelm von Steuben's time and has developed systems to train the force and monitor its progress, but its efforts to maintain combat readiness contain severe shortcomings. All the palaver over the regimental system and volunteer recruitment versus conscription, or adulation of the German Army of World War II does not help the combat company one iota. Units must train and fight in the established system. The responsibility for maintaining combat readiness must come from within the company. To expect or demand otherwise is foolish. Before examining how the company can develop a system to resolve this perennial problem, an examination of the salient factors that affect readiness is in order.

Personnel Turnover. Turnover is disruptive to the unit and represents a constant drain on the unit's continuity of experience. The company can expect to experience a 3 to 4 percent personnel rotation per month, which translates to a turnover of over one-third of the company in just one year. For key leader positions, personal turnover represents the greatest retardation to experience. Company commanders enjoy from 18 to 24 months in command; platoon leaders, 12 months; noncommissioned officers (NCOs), between 2 and 3 years. It takes officers about 3 months before they learn the fundamentals of their job in garrison and about 6 months for basic field tactics. NCOs generally enjoy an advantage of job experience even with rotation because they spend most of their career at the platoon level.

The average level of training for a platoon rarely matures beyond the basics of experience. The platoon leader determines the training objectives and can expect to participate in four to six field exercises where collective training at the platoon or higher level is conducted before his tenure is up. As a result, just as he gains the experience and confidence needed to progress to the next stage of proficiency, he rotates. Although the company commander's time in command is longer, the amount of experience gained at company-level or higher training exercises also averages between four and six. Worse, the commander must devote an extraordinary amount of energy to administrative, property accountability, and maintenance matters. If he doesn't, he likely will become a better tactician, but at the cost of his career.

In short, personnel turnover results in the loss of the unit's collective memory. Even units with recent combat experience suffer a rapid deterioration in proficiency after six months because proven tactical techniques and lessons (some learned at the price of blood) are not recorded in a practical way for the unit to use, and hence they fade.

Training Distracters. The commander is faced with many competing demands on his soldiers' training time. Even during major training events, a portion of his soldiers cannot participate because of schooling requirements (BSEP, GED, PLDC, BNCOC, and so forth), profiles, hospitalization, quarters, administrative matters (chapters), and judicial matters. The commander and first sergeant (1SG) must work hard to get more than 80 percent of their company into field training at any one time. The percentage of participation is even worse for lower-level training events. Some training (individual and minor collective tasks) is covered impromptu (hip pocket) or with little advanced preparation required (backyard), but a company needs to train collectively to operate effectively in combat, and that requires the company to train together for at least a week under the present system before it begins to operate as an entity instead of a collection of squads.

Infrequent training causes proficiency to deteriorate or perish outside of six months —sooner for difficult tasks. The result is that the company is forced to repeat basic, essential, collective tasks for nearly every field exercise simply to maintain rudimentary, collective training skills. Naturally, such repetition leads to boredom and inertia as soldiers practice the same tasks over and over again. It is very frustrating for the commander to devise interesting training scenarios without sacrificing the basics.

Military Education and Professional Development. Military education is

a career-long endeavor and is not confined to the classroom. Sometimes unit commanders forget that fact and expect the junior leader to seek professional growth without guidance.

Soldiers and leaders do not arrive to the unit completely trained. Generally, course curricula are very basic and are not necessarily pertinent to the needs of the leader's current or next job. No junior leader gains any appreciation of tactics by memorizing the definitions of such concepts as combat power, battle space, fundamentals of the defense and offense, and so forth. Instruction that generalizes principles and fundamentals such as security, maneuver, and priority of work on the defense is a waste of everyone's time. Of course, all of these themes are important. What the junior leader thirsts for are the details in the form of tactics, techniques, and procedures; the historical example; and weather and terrain effects. It is no wonder that units too often place listening and observation posts (LP/OPs) in irrelevant locations, make linear assaults on objectives without heed to the circumstances, or give undo focus to digging in the unit to the detriment of the other priorities of work in the defense. Although revisions to course curriculums are ongoing, limited time, resources, and support continue to impede the goal of producing qualified graduates. Moreover, instructor proficiency is a lottery. Some are excellent, others are quite poor. Although the chain of command of training schools attempts to weed out the poor instructors, this is no help to the affected students after the fact.

A curricular deficiency is the infrequent use of tactics, techniques, and procedures (TTPs) for company-level actions. In fact, practically no literature or instruction is devoted to detailed company operations. Small-unit operations require specific actions by specific individuals or elements. The only way I myself learned tactics was by trial and error and discussions with my peers. This is akin to learning about sex in back alleys—with about the same results. Few instructors are willing to discuss, let alone provide, specifics on how to approach certain tasks for fear of violating the sacrosanct tenet that students should be taught how to think rather than what to think. Such an academic approach is expected for advanced leaders, but junior leaders initially require more definitive examples. As they gain experience, they can then begin to formulate and experiment with new TTPs in the unit.

Unit professional development sessions (OPD/NCOPD) vary and can easily degenerate to being of little practical use except for general knowledge or entertainment. Often, good ideas lose their value over time because the company has no system in place to assimilate them. They wither on the vine because no one has figured out how to incorporate good ideas and techniques into unit training. Professional development should be devoted to enhancing military skills and not just be seen as an opportunity to conduct mandatory briefings (equal opportunity, law of land warfare, and so forth). Professional development sessions should focus on solving a particular tactical problem and then incorporate the solution into a TTP framework.

The quest for military knowledge about small-unit tactics is fraught with many false roads. Professional reading lists would appear to be an excellent method for guidance, and they would be if they were properly organized. Most are thrown together haphazardly. No thought is given to selecting books for their applicability to the intended reader. Junior leaders are given no guidance as to which books are helpful in their quest for tactical expertise. Instead, officers of all ranks are overwhelmed with a list of over a hundred books to read, ranging from the simplicity of Swinton's *Duffer's Drift* to the complexity of Clausewitz's *On War*. While many of these books are excellent in their own right, leaders are forced to wade haphazardly through them, searching for guidance for their own tactical problems. I am not suggesting that junior leaders can only read literature on tactics, but since platoon leaders and company commanders have such a brief tenure, they should focus on the pertinent literature first.

Another problem is that very little good literature is devoted to small-unit actions. The reader is forced to glean and to try to adapt applicable tactical lessons from literature on higher-level operations to the company level. Worse, the plethora of myths, misinterpretations of events, and hidden agendas are difficult to separate from facts.

The leader needs a reason for this investment of time and thought. Knowledge for the sake of knowledge is not pernicious, but it doesn't help the unit either. In this pursuit, the researcher needs a practical framework on which to apply the lessons of history or doctrine for his company.

Experience. Junior leaders who are new to the unit bring with them a repertoire of skills and knowledge that is potentially beneficial. But this repertoire can also be damaging or ineffective if not properly assimilated by the company. If a platoon's approach to tasks and subtasks is incompatible with that of other platoons, unity of effort suffers. This becomes critical for collective tasks that require close cooperation among company elements, such as passage of lines, occupation of an assembly area, and mission preparation.

Standardization is a proper step in ensuring unity of effort, but it is not the end state. Improvements to established methods and procedures are not possible if the unit has no means for the soldier to submit his ideas for review, evaluation, and adoption. The greatest danger for any military unit is not to remain current with changes in doctrine and technology. The unit needs to have a framework for looking at new ideas or changes in order to exploit their potential.

Inspirations that appear divine in the mind often prove to be infeasible, unsuitable, or unattainable once analyzed. The first step in this process is to put these ideas on paper. Writing down thoughts is difficult, and commanders cannot expect soldiers to be creative without a simple format to follow in order to help them organize their thoughts coherently.

Attitudes toward Training. Attitudes vary from unit to unit, depending on the agendas of the battalion, brigade, and division commanders. A pervasive attitude is that anyone can be a good tactician, but a good manager is prized. Granted, managerial skills are essential. But if these skills are held in higher esteem than tactical skills, junior leaders will pay less attention to tactics. It is human nature to pay more attention to those things that receive praise. In such an environment, junior leaders muddle through training exercises and evaluations because success is not measured on how well a task is executed but rather on whether all of the right boxes were checked on the training evaluation form. Unfortunately, higher command has great difficulty determining tactical proficiency on training evaluations because much of it is subjective (by evaluators), with somewhat inflated results. Maintenance, unit adherence to uniform directives, and property accountability are easier to gauge when evaluating commanders, and hence receive undue emphasis. It is difficult to fault a subordinate commander on tactics, because proper tactics are difficult to gauge. Commanders need to strike a balance between managerial and tactical skills. Otherwise, the upper echelons of the Army will consist of bureaucrats, the warriors having resigned long ago in disgust. As long as the company has no written procedures on how it does business, higher command will pay less attention to a unit's tactical proficiency.

Training Limitations. The Army has made great strides in the realm of training. Training devices (MILES), computer simulations, and the Combat Training Centers (CTCs) have all increased unit proficiency—to a degree. None of these can ensure long-term unit proficiency. The unit is responsible for capturing the lessons learned from training and finding the means of incorporating them. Without a means to document and assimilate TTPs, the unit's institutional memory is short lived.

Despite the great strides that have been made in training, technological limitations and artificial conditions deprive soldiers of combat realism. Of course, total realism is not possible because of safety considerations. We cannot subject soldiers to an artillery barrage or conduct unstructured live fire exercises, because these lead to casualties. Yet, it is just as wrong to give the soldier trainees the impression that training conditions approximate combat conditions. For instance, soldiers and leaders need to understand the psychological effects that the battlefield inflicts on soldiers. Leaders need to know what works and what doesn't when soldiers become gripped by fear. The leader must perform deep research to uncover all facets of this phenomenon. No doubt many leaders are knowledgeable on many invaluable subjects. But how many units benefit from this knowledge after the leader has departed? Without an obtainable reference, soldiers and junior leaders forget. Hence the expression, "Reinventing the wheel."

A unit is rarely able to go beyond the rudiments of collective training in peacetime. In the past, this degree of readiness was acceptable because it was enough to sustain a unit during the initial six months of combat, the theoretical amount of time required for the unit to gain enough experience before conducting more sophisticated operations—and ipso facto the United States never lost a war. World War II revealed the inherent weaknesses of U.S. peacetime training and readiness. Poor training, insufficient ground forces, and an execrable replacement system caused tremendous casualties and misery in the ranks of the infantry. Viewing the United States's victory from another angle, one can say that it is not the best army that wins wars, but rather the one that is the least fouled up. Simply because the United States was on the winning side in World War II is no cause to declare that its approach to training was correct. Many will say that the discrepancies have been corrected. Between the period of the Vietnam War and the conflicts following 9/11, no conflict lasted more than two months. Moreover, the United States is now engaged in counterinsurgency warfare, which often employs a different set of perishable war fighting skills. The inadequacies of training do not manifest themselves in short conflicts. Few nations are able to match the training and combat proficiency of the United States. Nonetheless, during crises, battalion and company-sized units often find themselves inserted rapidly into conflict regions where the local correlation of forces is not to their initial advantage. The enemy may not be as highly trained as we are, but he may be able to offset technological and qualitative superiority with superior numbers and strike before the United States or its allies can build up. At this point, victory rests on the shoulders of the frontline soldier. The Army has not figured out how to maintain peak readiness in each unit without keeping it in the field eternally. The Army is continually assessing how to provide the best-trained force possible—from the higher echelons of Army hierarchy (e.g., Training and Doctrine Command). Such an approach is needed to provide a comprehensive force for the nation. But it is the details that confront a unit in combat. Consequently, the development and progression of tactical proficiency must come from below.

Combat Experience. Depending on the type of conflict, soldiers gain a variety of tactical skills. Some skills can be applied to other forms of conflict, others cannot. Those skills not practiced in combat are perishable, and it is utter foolishness to train on those skills that have no application for an upcoming deployment. Yet, at some point, perishable skills can suddenly become germane in a new type of conflict. In view of this contingency, a unit needs a repository of knowledge.

Chapter 2

The Development of a Company Doctrine

The framework to which the first chapter alluded is the development and the cultivation of the company doctrine. A large measure of this development lies in the establishment of tactical standard operating procedures (TAC SOP). The cultivation results from establishing a repository of tactical thought. In short, company doctrine establishes the fusion of military science (organization, time and resource management, and equipment and weapon capabilities) and military art (tactical techniques, ingenuity, and field expediency) for the small unit.

> This book is intended for any small unit conducting tactical operations. The light infantry perspective is solely for simplicity and continuity of thought.

The company needs its own doctrine in order to formalize and standardize procedures and assist in the rapid orientation of new soldiers into their assignment. Not long ago, such a concept was impractical for company-sized units. Developing and updating such a detailed doctrine was beyond the line company's capabilities. It was not worth the effort. But with the advent of computers, companies and even platoons and squads can become more sophisticated. A company doctrine can now be produced, revised, and distributed without exorbitant time or effort.

THE TACTICAL SOP

As the name suggests, the tactical standard operating procedure (TAC SOP) establishes procedures for company field operations—the military science portion of the doctrine. It consists of the following:

- Missions broken down by supporting tasks, responsibilities, and priorities
- Coordination, planning, and orders templates
- Report formats

The identification of mission supporting tasks and the assignment of responsibility for each are essential for organizational efficiency, tactical effectiveness, and sustained operations. The commander, with input from his subordinates, establishes the tasks and subtasks. He decides who is responsible for each task or subtask, which have priority, and when each should be completed. These procedures are established in written form as a reference for himself and his subordinates, in a simple format, and packaged as a handbook for field use.

The contents of the TAC SOP are derived from the company Mission Essential Task Listing (METL). A good starting point for constructing the particulars of each task is mission training plans (MTP), field manuals, Army school student texts, and Department of the Army (DA) pamphlets. However, the product has very little practical use if it is not designed for the user and then field tested.

These procedures allow the company commander to focus his energy on planning and the exercise of command, confident that his subordinates are working on the myriad supporting tasks, thereby precluding the need to micromanage.

Because the TAC SOP specifies responsibility for each task, there is no ambiguity. When a task is not performed to standard, the commander can pinpoint the source and take corrective action. Lazy and incompetent leaders can no longer seek refuge in the common excuses of lack of communication, misunderstandings, and misinformation. It is therefore an effective tool for culling out ineffective leaders.

Newly assigned or appointed leaders refer to the TAC SOP not only to determine what is expected of them but also to describe how their job meshes with the larger framework of the company activity. It provides clarity of effort, allowing each soldier to see how planning, preparation, and execution of a mission unfold. In this sense, it serves as a continuity experience for orienting and acquainting new leaders to their job quickly.

Planning and orders templates and checklists are used to help the commander and his subordinates extract essential information quickly for mission planning and execution. They also serve as a memory device when the leader is fatigued and distracted. During intense periods of combat (and some training exercises), leaders become extremely fatigued and need a guide to refer to when planning. If the company commander does overlook something vital, his subordinates can remind him at any stage of planning, using the format as a quality control reference. In this manner, the company leadership ensures that the plan is nested and complete.

Coordination checklists, such as passage of lines, relief in place, and air assault planning, serve as frameworks for exchanging and disseminating information. If the subordinate making coordination is unable to extract the needed information because the other unit is not prepared, the commander knows to place more effort on this particular task (requiring a follow-on meeting, personal coordination, or higher headquarters involvement).

The company files away those tasks no longer needed in the TAC SOP because of a change in METL. In this manner, it is available for use when needed in the future without the company having to "reinvent the wheel."

> The caveat to following the TAC SOP is to avoid becoming too bureaucratic and pedantic in its application. It is a tool. If any portion is not applicable or is too time consuming for the current task at hand, ignore it. Part of your growth is the ability to use good judgment. If you become enthralled with the process and ignore the essence of effort, you are condemned to be a leader who is the master of inconsequence.

THE CULTIVATION OF MILITARY DISCOURSE

The formal establishment of military discourse is valuable to the unit, because these written thoughts serve as the repository of the unit's experience, knowledge, and intellectual development. It is not distributed as part of the TAC SOP for field use; its general, practical application in the field (to the unit as a whole) is minimal. Its usefulness lies in explaining approaches to tasks and also serves as a forum for new tactics, techniques, and procedures (TTPs).

The commander takes the lead in developing and maintaining this program. He is the major contributor and must promote the program if he expects subordinates to follow suit.

The commander establishes many goals as he formulates the format of the repository of thought. First, the repository addresses the TAC SOP itself. It provides explanations for procedures, priorities, and delegation of effort. These thoughts are really essential, because they give subordinates insight into the commander's problem-solving approach. Later, as subordinates contribute to the discourse or written product the corporate effort establishes a pride of authorship, which induces subordinates to follow the SOP out of professionalism rather than because of command directives. Second, the discourse or writing program is an excellent tool for professional development. Bearing in mind that the ultimate goal is to incorporate lessons learned into the TAC SOP, the mentor or instructor can address techniques and procedures logically.

Military prose is an oft-neglected discipline. Subordinates, particularly officers, need to express their ideas on paper as part of their military education. The soldier does not operate in the abstract—he is a practitioner. Many ideas appear brilliant in our heads until we are forced to write them down, view their logic and practicality, and defend them before our peers. If we cannot translate a thought into clear written words and defend its logic, it is of no practical use. We draw our ideas foremost from our experiences and education. Largely, our thoughts are the result of a solution to a problem encountered or of a better way to perform a task.

Writing forces the leader to translate doctrine, theory, and the experiences of others to his level of practice. These thoughts must be tailored to the unit's needs if they are to be practical. The researcher soon discovers that many of these writings do not stand up well under scrutiny. Pundits, intellectuals, and other "professional" military theorists who have never heard a "blank fired in anger"[1] cannot provide us with the realities and insights of combat experience. Armchair strategists possess the greatest army in the world—in their heads—because their units never get lost; never misinterpret orders; never need sleep, food, or water; are not exposed to normal emotions, such as fear, depression, or panic; never have exhaustion and fatigue affect their decisions; and never have an unclear picture of the enemy and themselves. What a perfect world they live in. Others carry hidden agendas for promoting pet theories. Combat leaders must be circumspect of others' findings. Even authors with military experience and distinction require scrutiny. The bottom line is not to parrot the experiences or ideas of popular theorists and soldiers or to disregard those currently not in vogue without first analyzing their work. Many nations or leaders with poor reputations in particular conflicts did many things right and should not be ignored. The bottom line is to extract what makes sense and can be applied.

> Military writers should not allow the echoes of the past to smother innovative ideas and thoughts.

There are many military myths floating about that claim to be panaceas to tactical problems. We have all been bitten by this bug because it excites the imagination to have a simple explanation for historical successes or failures. A little common sense is a simple cure for this ailment. We should avoid foreign words and phrases whose meaning is open to interpretation. Few know the exact meaning of *Auftragstaktik, Schwerpunkt, Coup d'oeil*, and *Fingerspitzegefuehle*. Other than demonstrating mastery of arcane concepts, the intent is lost— so be specific, even if it requires more words. Avoid the urge to quote famous personages and historical anecdotes. This detracts from the intent of the repository, turning it into a discourse better appreciated in the classroom. Generally, most quotations and anecdotes are too general or address tactical issues not applicable at the company level. Use historical references to address a specific battlefield problem, to support an idea, or to document research for others. If you read an idea that sounds good, analyze it in writing, war game it, and test it out in training. If it works, adopt it—there is no plagiarism on the battlefield.[2] Once a concept is proven in training, it belongs to the unit. Besides, soldiers need to dispel the belief that ideas must be confirmed by "experts" rather than being self-evident.[3]

The repository of thought also serves as an excellent place for exceptional TTPs, which the leader can insert within the text of the tactical thoughts. Such TTPs range from health and hygiene to individual combat techniques.

> **TTP**
>
> In order to draw the attention of the reader to an exceptional TTP, use of a box, as illustrated here, is a useful tool.

THE END PRODUCT

The company doctrine is a living document. It is a corporate effort, meaning that anyone with a good idea is encouraged to contribute to it.[4] It becomes a tool for sharing information and building a bond of mutual trust and cooperation throughout the company. It serves as a type of playbook, allowing leaders and soldiers to discuss or review a task first, using the SOP as reference. It rapidly accelerates the time required for the unit to gain tactical proficiency in the basic tasks in order for it to progress to more complex tasks smoothly.

Company doctrine also serves as an excellent format for instruction before hands-on training. The TAC SOP serves as the vehicle for instruction, while the repository of thought serves as the lesson plan. This allows the common soldier to understand the reasoning behind certain procedures. This method of communication surpasses by far the common experience of muddling through from one field problem to another, shrouded in a cloak of ignorance, myth, rumor, and inertia. Despite an almost perverse pride in eschewing written thought, soldiers need something tangible, concrete—something they can sink their teeth into— that they can refer to in small blocks, in order to cultivate their craft. It gives real substance to the after-action report. Leaders and subordinates can take out the TAC SOP and address those aspects that worked and those that did not and can then recommend changes to the SOP to make it better. Because computers make changes to the TAC SOP relatively painless, the company can have the revised chapters completed within a week of the after-action review (AAR). This is the process of capturing lessons learned, measuring progress, and creating an elite unit.

THE ORGANIZATION OF THE PRODUCT

The book is organized like a company commander's field book. Chapters are organized to reflect the sequential events common to most military operations as with a METL. Each chapter begins with the tactical thoughts followed by the TAC SOP. Each leader should also include blank pages in each chapter for notes on relevant subjects. As a rule, subordinates should only bring the TAC SOP portion to the field in order to keep the size down.

The establishment of the company doctrine is unique to each company. The example TAC SOP and repository of thought, which make up this book, are my development and should not be blindly copied. Otherwise, this book will be of little use to the reader—the practitioner of small unit tactics. My research, organizational format of the book, and thoughts should only serve as a starting point and not as ends in themselves. Besides, the sheer joy of discovering one's own creative process is a reward in itself. The junior leader has limited troop time; he shouldn't squander it.

NOTES

1. A phrase by Colonel Timothy Wray, military history professor at West Point, during an interview with him in 1991 concerning tactics and civilian pundits. This highlights that junior leaders should not accept the conclusions of civilian military writers without question.

2. Let there be no misinterpretation of my intent. Plagiarism is the stealing of others' ideas in the scholarly world, and hence their ideas should be acknowledged. Besides, documentation allows others to review the primary sources and draw their own conclusions—a very healthy process.

3. The danger of using foreign phrases is that the audience will not acknowledge that they do not understand out of fear of embarrassing themselves. Also, foreign terms can change meaning over time. For example, *Coup d'oeil* meant one thing to Frederick the Great and quite another to Napoleon. Many of the terms so familiar with the Wehrmacht are no longer in use with the Bundeswehr. Even in the U.S. Army, where doctrine seems to change every three years, it is difficult to keep up with the proliferation of new terms let alone the evolution of old terms. Take for example the concept of those activities at the lower spectrum of conflict—guerrilla warfare, counterinsurgency operations, pacification operations, low-intensity warfare, operations other than war, and support and sustainment operations.

Part of this evolution of ideas is no doubt an attempt to gain clarification. But sometimes, I wonder if it is also used to support a better Officer Evaluation Report or end of tour award.

4. Timothy Lupfer, *The Dynamics of Doctrine: The Changes in German Tactical Doctrine during the First World War*, Combat Studies Institute (Fort Leavenworth, Kans.: U.S. Army Command and General Staff College, 1981), 8–11. The concept of corporate effort is nothing new. German tactical lessons from World War I illustrate the power of ideas from the collective thoughts and experiences of frontline soldiers rather than from a committee from above.

However, the Germans were not the only ones to exploit the corporate effort in solving battlefield problems. Both the British Army during World War I and the U.S. Army during World War II excelled in harnessing the ingenuity of frontline soldiers to break the deadlock of the trenches or the hedgerows. Paddy Griffith's *Battle Tactics of the Western Front: The British Army's Art of Attack 1916–1918*, and Michael Doubler's *Closing with the Enemy: How GIs Fought the War in Europe, 1944–1945* are both excellent sources for study in this regard.

Chapter 3

Standardized Identification

Standardized identification procedures pay big dividends and have numerous practical uses. The use of symbols and colors gives flexibility to planning. Colors are generally more effective at night and symbols during the day. Symbols can be used to mark the *A* and *B* bags (duffel bags) for quick identification. This procedure makes the building of pallets for deployments and distribution of baggage faster and more organized. Painting a white square with the unit symbol in black paint and the soldier's last name and Social Security Number ensures that unit baggage remains together. Sewing a luminescent square with embroidered symbol and last name on the rucksack also pays big dividends in all situations.

In tactical situations, illuminated colors (directional chemlights) are excellent control measures. They can identify the left and right flanks of an element occupying a position at night, unit checkpoints, release points, and rally points. Guides with colored lens on flashlights are an excellent method for assisting the linkup with their respective elements. Permanent alcohol pens or magic markers on a clear lens work well in making multiple colored lenses. With a little thought and enterprise, other techniques from the soldiers will help alleviate commonplace problems. With the introduction of frequency-hopping radios, the use of code names eliminates a host of problems for leaders. Habitual use of code names immediately alerts radio-telephone operators (RTO) of a radio contact attempt. Ingrained code names—meaning everyone recognizes the association between code name and unit or person—virtually eliminate noncontact radio problems caused by RTO daydreaming. Use of code names (for example, "Mongoose 6" refers to the commander, "Cobra 1" refers to the AT SECTION Leader) allows quick and intelligible recognition of leaders and company elements (subunits) on the radio. Since radio transmissions tend to become garbled, use of personal names can cause confusion. Code names can still be used if frequency-hopping communications are not possible. Although it risks compromising communications security (COMSEC), company-level intelligence is probably the lowest priority in the enemy's collection effort.

TABLE 3.1 STANDARDIZED IDENTIFICATION

Element	Symbol	Color	Code Name
HQ	△	Blue/Green/Red	MONGOOSE
FIST	●	N/A	SMOKE
MEDIC	✚	N/A	DOC
SUPPLY	—	N/A	LOG
COMMO	ꙅ	N/A	FREAK
NBC	✕	N/A	BUGS
AT	↑	Blue/Green	COBRA
MORT	↕	Blue/Red	SCORPION
1 PLT	I	Blue	REDSKIN
2 PLT	II	Green	BUSHMASTER
3 PLT	III	Red	HEADHUNTER

PERSONAL IDENTIFIER

Commander	6
XO	5
Platoon/Section Leader	1
1SG/PSG	2

Chapter 4

Occupy Assembly Area and Priority of Work in Assembly Area

Occupying an assembly area during the day is not too difficult, but at night it can be a nightmare. Within minutes, the plan, which once looked so simple and logical, is transformed into a morass of confusion and chaos. Since operational security (OPSEC) considerations require night occupation, the tactical standing operating procedures (TAC SOP) permits secure and organized occupation. Only by practicing occupation by day, then by night, will a company perform it to standard without divulging its presence. If the company is occupying an independent assembly area, the procedures used in **Occupy Simple Perimeter** (chapter 15) are sufficient to allow for an organized occupation. If the company is participating in a battalion-sized assembly area, the company commander coordinates the use of its organic company guides. Battalion guides are not attuned to company procedures and are not accountable to the company commander for substandard performance. Success depends on centralized planning and decentralized execution.

Once the company establishes a set pattern for occupying an assembly area and practices it both day and night, this task becomes a minor operation instead of consuming the minds of the higher leadership and requiring constant supervision.

Security Halt. The purpose of the security halt is to place the company in a secure place while the quartering party prepares the assembly area for occupation. In this manner, the company remains in the most secure position possible with a ready defense. Even if occupying a battalion assembly area, it is best for the company to conduct a security halt short of its designated sector and to wait for the quartering party (Q-party) guides to lead it into position.

Quartering Party Activities. For battalion assembly areas, the executive officer (XO) needs to coordinate with the S3 Battalion Operations Section on the size allowed for the Q-party. Generally, there will be only enough room for 10 soldiers in each Q-party, so the XO will need to plan accordingly. The only reasons to use two squads to clear and prepare the assembly area are speed and unit cohesion. Work details and other tasks should be performed by established elements (fire teams, squads, platoons). If the Q-party has contact with the enemy during the course of a task, it will fight more cohesively. Assigning a fire team per platoon sector with a fire team in reserve makes good tactical sense. The quartering party clears and marks the company sectors quickly. The battalion quartering party may not mark company boundaries, so the company Q-party will need to coordinate or reach agreement on boundaries with the other company Q-parties. Marking of platoon and section boundaries is in accordance with (IAW) **Standardized Identification** (chapter 3) symbols or colors.

The listening and observation posts (LP/OPs) provide early warning and screening from enemy units. The LP/OPs react immediately and aggressively to enemy incursions. If contact with the enemy occurs, the best way for the LP to alert the company is to use its weapons without regard to compromising the assembly area. Alerted, the company can maintain its hasty perimeter for immediate defense or react to the incursion. Any disadvantage resulting from a false alarm is countered by the assured security and alertness of the unit. Better to explain to the battalion commander why the assembly area was compromised rather than have an enemy force rampage through the area at will.

Occupation of the Assembly Area. Once the XO has conducted a back brief with the guides about the occupation plan, he sends squad 1 to the main body to lead it to the assembly area company command post (CP)—this becomes the platoon release point. To preclude confusion, each platoon guide uses the appropriate ID color marker (flashlight or chemlight) so that the appropriate squad links up with the right guide quickly and moves on to its sector. The platoon guide leads the platoon to the tentative platoon CP (normally center of sector), where the platoon sergeant (PSG) will establish a hasty perimeter. The guide shows the PSG the right and left flank markings and the LP/OP. Once this task is accomplished, the guide and the Q-party LP/OP report to their parent platoon. The PSG may have his squad leaders and a designated LP/OP accompany him and the guide during the recon. At this point, the company begins working on the **Priority of Work in the Assembly Area** (this chapter).

PRIORITY OF WORK IN THE ASSEMBLY AREA

The primary consideration of the commander is establishing an effective defense.

Assembly areas can be death traps for units. They give soldiers a false sense of security, because they are normally located behind friendly lines, diminishing the likelihood of enemy contact. It is too easy to blame lax discipline on the unit when priority of work in the assembly area is not tested during training exercises. Units tend to regard assembly areas as administrative areas for deployment to and from training events. So, the priority of work in the assembly area is rarely given due consideration; the leadership is focused on preparing for the first mission.

The enemy is not obliged to allow preparation for the mission in assured security. During World War II, German units often conducted raids and spoiling attacks against Soviet rear areas for the purpose of disrupting offensive preparations.[1] Regardless of the level of conflict, an assembly area is prey to enemy reconnaissance, raids, and harassing fire from artillery. Because of the threat, the priority of work in both the assembly area and the defense is similar. Only command emphasis differs.

Establishing Security. Depending on the time of occupation, the first sergeant (1SG) and PSGs establish the permanent LPs (night) or OPs (day). The LP/OP has an active job of providing early warning, screening the company from enemy reconnaissance, and so forth. The PSG develops the LP/OP occupation schedule and alert plan (see the section on LP/OP activities in **Priority of Work in the Defense**). The battalion assembly area is a busy place, with numerous deploying combat support (CS), combat service support (CSS), and combat units moving about. LP/OPs may need to direct these units away from the company sector of the assembly area, particularly at night. Nuclear, biological, and chemical (NBC) alarms are only employed if directed from the Battalion S3. Do not make busywork for the soldiers (checking the blocks and painting rocks) when priority tasks need to be done. If the NBC noncommissioned officer (NCO) wants to train on this, authorize it after the other tasks are completed.

Clearing patrols are simply a precaution taken during periods when the enemy is most likely to attack. This active security measure is not only a prudent precautionary measure but also a psychological measure for the soldiers. Patrolling gives soldiers a feeling of control over their fate and challenges the enemy in a test of wills regarding the ownership of the area to the unit's front. It is practiced in training to accustom soldiers to this activity. If not practiced in peacetime, everyone (soldiers and leaders alike) will be reluctant to employ the practice in combat for fear of fratricide. Establishing security even for the relatively safe assembly areas enforces in the soldier's mind the necessity for security at all times. Stressing this habit early in a soldier's career pays dividends in the future.

Establish Defense. Realistically, establishing a hasty defense takes about one hour to establish. Even though each squad occupies its sector immediately, troops do not begin preparing their hasty fighting positions until the platoon leader emplaces the crew-served weapons, and the squad leader recons his sector and selects the location of each position. Once the platoon leader positions his machine guns (and other crew-served weapons), he determines what type of fire (grazing fire or principal direction of fire [PDF]) is most effective from each position. The mindset in the Army is to establish grazing fire for machine guns (MG) regardless of terrain. Grazing fire is the trajectory of rounds, one meter from the ground and out to 600 meters. The platoon leader is cautioned not to select an MG position solely because it allows good grazing fire. As a rule for most defenses, if the tentative MG position causes the unit to overextend itself to protect it or places any part of the unit in an untenable position, select a position more easily protected even if its fields of fire are degraded. The defense is built around the crew-served weapons, and they must be protected.

Platoon leaders can emplace machine gun positions and establish squad sectors simultaneously once they become experienced. Squad leaders establish hasty positions in pairs in case two-man fighting positions are required. For two right-handed firers, the right position should be angled to the right in order to account for the natural angling of the weapon to the left. (See figure 4-1.) In this regard, the space between the pair of positions is not more than two M-16 rifles. The defense requires only hasty positions (18 inches deep times the body width with slight berm on three sides). From the prone position, the bottom of the hasty position needs to slope gradually upward from the waist forward in order to conform to the body's range of motion, making the position ergonomic. (See figure 4-2.) To compensate for the extra exposure caused by the upward angle, the berm is built up slightly toward the front for added protection for the

FIGURE 4-1

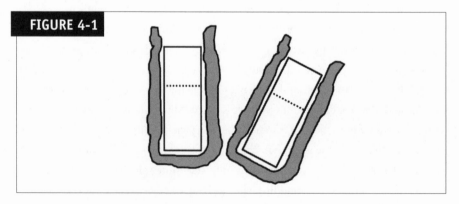

Hasty Positions in Vee Pattern

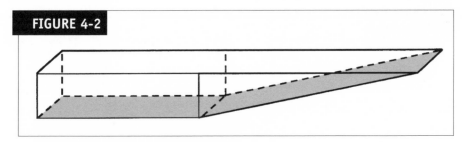

FIGURE 4-2

Ergonomic Construction of Hasty Position

soldier. Such positions should take about fifteen minutes to dig when the soldier uses his entrenching tool (E-tool) as a pick (90-degree angle). This is really the only way to use the E-tool, yet too often soldiers attempt to use it as a regular shovel to break ground; in this mode, the soldier comes to the conclusion that the E-tool is useless and waits for the squad pioneer tools to make the rounds, wasting time.

PSGs and antitank (AT) section leaders actively supervise the production of range cards, because crews are apt to complete them haphazardly, regarding it as just a paper drill. The accuracy of range cards becomes important when the platoon leaders and company commander assimilate them into their own sector sketches.

Securing unit flanks is an essential task, because enemy reconnaissance seeks to locate seams for main force exploitation. Granted, in an assembly area the odds are remote that the enemy forces will be able to exploit a seam, but platoon leaders are tasked with securing to flanks to emphasize the importance of this procedure and to develop it into a habit. Besides, tying in the flanks takes little time in the assembly area.

Develop Defensive Plan. Sector sketches serve one purpose: they allow leaders at each level to visualize the defense as it actually exists. They alert leaders to natural weaknesses in the defense, such as uncovered gaps, dead space, and untenable positions. For this reason, sector sketches and range cards at all levels are standardized, using a point of reference system (grid lines and azimuth tick marks) to assist the commander in his company sector sketch. This is covered in more detail under **Sector Sketch** (chapter 20).

The executive officer plots at least three starting points (SP) in anticipation of departure from any direction. (See figure 4-3.) He locates these far enough away from the assembly area to allow the company elements to deploy without congestion. In consultation with the commander, he determines the order of march for departure along any of the egress routes for swift departure of the assembly area. The company can always modify the order of march during a security halt on the move. This method gives the

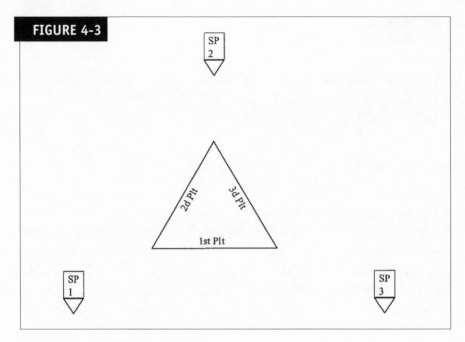

FIGURE 4-3

Establishment of Starting Points

commander great flexibility in moving his unit quickly and without congestion. The order to move the company can be either a part of the operation order (OPORD) or a fragmentary order (FRAGO), but brevity is preferred: "Alpha Company departs at SP 2 at 2315 hours. Order of march is 2d, HQ, 3d, Mortars, AT, 1st. Time now is 2245." The commander and his radio-telephone operators (RTOs) can move to the SP immediately and await the main body. This technique allows him to evaluate the timeliness of the movement, march, light, and noise discipline on the spot. Once the lead platoon passes through, the commander's CP falls into place.

Reconnaissance and security (R&S) teams provide the commander with current information of the surrounding terrain and enemy activity. They also identify covered and concealed routes to the line of departure and line of contact (LD/LC). The commander weighs this information with the amount of friendly activity in the area and uses his judgment accordingly. Even if not required, the company R&S plan is sent to the S3 for consideration. The mortar section leader assigns target reference points (TRPs) along the route for fire support to the R&S team.

The fire support team (FIST) and mortar section leader plan TRPs forward of and within the perimeter in order to add depth to the defense plan. This configuration allows quick artillery response from enemy attack and assists the unit in breaking contact with the enemy forces if pushed out of the assembly area. The FIST officer does not plan variable time (VT) fuses for TRPs within the assembly area because no positions have overhead cover.

Team leaders back-brief their soldiers on the contingency, security, and alert plans once hasty positions are complete. Knowledge of the alternate assembly area is crucial in case the enemy forces overrun the assembly area. Team leaders should show this area's position on the map and point it out to their soldiers on the ground (at least the direction and distance). The soldiers should record contingency plan information for reference at each position (laminated cards and grease pencils are a possible technique). Running passwords are reserved for R&S patrols and clearing teams entering the perimeter in an emergency. They are rarely used, but they come in handy when needed. An additional task for a selected R&S team is to recon the alternate assembly area to verify its location and suitability for the unit.

The security and alert plans establish the alert posture of the company at any given time. LP/OPs are counted when determining the level of security. The PSG manages the LP/OPs, but manning comes from the squad directly behind the LP/OP. The affected squad leader includes his personnel contribution in his percent security. The squad leader manages the security and alert posture within his sector and may use any technique that will ensure proper security: he may consolidate squad personnel not pulling sentry duty in hasty fighting positions around the squad CP so that they can sleep; he may keep everyone in fighting positions and rotate sentry duty by fighting position on a time schedule; or he may select a position slightly forward of the perimeter (a quasi LP/OP) for sentries to be located. The squad leader and his two team leaders share supervisory responsibility for checking on the sentries. Stand-to for BMNT is a particularly unpleasant task, especially when it is cold. An effective technique for getting the unit 100 percent alert, dressed, and in fighting positions is to have each soldier consume a sugar packet and perform 20 to 30 push-ups to help them wake up. Leaders rise before stand-to and begin checking the perimeter as stand-to begins. The first sergeant (1SG) is in charge of supervising compliance with stand-to. The commander and platoon leaders also inspect the security and alert posture from time to time. If discrepancies are noted, officers bring them up with the PSG and 1SG to correct.

TTP

Soldiers can record recurrent information using grease pencils on laminated 3 X 5 cards.

Very rarely will there be a need for obstacle emplacement to support the assembly area. But the commander at least considers likely places for obstacles in case the assembly area becomes long-term emplacement.

Establish Communications. Communications appear to be a never-ending problem. It could be gremlins; if so, they only affect communications from higher to lower, never vice versa. Generally, the RTO maintains his radios by frequently cleaning the hand mike terminals and antenna, using an erasure. Waterproofing the mike using a plastic bag and a rubber band extends the span of continuous radio contact. If a problem persists, changing batteries is necessary, even if the expected battery life is not due. If communications fail, dispatch a runner to the next level of command. Assembly areas are small enough to make this an easy and fast option. Commo wire should be used whenever the term of occupation is longer than 12 hours (otherwise setting up communications is not really worth the effort, because the laying and recovery of wire take a couple of hours each).

Conduct Sustainment. Sustainment is the first sergeant's main responsibility, because it affects the well-being of the unit, and his association with that task should be in the minds of the soldiers. The commander checks sustainment but directs his concerns to the 1SG.

The 1SG gives the commander the unit physical security report at the end of BMNT and early evening nautical twilight (EENT). After the morning meal, he provides the commander with the personnel, equipment, and essential supply request status reports. If the 1SG is busy with other duties, the XO collects this information. With this information, the commander voices any concerns with the battalion staff or battalion commander, or both together, as a matter of business. The 1SG turns in these reports to the battalion staff after the breakfast and dinner meals.

The assembly area supply point is the distribution point for resupply. Because of the proximity of company positions, platoon details conduct resupply activities at this point without significant problems. The supply sergeant is responsible for maintaining tools for opening crates, and so forth. If a company element needs to borrow a special tool, the PSG signs for it to maintain the chain of custody. Failure to do so results in lost accountability of the tools most needed.

Soldiers sometimes fail to load their magazines as a result of the rush of activity. Often, ammunition boxes are quickly thrown in the rucksack immediately, with the intention of filling their magazines later. Usually they do, but it is disconcerting to see soldiers frantically filling magazines during a fire fight because it slipped their mind to fill them earlier. NCOs ensure that the soldier is ready to fight fully armed. Another point for the NCOs to check is the magazine spring, which loses its effectiveness over time. A tired spring often results in a misfeed of ammunition into the chamber, so NCOs make this spring part of their checks. The supply sergeant maintains a small stock of magazines for a fast exchange.

Health and hygiene normally becomes important in a sustained conflict or deployment in the field. Training exercises are sufficiently short that health and hygiene does not become a problem and is generally ignored. To prevent laxness, the commander makes a point to check specific preventive measures such as food and water consumption, water purification, the sleep plan, and field latrines.

The 1SG enforces the tactical feed plan to prevent soldiers from clustering around the hot meal distribution point and assuming the picnic posture. The 1SG manages the feed order and schedule for each element to line up for chow. The PSG/section leader enforces the five-meter-line interval between soldiers and ensures that his soldiers eat in a covered or concealed position. As an added precaution, soldiers do not eat in groups larger than a squad. Even the consumption of MRE (meal, ready to eat) rations requires an alert posture and trash discipline. The security posture is maintained during distribution of hot meals with sentries remaining on duty. If consuming MREs, soldiers do so at the squad CP while sentries maintain the security posture. The PSG ensures that the sentries have been relieved for their meal before he eats.

Test Weapons and Equipment. If time and a safe locale exist, soldiers need the opportunity to zero and test fire their weapons, especially machine guns. The zeroing of night observation devices on weapons is also accomplished during test fire.

In-process Replacements and Attachments. The 1SG welcomes new soldiers and attached units to the company. A quick briefing on the tactical situation and the TAC SOP helps assimilate them quickly and heightens their confidence in the unit. For incoming leaders, a copy of the TAC SOP apprises them of their responsibilities.

Prepare for Mission. The officer chain of command will likely be immersed in this activity, leaving the NCOs to manage the major aspects of the assembly area. Since some NCOs will be involved in the planning process, they may need to delegate some of their assembly area duties to subordinate NCOs. The important thing is that the duties get accomplished without higher leadership micromanaging everything.

Departure of the Assembly Area. The 1SG does a walk-through of the assembly area to ensure that it is sanitized. This is not only a common courtesy for subsequent units, but also a good operational security (OPSEC) measure. A handy technique for ensuring compliance is for the 1SG to bring the PSGs and section NCOs along to clean up the assembly area themselves if it is not sanitized. This instruction will probably only need to be done once before PSGs take this duty seriously.

NOTES

1. U.S. Department of the Army, *German Defense Tactics against Russian Breakthroughs*, Department of the Army Pamphlet No. 20-233 (Washington, D.C.: October 1951), 15–19. The company cannot get into the mindset that the enemy forces will not attack because they are on the defensive. A good spoiling attack will inflict severe losses on any unit that makes this assumption.

TABLE 4.1 OCCUPY ASSEMBLY AREA	
Task	**Responsibility**
1. Security Halt:	
a. Halt company 200–300 meters from tentative assembly area and form hasty perimeter.	CO CDR
b. Organize QP: two squads, AT section leader, mortar section leader, and NBC NCO (if required). Exchange **Five-Point Contingency Plan** with CDR: 1) The task. 2) Return time.	XO

TABLE 4.1 *(continued)*	
3) If Q-party does not return by appointed time, attempt to establish contact with it. If contact not made, company occupies assembly area in force. 4) If Q-party is attacked, it should return to main body. 5) If company is attacked, it should move to assembly area.	XO
c. Upon departure of Q-party, adjust perimeter to establish 360-degree tactical posture.	1SG
2. Quartering Party Activities:	XO
a. Establish initial security at tentative Co CP.	SL I
b. Organize clearing plan: Designate platoon sectors for squad 2 (3 fire teams) to clear and time allotted for clearing perimeter sector.	XO
c. Check area for NBC contamination if applicable.	NBC NCO/NBC TM
d. Clear designated sector (box, zigzag, cloverleaf, etc.).	SL II
e. Check for and remove or mark obstacles and mines.	TM LDR
f. Establish LP/OP per platoon sector.	TM LDR
g. Mark PLT left and right flank limits **Standardized Identification.**	IAW TM LDR
h. Select Co mortar position.	MORTAR SEC LDR
i. Select AT firing positions.	AT SEC LDR
j. Verify Co CP location and select Co trains location.	XO
k. Conduct brief-back with clearing teams for occupation plan.	XO
3. Occupy Assembly Area:	CDR
a. Dispatch squad I to guide main body to assembly area Co CP.	XO
b. Link up with designated platoon leaders at assembly area Co CP (Plt RP). 1) Guide to Plt CP. 2) Show PSG left/right limits and LP/OP. 3) Pick-up QP LP/OP, report to parent platoon sector.	Squad II Guides
4. Begin **Priority of Work in the Assembly Area.**	ALL

TABLE 4.2 PRIORITY OF WORK IN THE ASSEMBLY AREA

Task	Responsibility
1. Establish Security:	1SG
a. Select permanent OPs [Day] with covered and/or concealed route back to defensive positions. (N / 30 minutes)	PSG
b. Select permanent LPs [Night] and emplacement of early warning devices (PEWS, etc.) before EENT.	PSG
c. Emplace NBC alarms if danger exists. (N / 30 mins)	NBC NCO/PLT NBC TM
d. Identify and brief BMNT/EENT clearing patrols— clear within 300 meters of positions to disrupt enemy in surveillance or assault positions. (Before EENT/BBMT)	PLT LDR
2. Establish Defense:	CDR
a. Select crew-served weapon positions and sectors of fire. (N / 15 mins)	PLT LDR/SEC LDR
b. Verify platoon sector and select squad sectors. (N / 30 mins)	PLT LDR
c. Designate company reserve and position. (N / 30 mins)	CDR
d. Hasty crew-served weapon positions prepared: waist deep with kneeling or sitting position for AT weapon or body width X 18 inch deep for MG; firing platform for AT and MG weapons; MG grazing fire or PDF established; immediate camouflage. (N / 45 mins)	PSG/AT TM LDR
e. Select hasty fighting positions and sectors of fire. Ensure positions are body width and 18 inch deep, with sector stakes, fields of fire cleared, and immediate camouflage. (N / 45 mins)	SL
f. Verify and submit **Range Card** to platoon/AT CP. (N / 1 hr)	PSG/AT TM LDR
g. Coordinate flank security with units on left and right: exchange flank position's sector of fire, LP/OP positions; ensure avenue of each approach is covered. (N / 1 hr)	PLT LDR

TABLE 4.2 *(continued)*	
h. Plot at least three departure-starting points (SP), each at least 100 meters from assembly area. Disseminate to company elements. (N / 1 hr)	XO
3. Develop Defensive Plan:	CDR
a. Submit copy of squad **Sector Sketch** to platoon CP. (N / 2 hrs)	SL
b. Submit reconnaissance and security (R&S) plan to Co CP (within 1 km of perimeter). (N / 2 hrs)	PLT LDR
c. Submit company fire support plan to commander. (N / 2 hrs)	FIST OFF/MTR SEC LDR
d. Submit one copy of each AT **Range Card** to company CP. (N / 2 hrs)	AT SEC LDR
e. Submit one copy of platoon **Sector Sketch** to company CP. (N / 3 hrs)	PLT LDR
f. Back-brief soldiers on contingency plan (from OPORD): Challenge and password; location of alternate assembly area on map—point out the actual direction and distance from current location. (N / 2 hrs)	TM LDR
g. Back-brief soldiers on security plan (from OPORD): (N / 2 hrs)	TM LDR
1) % Security during daylight.	
2) 100% Security during stand-to.	
3) % Security during night.	
h. Back-brief soldiers on alert plan: (N / 2 hrs)	TM LDR
1) Stand-to for BMNT: 30 mins prior to BMNT until 30 mins past BMNT. Advance LP to OP position.	
2) Soldiers are in full uniform with all equipment donned throughout stand-to. For BMNT, soldiers eat one packet of sugar and perform 20 push-ups to heighten their alertness.	
3) Stand-to for EENT: 30 mins before EENT until 30 minutes past EENT. Inspect all (NODs) and thermal devices. Withdraw OP to LP positions.	

(continued)

TABLE 4.2 *(continued)*

i. Produce obstacle plan on key avenues of approach and include in company **Sector Sketch** [if directed by Bn HQ]. (N / 2 hrs)	CDR/ENG LDR
j. Submit one copy of company **Sector Sketch** to battalion CP. (N / 5 hrs)	CDR
4. Establish Communications:	COMMO CHIEF
a. If no radio commo or if radio listening silence, platoon/section sends one runner to Co CP until wire commo is established. (N / 0)	PSG/AT TM LDR/ MORT SQD LDR
b. Lay commo wire to PLT/MORT/AT CP and Co Trains. (N / 1 hr)	CO RTO
c. Lay commo wire to LP/OPs. (N / 2 hrs)	PLT RTO
5. Conduct Sustainment (times are subject to higher HQ tactical SOP; otherwise sustainment should begin once the defense is established):	1SG
a. Conduct physical security checks on all sensitive items by serial number and submit report to 1SG at end of BMNT and EENT.	NCO CHAIN
b. Submit personnel strength, casualty reports, and witness statements to 1SG during morning meal.	PSG/SEC LDR
c. Submit equipment status report and maintenance requests to supply sergeant during morning meal.	PSG/SEC LDR
d. Submit supply status report and resupply requests to supply sergeant during morning meal.	PSG/SEC LDR
e. Collect above reports for submission to pertinent Bn staff sections.	1SG
f. Supervise maintenance of equipment and weapons with priority to weapons and radios.	NCO CHAIN
g. Coordinate for supply and maintenance support with Bn S-4.	SUPPLY SGT
h. Establish resupply plan with PSG and SEC LDR:	SUPPLY SGT
1) Establish company resupply point:	
a) Aerial: prepare, mark, and secure LZ to the rear of unit.	

TABLE 4.2 *(continued)*	
b) Vehicle: covered and concealed unloading site to the rear of unit.	SUPPLY SGT
2) Resupply for platoons and sections conducted at company supply point.	
3) Proper tools for opening crates, cutting wire bands maintained at Co trains (wire cutters, crowbars, etc.).	
i. Ensure each rifleman has his allotment of rifle magazines and that each magazine is filled to capacity. Excess ammo is stowed for quick access without compromising the position.	NCO CHAIN
j. Health and Hygiene:	1SG
1) Monitor food and water consumption (8 qt. per day minimum) of soldiers.	NCO CHAIN
2) Establish sleep plan: 4 to 6 hours of sleep per soldier are required. Maintain same sleep schedule per soldier to maintain circadian rhythm.	OFF CHAIN
3) Random check of soldiers' feet, hydration, and general health.	CO and PLT MEDICS
4) Disseminate first sergeant's sick call procedures.	NCO CHAIN
5) Monitor personal hygiene of soldiers.	NCO CHAIN
6) Supervise field sanitation team.	PSG
7) Check water for potability. Water purification tablets on hand for each soldier. Purify water using tablets or by boiling IAW FM 21-10, paragraph 46–48, 51–54.	FLD SAN TM
8) Establish and supervise construction of one field latrine per platoon located at least 30 meters behind defensive positions, away from water sources, and from feeding area. Inspect inside and outside of assembly area for fecal and urine contamination.	PSG/FLD SAN TM
9) Use field sanitation kit to ward off insects/ rodents: repellent, traps, and poison.	FLD SAN TM
k. Establish tactical feed plan:	SUP SGT

(continued)

TABLE 4.2 *(continued)*	
1) Maintain local noise and light discipline.	NCO CHAIN
2) Enforce five-meter interval between soldiers in serving line and meal site. Ensure soldiers use available cover and concealment.	NCO CHAIN
3) Maintain security of perimeter during meals IAW security plan. Enforce consumption of rations behind fighting positions.	NCO CHAIN
6. Test Fire Weapons. Mount and zero NODs on machine guns and AT weapons before EENT. CO CDR approval required.	
7. In-process replacements/attachments: brief on situation (friendly and enemy) and TAC SOP.	1SG
8. **Mission Preparation.**	CDR
9. Upon departure of assembly area, inspect area for trash, open latrines, mines, obstacles, commo wire, aiming stakes, etc., ensuring the area is sanitized.	NCO CHAIN

Source: U.S. Department of the Army, *Mission Training Plan for the Infantry Rifle Company,* Army Training and Evaluation Program No. 7-10-MTP (Washington, D.C.: 1988), 5-49–5-52, 5-172–5-175, 5-204–5-208, 5-209–5-212. "Occupy Assembly Area" (7-2-1022), "Maintain Operation Security (7-2-1057), "Perform Personnel Actions" (7-2-1037), and "Perform Logistical Support" (7-2-1048) form the foundation document for this portion of the SOP.

Chapter 5

Mission Preparation

Preparation for any mission encompasses both the decision-making process (DMP) and troop-leading procedures (TLP). As with any process, the DMP is a tool for producing an adequate, tentative plan for the operations order (OPORD)—the 80 percent solution. Leaders should not attempt to develop the perfect plan, because there usually isn't enough time and information. Leaders must avoid becoming enthralled with the process to the detriment of the mission. Traditionally, leaders spend too much time on planning and not enough on preparation, brief-backs, and rehearsals. More often than not, failure is traced to brilliant plans but poor execution. Leaders should spend minimum time on developing the tentative plan, because so much critical information is unknown. As with any process, the plan will develop as a result of new intelligence, back-briefs, and rehearsals. Both DMP and TLP are analytical and time-management tools to help leaders focus combat power at the decisive point. Leaders waste time and resources attempting to address inapplicable matters in a blind attempt to follow worksheets, formats, and checklists. Leaders are expected to think and come up with solutions to problems quickly, not fill out charts in an effort to please superiors. The bottom line: if an analytical tool or worksheet is not applicable or does not address the problem, leaders should not use it.

Because time is likely to be the most critical factor, the commander seeks every opportunity and stratagem to exploit the little time provided. He never knows when the mission schedule will be accelerated. The commander refines the plan to a 100 percent solution as a result of reconnaissance, intelligence updates, back-briefs, and rehearsals. It is not unusual for the commander to finalize the plan at the objective rally point (ORP).

The commander should use all information to adapt the theoretical (tentative) plan to a practical plan. Conversely, he must balance any changes he makes with the time available. His changes cannot be so radical that the company cannot adapt in time. With practice, the process of planning and preparing for a mission becomes seamless and fast. The amalgamation of the DMP and TLP comprises the following: The alert, receipt of the battalion operations order (OPORD), mission analysis and issuance of warning order, make the tentative plan, start movement, issue OPORD, conduct reconnaissance, conduct inspections, conduct rehearsals, complete the plan, and talk to soldiers.

THE ALERT

Upon receipt of the battalion warning order (or notification that a mission is pending), the commander immediately gives his subordinates warning order #1, comprising the nature of the mission, the area of operation, and likely tasks in order to energize the company. Most preparation activities are routine regardless of the mission (zeroing weapons, soldiers' load, and so forth), and some require early attention because they are time-consuming (PZ preparation for air assault, overhead cover packages for fighting positions, and obstacle packages). The company cannot afford to be idle while awaiting the mission, because this would put it behind the power curve.

RECEIPT OF THE BATTALION OPORD

The company radio-telephone operator (RTO) and fire support team (FIST) officer accompany the commander to the OPORD. The RTO copies the operational graphics overlay and makes multiple copies for the company elements while the commander receives the OPORD briefing. He also coordinates changes in communications (new encryption fills, signal operation instructions [SOI] change out, and so forth). The FIST officer receives the fire support overlay and performs initial coordination with the battalion fire support officer (FSO). The commander prepares the company warning order during any lulls. Upon notification, the RTO radios the company leadership (while en route or upon return) of the time and place of the company warning order.

MISSION ANALYSIS AND ISSUE WARNING ORDER

In conducting the mission analysis, the commander's main concerns are the restated mission (although rarely different from the assigned task from the battalion OPORD), the time schedule, and the identification of critical preparation tasks. The restated mission must clearly identify the company's task and purpose (with emphasis on purpose). The company mission is normally clearly identified and straightforward from the battalion OPORD, requiring little need for listing specified, implied, and critical tasks for a restated mission. These tasks are important for delegation during the warning order and development of the plan though. Within 30 minutes of his return, the commander issues warning order #2, comprising the situation relevant to the company, the restated mission, the preliminary analysis of time for mission preparation, and delegation of tasks. The first sergeant (1SG) checks attendance and informs the commander hen all are present. The RTO distributes the battalion map overlay to orient the attendees to the area of operation. The commander informs the participants that this is the warning order. (Attachment leaders may assume that it is the OPORD.) After giving the situation and mission, he gives the time schedule up to the time of departure in accordance with the ⅓, ⅔ time rule. The commander has no formal staff to help him plan and coordinate the mission. He therefore assigns tasks to subordinates to assist in OPORD and mission preparation.

DELEGATION OF TASKS

The commander assigns portions of the OPORD preparation to specific subordinates while he focuses on developing the "tentative plan" for the OPORD. This allows him to concentrate on planning the battle at the decisive point, while subordinates work on the more routine aspects of the OPORD. For other assigned tasks, subordinates refer to pertinent tactical standard operating procedures (TAC SOP) checklists/planning formats and worksheets. The commander checks on the progress of preparation by having subordinates back-brief him before the OPORD.

OPORD paragraphs I-III. The commander, FIST officer, and executive officer (XO) form the S3/S2 planning cell. The commander and FIST officer analyze the friendly side, while the XO tries to ascertain what the enemy will do. Each considers mission, enemy, terrain and weather, troops, and time available (METT-T) and observation and fields of fire, cover and concealment, key terrain, and avenues of approach (OCOKA) as they apply to each side and refers to the planning considerations of the offense or defense. At first glance, analyzing terrain and weather appears to be quite daunting and time-intensive. But with practice, the commander can glance at the aspects of terrain and weather quickly and extract the important from the mundane, developing a keen appreciation of the tactical uses of terrain and weather.[1]

The commander and XO explore and discuss the strengths and weaknesses in the course of action (COA) as it applies to each side. The commander develops only one course of action, because the limited time, area of operation, and resources on hand do not warrant the development of more than one. Again, few operations fail because of a poor plan; most fail as a result of a poor execution. The commander should therefore focus appropriately.

The commander and XO then war game the course of action with the XO playing the thinking enemy as opposed to a doctrinal automaton. As a word of caution, it is not enough to consider the enemy most likely course of action. The XO must identify every course of action the enemy will find suitable, feasible, and acceptable, because he may switch to another COA as the engagement develops. He should not discount an enemy course of action because the war game reveals a weakness. The enemy may choose that course of action for whatever reason. The company plan should be flexible enough to anticipate even a foolish enemy course of action and make him pay for his mistake. The XO should always consult the battalion S2 for the enemy most likely and most dangerous COA, as well as his latest doctrinal changes and tactics. To be a thinking enemy, he must understand that enemy. It is the XO's job to help the commander focus on fighting the enemy and not the plan. The FIST officer assists the commander in selecting the engagement areas and kill zones and the proper type of weapons and munitions for each critical event. In this manner, he helps the commander set the conditions for defeating the enemy forces at each stage of the engagement.

Other OPORD paragraphs and annexes. The commander assigns completion of other OPORD paragraphs to subordinates whose positions make them suitable for the task. During the OPORD, they brief only critical information. Information such as radio frequencies, call signs, and command post (CP) locations should be recorded on 3 X 5 cards and handed out to the leadership. Subordinate officers and noncommissioned officers (NCOs) assigned other supporting tasks refer to the relevant checklists and procedures for preparation or coordination. The end product becomes an annex to the OPORD. During the OPORD, they brief their portion and use visual aids (sketches) as much as possible to disseminate information.

Reconnaissance. As the critical piece of any mission, the commander must deploy a reconnaissance element to the objective or defensive sector. As there is no organic recon element, the commander must make one—normally the antitank (AT) section if the armor threat is low. Since the company recon element must deploy early, the rest of the company must provide supplies to the recon element from existing stocks in order to ensure that it has plenty of ammunition, water, and food. The company replenishes its stock of supplies during the normal preparation period.

THE TENTATIVE PLAN

To reiterate, planners do not attempt to make the perfect plan. They should concentrate on a plan that affords the greatest range of opportunities to defeat the enemy in detail. This is the concept of the Central Position (although actually a term associated with higher-level tactics). In the offense, the commander disposes his company so that it threatens several points in the enemy defense and, once committed in the attack, can defeat the defender in detail through overwhelming combat power against a portion of the enemy while maintaining the capability to respond to the enemy's actions and reactions. In defense, the commander attempts to destroy the attacker piecemeal. Using terrain and obstacles, the defender engages and destroys a portion of the enemy without suffering mutual attrition. The defender maintains freedom of maneuver to shift forces and fire against an enemy threat and then realign its forces before the next threat develops. The success of the central position depends on freedom of maneuver as well as time and space considerations. Development of the tentative plan comprises METT-T analysis, courses of action analysis for the defense of attack and course of action selection.

METT-T

The process of planning can become quite laborious and perhaps more detailed than is warranted. Keep in mind that the plan changes as the commander is provided with more information from the S2, S3, the company scouts, feedback from back-briefs and rehearsals, and the commander's own battlefield intuition. At the company level, the intelligence preparation of the battlefield (IPB) consists of METT-T and OCOKA, which help focus the planner's analysis. METT-T aligns the task to be accomplished with opposing force, the environment, and the time and resources available. Because time is of the essence, the commander and XO consider these factors very rapidly and do not become bogged down on minutiae. These factors help the commander and XO focus on locating the decisive point and setting the conditions for applying overwhelming combat power at that point.

Mission. The planner concerns himself with only those factors that influence the mission. The restated mission and higher command's mission and intent provide the purpose of the company operation and are the guiding hands for all planning. Understanding the purpose of the company operation is critical in establishing initiative. Purpose is the essence of mission orders.

Terrain. At the company level, the modified combined obstacle overlay (MCOO) translates to OCOKA—Observation and fields of fire, Cover and concealment, Obstacles, Key terrain, and Avenues of approach. Whether on the offense or defense, OCOKA applies to both sides. Analyzing the effects of terrain and weather on the operation applies to both the friendly and enemy COAs. Terrain and weather effects help the commander understand how his and the enemy's forces are degraded as well as eliminating COAs that become unfeasible (beyond a unit's capabilities).

Observation and fields of fire. This concerns the unobstructed range for visual observation and weapon effectiveness as they relate to the terrain. High ground is particularly important, because it allows the observer early detection of approaching enemy forces. Additionally, high ground provides greater weapon accuracy and coverage on the enemy located on lower ground. Naturally, this capability is contingent on the amount of vegetation available. If a hill is heavily vegetated, no one can exploit its natural advantage. Hence, the planner must consider height and vegetation together as a package when considering observation and fields of fire. It was for these reasons that Little Round Top at Gettysburg was more tactically valuable than Big Round Top. Even though Big Round Top enjoyed greater height, its heavy vegetation nullified this advantage. Little Round Top, however, provided excellent fields of fire to its front. Unfortunately, a map reconnaissance cannot determine the effects of vegetation on observation—this certainty requires ground reconnaissance. On the offense, observation considerations point to the best place to emplace the fire support element. In the defense, observation and fields of fire determine the number and type of weapons to use in relation to the terrain and vegetation—open, flat terrain favors the use of anti-tank weapons and a few small arms weapons in dispersed defensive positions, because this type of terrain favors a mechanized attack; increased vegetation and broken terrain require a greater density of defensive positions and more small arms weapons, because this type of terrain favors a dismounted infantry assault. The commander identifies surrounding terrain (which the attackers may use as a support by fire position) to either deny this position with mines or indirect fire, or to establish his position where cover and concealment provide protection.

Cover and concealment. In the offense, the attacker seeks a route that conceals his approach to the objective. Cover is incidental if it contributes to this end. Cover and concealment are factors in the selection of the fire support position, because they minimize the amount of defensive counterfire. Ideally, the position allows the fire support element to target a specific portion of the enemy defensive sector without being exposed to unsuppressed enemy weapons. A covered and concealed assault position also conceals and protects the assault element from enemy defensive fire until the assault begins. In the defense, the commander selects areas within the defensive sector that allow defensive positions to engage exposed attackers without revealing themselves to enemy units on the periphery (those units not in the defensive fields of fire or engagement areas). With woodline defenses, defensive positions are located at least 100 meters from the edge in order to avoid the effects of enemy preparatory fires and to make it more difficult for the enemy to pinpoint fighting positions. Likewise, reverse slope defensive positions are protected from the massed offensive fires of the attacker. Sectors that lack cover and concealment require dispersed defensive positions and exploitation of natural swells and hollows in the terrain. The commander identifies surrounding terrain that masks the attacker's approach and plans countermeasures (additional obstacles and a greater density of defensive positions).

Obstacles. Obstacles inhibit freedom of maneuver and time-space synchronization for the attack. In the offense, identification of natural obstacles divulges possible locations for enemy defensive positions, particularly if the obstacles form a choke point (defile, bridge). One of the crucial tasks of the reconnaissance element is to verify the true nature of the natural obstacles and the reinforcing manmade obstacles. Once the extent of the obstacles is identified, the planner can request or organize breaching assets. In the defense, the commander incorporates existing obstacles into the defense. They help the commander identify possible locations for the decisive point as well as a possible economy of force. They also indicate the extent of reinforcing obstacles required to strengthen the defensive effort. The commander does not develop an obstacle plan until he has chosen a course of action.

Key terrain. In both the offense and defense, the commander identifies terrain that allows him to place effective fire against the enemy, dominates surrounding terrain, controls a choke point, or forms the keystone to the defense. Normally, key terrain is associated with the decisive point. As already touched upon, elevated features with good observation and fields of fire are generally key terrain. We make a further distinction by labeling key terrain that is critical to mission success as decisive terrain. In the offense, key terrain can determine intermediate objectives if their seizure allows us to place effective fire on enemy positions or if their neutralization (indirect fire or smoke) prevents the enemy from placing effective fire against our advancing troops. They are progressive objectives, whose control or neutralization promote the continuation of the advance (successive support by fire positions for instance). Selection of key terrain for the security element is important. Such terrain aids in the isolation of the objective, control of the main entry and egress routes, early warning of reinforcements or counterattacking forces, and establishment of a defense or delay with few forces. The seizure of the decisive terrain normally has operational implications for higher headquarters. It represents the immediate objective of brigade headquarters. However, at the company level, terrain is a means to an end—the destruction or rupture of the enemy defense and destruction of organized resistance. Key and decisive terrain is normally the keystone for the defense. The defender turns surrounding key terrain into redoubts, protecting the citadel (decisive terrain) from immediate fire support and assaulting forces. If forced to, the defense collapses on itself around the citadel. If forced to withdraw, the objective of the counterattack is the recapture of the decisive terrain. Often, the psychological impact of key terrain is such that its mere seizure or retention decides the course of the engagement.

Avenues of approach. Control of transportation networks is important for sustainment. Eventually, the attacker must gain control of these networks to continue the offensive. As such, a major avenue of approach may have operational importance. No matter where the attacker launches his initial thrusts, he will eventually need to converge on these networks. In the offense, the commander identifies avenues of approach for counterattacking forces into the objective area. He also considers which terrain controls the avenues of approach from the defender's perspective and addresses this terrain in his plan. In the defense, the type of avenues of approach determines weapon systems coverage. Roads are ideal for vehicles and require coverage by heavy weapons. Because roads offer the enemy the fastest avenue for attack, the defender focuses its initial effort in covering these immediately and in preparation of defensive positions. Because attacking dismounted infantry is not limited to a specific avenue of approach, the defender's most prudent tactical technique is the perimeter defense. The problem for the defender is that a perimeter defense cannot cover the same area as a linear defense. One technique is to have the initial defense collapse on itself by disposing supplementary and alternate positions in such a manner that the company ultimately forms a perimeter defense as the engagement progresses.

Weather. Now the planner examines the effect weather has on weapon effectiveness and mobility for both the enemy and friendly forces. Generally, inclement weather favors the defenders and makes planning a little easier. At this point in the process, the commander makes an assumption regarding the effect weather will have on each side's mobility and weapon effectiveness regarding visibility.

Mobility. For the attacker, mobility is critical. The old adage that weather applies equally to both sides is bunk.[2] If the attacker's mobility is restricted, he cannot quickly mass forces and weapon systems at the point of his choosing. If the defender has a defense with superior interior lines relative to the attacker, it can shift forces far more rapidly to a point that the attacker threatens. For the defense, anything that delays the attacker gains time, and time is what the defender always seeks. If the attacker requires more time to form an assault, the defender has time to bolster its defenses. Rain, mud, deep snow, and periodic thaws force the mechanized and motorized forces to advance along or near improved roads. Rain and periodic thaws can cause minor streams to swell and gentle slopes to become slick, thereby prohibiting mechanized transit. Likewise, these conditions bring dismounted infantry advances to a crawl. The attacker cannot take full advantage of mobility corridors or presuppose that vantage points for key weapons are accessible in support of the attack. As it is more difficult to assess how much inclement weather will slow the constituent forces of an attack, planners have great difficulty synchronizing the attack. Practically, the only advantage afforded the attacker is that the objective is easier to isolate against enemy reinforcements and counterattacking forces. Hence, any effect on mobility is to the defender's advantage. He can concentrate his forces and key weapons on major avenues of approach and make better use of economy of force without undue risk. Nevertheless, the defender must contend with the fact that the weather will eventually allow for greater mobility. The defender should plan the defense to adapt to such improvements and take advantage of any delays to add depth to the defense.

Visibility. Again, lack of visibility places the attacker at a disadvantage. Fog, heavy rain, haze, and mist, as well as smoke, dust, and smog, reduce visibility. With poor visibility, the effectiveness of support by fire positions is degraded. It is more difficult for the attacker to place and observe indirect fire on specific points. The attacker also finds it more difficult to pinpoint and destroy the defender's weapons, which, relying on range cards and sector stakes, can bring effective fire on the attacker without actually seeing him. Besides the obvious effect on land navigation, an advance is more apt to bog down even against light enemy contact. Each individual moves more slowly (especially at night) because he cannot see very far forward and recognize likely dangerous terrain (engagement areas, good fields of fire, and likely places for enemy positions). He feels isolated because he sees few of his comrades. He is apt to imagine that his small group has broken contact with the main body and is hence on its own. He anticipates danger with every step, exhausting him and eroding his will. Fog intensifies this sense of isolation, because it dampens sound. Once enemy contact is made, each soldier is likely to go to ground and remain there because he cannot see progressive action. This is a real challenge for leaders to get control and regain forward movement. Unfortunately, this sense of isolation is nigh impossible to replicate in training.

TTP

During training, a 20-mile approach march to the attack objective or defensive sector, performed at night with little rest, inures soldiers to the hardships of combat. The attendant numbing effects of fatigue, sleep deprivation, darkness, and exposure to the elements help prepare the soldier for the realities of combat. The goal is to break the cycle that surprise and danger have on fatigue at the moment of reckoning.

To mitigate this effect of isolation, the company must devote a substantial amount of training at night over long distances and in all kinds of weather. The opposing forces (OPFOR) contest the approach march up to the final objective. In this manner, both friendly forces and OPFOR get the same benefits from this training. Soldiers will not enjoy such training and will experience a decline in effectiveness. Leaders will make errors in judgment as a result of fatigue. In short, the commander will have a more realistic appraisal of his leaders, soldiers, and himself under stressful conditions. The commander can account for this knowledge in his plan.

Another factor to address is the natural tendency of soldiers to become paralyzed upon contact. Training must require each leader to automatically inform those around him of the nature of the enemy contact (even if the leader feels that the nature is obvious) and require that each soldier relay shouted commands and information forward and rearward. The common fear is that shouting will give one's position away to the enemy. This fear is unsubstantiated, because the enemy usually already has a good idea where the force is. Moreover, voice commands slice through the soldier's sense of isolation. Leaders need to understand that getting soldiers to return fire is not automatic, and that they must prepare for such a contingency. Often, leaders can break this paralysis by getting the soldier to perform some positive act—firing his weapon, repeating commands, bringing up ammunition, or applying first aid. Another problem is the reluctance of soldiers to fire their weapons (or have an excuse for not firing it) because they could acquire no target. Accurate marksmanship is critical during training, because it heightens the soldier's confidence. However, soldiers must understand that when no target is clearly pinpointed, then they must fire suppressive fire in the general area of the enemy fire (using sound as the basis for line of fire) to suppress the enemy, which in turn lessens the enemy's rate of fire and accuracy. Firing automatically upon contact is a positive action that gives the soldier courage and a psychological boost to maintain forward movement. Conversely, firing back on the enemy forces quickly and resolutely may cause them to quit their positions posthaste.

Paradoxically, the problems associated with the attack do not plague the defense. Provided the defense adheres to the security and alert plans, key weapons have range cards, and positions have interlocking fires with sector stakes, its fires even under poor visibility can still inflict heavy casualties on the attacker.

Lastly, the nuclear, biological, and chemical noncommissioned officer (NBC NCO) provides the commander with the effects of weather on the use of chemical weapons and smoke. At a minimum, he must address wind direction, the amount of smoke needed to achieve the commander's desired effect, and the munitions and equipment available. Conversely, he addresses the enemy's use of chemicals in the same manner.

Enemy. Of the enemy forces facing him, the commander will know little. Initial information on the enemy will be general (likely mission and course of action, disposition through doctrinal templates, projected strength and composition). If the enemy does not have an established doctrine for the purpose of a template, the commander must analyze the terrain in the objective area first and determine the most likely places for the enemy to establish a defense or the most likely scheme of maneuver for an attack. Determining the strength of the enemy force, his organization and disposition of forces, his probable fire support within range, and NBC capability (a necessary precaution) are essential for a successful attack. Over time, the company constantly assesses the enemy's pertinent peculiarities; his strengths and weaknesses, to include his command and control system (how quickly can the enemy effectively react to our actions); his weapons and equipment (type and range, night vision equipment, and maintenance discipline); his general discipline, his mobility (this helps us focus on probable avenues of approach and the speed in which the enemy forces can launch an attack or counterattack), and the general efficiency of his combat service support (how long can he sustain an engagement before he is forced to retire). The commander can only guess at the enemy's personnel and equipment strength (the effects of recent combat) and his morale (the effects of recent success or failure; the elite status and training; the belief in his cause). All this information helps the commander determine which courses of action are within the enemy's capabilities and where the commander can take acceptable risk.

Since risk is inherent in all combat operations, the commander must address it in his planning and accept risk based on COAs within the enemy's capabilities. The commander employs economy of force with some of his forces in order to achieve mass against the enemy with the rest of his forces. Risk is the probability of the enemy striking against the economy of force in sufficient force to upset the success of the plan. The commander cannot afford to wait for intelligence from higher HQ to provide more details on the enemy forces before starting. If they arrive, intelligence updates will be late and have significant intelligence gaps. It is better to adjust the plan from higher HQ and the company scouts as the mission progresses than to wait for all the facts to come in.

Troops available. Troops available describes all friendly assets allocated for the mission. The commander task organizes these assets for the mission using the concept of tailoring the force. The commander maintains his own **Unit Manning Roster** (see chapter 10) in order to have a good idea of his platoon or section available strength. This helps him identify which platoon will be the main effort and whether augmentation is required. It also gives him an idea of which weapons are available and who is operating them. The commander and each subordinate leader determine the size, allocated weapons and special equipment, and organization of his unit for each task identified or assigned (explicit or implied) Subordinate leaders provide feedback regarding the sufficiency of the force package as it relates to the task. Regarding size, platoon integrity is not sacrosanct. Subtasks may require a reduced platoon, an augmented platoon, or even a composite force in which the platoon does not form the base.

For example, the organization for an attack generally requires an assault, a security, and a fire support element. The assault element may be composed of an augmented platoon or two reduced platoons for the penetration of the defensive perimeter followed by a reduced platoon for exploitation. The security element generally is composed of the equivalent of one or two squads, perhaps a rifle squad augmented by a portion of the AT section. The fire support element is best used as a composite force with medium and light machine guns, antitank rockets, mortars, and grenade launchers. There are limits to the tailoring of force, though. Squad integrity is maintained as a matter of course.

Lastly, subordinate leaders must be proactive in the preparation of assigned and implied tasks. The commander may assign a platoon as the assault element with no further details. The platoon leader considers that preparing a breach may be required and organizes a breaching team. The squad leader of the breaching team assesses that most obstacles are composed of wire or mines, or both. He organizes his team into smaller elements, such as the wire-cutting team and the mine-probing team. The squad leader does not hesitate to request more support such as smoke for concealment, special equipment (a bangalore torpedo, demolition charges), or specialized personnel (engineer squad), if extensive breaching is foreseen. Thus, the concept of aligning the appropriate force for the specific task flows from top to bottom and back up again.

The commander takes advantage of attachments, particularly tanks. (See figure 5-1.)[3] At every opportunity, the commander should request a tank attachment, because it increases the firepower of the company significantly. For deliberate attacks, the company must protect these assets by scouting ahead, suppressing enemy AT weapons, and guiding the tanks into good firing positions (preferably hull or turret defilade). The tank-infantry team can eliminate practically every enemy position and weapon before it. Of course, the infantry needs to curb the tankers' blitzkrieg mentality and convince them that close armor-infantry teamwork is essential to success. Since tank crews have diminished senses when buttoned up, the infantry acts as the eyes and ears of the tanks—human senses still have an edge over technology in this regard. To supplement communications, the infantry can affix field phones to the rear of tanks in order for all infantrymen to be able to talk directly with the crew during contact. Infantry soldiers should ride on the tanks during the approach march, using the turret as protection. The lead tank should have no more than two or three soldiers located directly behind the turret for the purpose of spotting enemy positions to the front and flanks.

German combat experience with infantry-armored teamwork suggests some tactical insights: In order to ensure shock and surprise, armored vehicles approach the front at night while artillery or mortar fires mask the sound; in other sectors trucks with loud speakers transmitting recordings of tracked movement or tractor movement can deceive the enemy. Infantry riding on top of armored vehicles have the task of alerting the vehicle commander of poor soil conditions, mines, enemy anti-tank guns, and aircraft. If possible, infantry and armored units should form a habitual relationship because this develops mutual trust and support. During combined arms operations, the infantry provides security at all times. In the attack, each armored vehicle remains with or slightly to the rear of the infantry line and engages the enemy. When the infantry assaults, the armored vehicle passes forward in open order to close with the enemy. Infantrymen do not cluster directly behind armored vehicles because armored vehicles usually draw enemy artillery fire. During an advance over open terrain, armor vehicles precede the infantry. In complex terrain, the infantry moves ahead of the armored vehicles. Upon contact, the armored vehicles suppress enemy weapons while the infantry closes with the enemy. An Air Force liaison officer at the battalion TOC or TAC coordinates close air support. Orange panels and smoke indicate friendly units; armored vehicles indicate enemy positions with tracer ammunition or lasers for aerial munitions. Armored vehicles are best held back for counterattacks rather than being incorporated in the front line defenses. Although incorporating them into the defense greatly strengthens defensive sectors and provides a significant boost to morale for the ground troops, the resultant high losses can exceed the benefits.[4]

The tank-infantry team conducts battle drills for enemy contact in order to ensure that the infantry can dismount quickly without fratricide. As a team, the infantry identifies enemy positions and suppresses enemy antitank weapons while the tanks destroy pinpointed enemy strong points, machine gun bunkers, and entrenchments. One infantryman per tank guides the tank into a support by fire (SBF) position, which allows the tank to engage the intended target without being exposed to other fires, and helps the tank commander identify targets using the field phone. This infantryman is located either on top of the tank behind the turret or beneath the forward portion of the tank, where he is protected from shrapnel and enemy fire and guides the tank commander onto the target using the clock method of direction or compass bearing and distance. The forward observer (FO) rides with the platoon leader and uses the tank as a good platform for extending his vision for calling in fires. Engineers strengthen this team by clearing obstacles, destroying hardened fortifications, and strengthening trails and bridges.[5] Engineers can load heavy demolitions on the rear of their assigned tank. While the infantry and tanks suppress enemy positions, the engineers can come forward to reduce obstacles or destroy enemy fortified positions.

With indirect fire support, the infantry-tank-engineer team represents one of the most versatile and powerful forces on the battlefield. During extended halts, the company forms a hedgehog defense, forming two concentric perimeters (see figure 5-1a).

The infantry establishes the outer ring with hasty or fighting positions and listening and observation posts (LP/OPs). The tanks are well camouflaged (nets or foliage) and form the inner perimeter, supplementing the infantry sectors of fire. In this manner, the infantry protects the armor from enemy close assaults and the armor protects the infantry with heavy firepower.[6]

Time available. Time figures prominently in the execution of every plan. The commander develops and manages his time schedule carefully, attempting to gain as

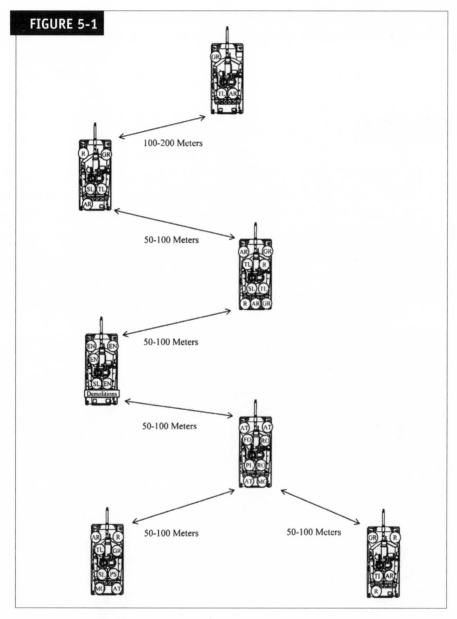

FIGURE 5-1

Infantry-Tank-Engineer Team Mounted Configuration (Platoon-Size)

much slack time as possible to keep him ahead of the time schedule. Subordinate leaders acting proactively during the preparation and planning phase are helpful in this regard. In the attack, however, the greatest problem for the commander is the time expended moving from the line of departure to the objective rally point. The *Land*

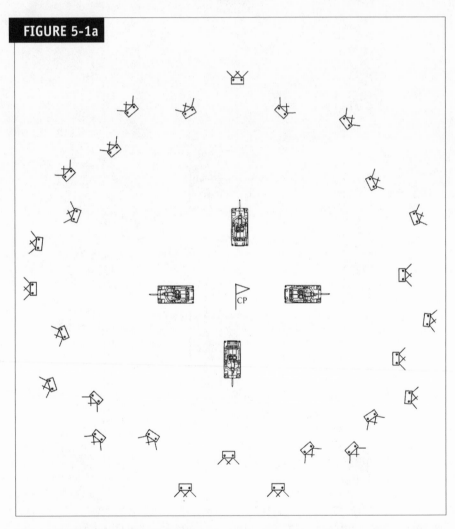

FIGURE 5-1a

Infantry-Tank Team Hasty Defense

Navigation Worksheet is essential because it gives the commander a good estimate of the time expenditure during this phase of the operation and allows him to take actions to ensure that the operation remains on schedule. The commander wants to use all his slack time for the leaders' recon and the movement of the company into position before the assault. For the defense, he will rarely have enough time to complete fortifications before the enemy attack. The **Priority of Work in the Defense** is the best method for managing the effort.

COURSES OF ACTION ANALYSIS

The development and selection of courses of action is the product of METT-T analysis. Ideally, the commander selects two or more courses of action for analysis and comparison. Rarely will the commander have the luxury of time to do this. Often, he chooses one COA for analysis and war gaming. Junior leaders feel uneasy focusing on only one COA, because it contradicts the Armys decision-making process (DMP) doctrine. Such standards are unrealistic at the company (and even battalion) level. Unlike corps and division staffs, which have days to perform the DMP, company commanders (without a formal staff) have only a few hours to develop a plan and prepare for a mission. The commander should not become a slave to the process, but instead should focus on developing an adequate plan. General George Patton's observation of planning and execution remains valid: "A good plan violently executed now is better than a perfect plan next week."[7] The fine-tuning and rehearsal of an average plan will more often result in success on the battlefield. And success is the bottom line.

At the company level, the tactical goal of the defense and attack is to defeat the enemy while avoiding mutual attrition. The use of terrain, firepower, and maneuver are the means to that end. Defeat does not occur instantaneously, so the commander should not delude himself into thinking that the fight will be decided in the opening phase of the engagement. He must analyze the plan through to the transition point (potential branches and sequels). The plan seeks to destroy the enemy forces in detail by fixing their main effort, isolating a portion, and eliminating it. Expecting that the enemy forces are attempting to do likewise, the plan uses terrain, weather, and selected assets as a counter. Because avoiding mutual destruction is a higher goal, the commander may select positions that are not optimum in terms of observation and fields of fire because of enemy capabilities. In all probability, the enemy forces will not remain engaged to the very end. The commander or his soldiers generally realize that the engagement is going against them and will break contact or surrender. The company commander does not consider force ratios or Krasnovian equations. What is the force ratio of a hundred men armed with only rifles against a machine gun bunker with panoramic fields of fire? It is irrelevant. Commanders think in terms of task-organizing specialized teams and weapons to eliminate specific targets and execute specific tasks in pursuit of the mission's purpose. A brief analysis of tactical techniques for the defense and the attack addresses how best to destroy the enemy without suffering mutual attrition.

THE DEFENSE

Disposition of the Defense: Where to place the main line of defense.

Forward-edge dispositions. Placing the main line of defense on the forward edge of a terrain feature (military crest of hill or ridge, woods, urban area, and swamp) permits greater detection and engagement of the attacker. The more open and flat the intervening terrain, the more opportunity available to maximize weapon capabilities. There are several advantages to this disposition. The defender engages the attacker at maximum effective range for each weapon system, eroding the attacker as he advances. The engagement reaches a crescendo as the attacker closes to 300 meters of defending positions. At this point, the attacker is repulsed, because all defensive fires, including the final protective line (FPL) and final protective fires (FPF), and the well-timed counterattack crush him. Unfortunately, such favorable terrain is rare. Generally, most terrain limits observation and fields of fire to 500 meters. This possibility does not negate its use, however. The chief advantage of this defense lies in the exploitation of firepower. If the attacker moves into the engagement areas either in haste or without an alternative avenue of approach, defensive fires can be devastating.

There are other advantages to the forward-edge defense. It permits the defender to shift forces under cover or concealment to the threatened point of the defense during an attack. Logistical activity is likewise protected from enemy interdiction. If supported by the terrain, the defense can establish a defense in depth to greater effect. During the course of the engagement, the defense can withdraw along covered or concealed routes to alternate and supplementary positions. A correct timing of the withdrawal can take the punch out of the attack, placing intervening terrain between the attacker's support by fire position(s) and the assault element. The assault element is forced to advance without the benefit of the fire support element or wait for it to displace forward, thereby losing momentum. I must emphasize that the terrain must support such an option since a withdrawal under pressure is not only difficult but also hazardous.

The risks. A forward-edge disposition is easy for the enemy to detect, degrading the use of surprise. Forward-edged defenses allow the attackers to mass firepower against defensive positions too. Generally the attacker has the edge in such contests. Once detected, the enemy will attempt to circumvent kill zones and strike in the flank or rear. This is another major disadvantage—the attacker retains his mobility when confronting a forward-edge defense and can shift his forces at points of his choosing before becoming decisively engaged. Even if the defense is successful, such actions tend to deteriorate into slugfests with high attrition on both sides.

Applicable techniques. Generally, the company has little leeway in deciding its dispositions. Forward-edge defenses may be the only option if the terrain allows this type of defense or if the mission requires it (part of a bridgehead or lodgment area). To reduce mutual attrition and maintain a degree of surprise, the commander can position the main line of defense about 100 meters inside the forward edge of the tree line or one to two buildings from the edge of a town. Since the wood line and the first row of buildings are common targets for offensive preparatory fires, the company avoids unnecessary casualties and degradation of the main defense. Emplacement of dummy positions and obstacles forward of the main line prompts the enemy to deploy prematurely and to waste firepower. In the case of woods, an overaggressive enemy may find himself in a dilemma as his lead element plunges into the wood line and finds itself decisively engaged at close range while his follow-on forces and support by fire elements are unable to assist because of the masking effects of the terrain and the engaged lead element.

The attacker risks fratricide by calling for fire because of the proximity of both sides. Withdrawal of the lead element requires movement into open terrain, where the defender's fires extend with little degradation. Defending a town is even more devastating, because the attackers' synchronization falls apart rapidly upon entering the initial buildings. The attacker is subject to high casualties clearing these buildings, because obvious points of entry (windows and doors) are mined and within the fields of fire of defensive weapons in echelon, which are activated whenever the enemy soldiers silhouette themselves when passing or opening a window or door.

Except for concealment, interior walls are just as lethal, because they give the enemy soldiers a false sense of security and are no protection from the majority of munitions. Offensive fires tend to rubble buildings, making an advance even more difficult and transforming buildings into redoubts. Properly positioned weapons can fire on the channeled attacker with little chance of detection. Is it any wonder ground forces hate attacking urban areas? Although the defense sacrifices greater weapon ranges with this technique, the payoff is fewer defensive casualties and a good chance of surprising and dealing the enemy a heavy blow.

A corollary disposition is a defense along the topographical crest of a ridge or hill. This not only enjoys the advantages of the forward-slope defense but also allows the defenders to occupy defensive positions, reinforce threatened sectors, and withdraw to supplementary or alternate positions under greater cover and concealment. The major risk is the ease with which the enemy can place indirect fire along the crest once the attacker ascertains this type of disposition. This defense has little depth for the main line of defense, reducing the ability to stagger defensive positions (use of the lazy W) and risks exposing supplementary and alternate positions if the attacker gains a foothold on the crest. The defensive counter is greater dispersion of positions, which may degrade the amount of firepower into the kill zones.

Reverse-edge dispositions. The rear-edge defense is the antithesis of the forward-edge defense. Locating a defense at the rear of a terrain feature (a defile, woods, swamp, town, or slope) means that the defender sacrifices greater weapons range and effectiveness for better protection, concealment, and surprise. By the nature of the terrain, the attacker is naturally channeled along few avenues. Rear-edge defenses are a prudent counter to an enemy that uses mechanization and firepower to great effect. Whether higher command possesses local fire superiority or not is irrelevant to the company receiving enemy fire. After all, the enemy's capability to fire 50 shells or to strafe company positions does not require fire superiority. This is simply a statistical term reflecting a trend, which is of more interest to historians than to the poor soldier living through a bombardment or air attack.

The advantages to a reverse-edge disposition are as follows. Soldiers can prepare their positions and obstacles under cover and concealment. The maintenance of surprise is easier and requires the enemy to devote considerable reconnaissance resources or accept heavier casualties by moving to contact. The defense enhances surprise by incorporating the deception plan—using dummy positions and obstacles along the forward edge. (Refer to the **Defense: Deception Plan**, chapter 19, for details.) To be believable, there must be activity among the dummy positions during the day. Except for the LP/OPs, soldiers withdraw to their main positions at night. There is a good chance that the attacker will have degraded mobility as a result of passing through restricted terrain before making contact with the main line of defense. Once the main body is committed into a defile, deep woods, or an urban area, both command and control and mobility become degraded. Mechanized forces must dismount their infantry to clear the terrain ahead. Likewise, the attacker finds it more difficult to position heavy weapons forward to support the engagement. The defense, however, possesses relatively good firepower and mobility, retaining the capability to shift forces and reinforcements forward quickly.

The risks. The defense has limited depth if the terrain to the rear possesses little cover and concealment. Once the attacker secures a key foothold along the rear edge of the terrain feature, he can control the rear area, intercepting anything moving forward and rearward. Defensive fires into the depths of the terrain feature are degraded, whereas the attacker's fire may extend into the rear of the defense and intercept logistical support. Unless there is a successful counterattack, this type of defense tends to degenerate into battles of attrition, because the attacker can entrench close to the defense and reinforce its forces under cover and concealment. Moreover, the enhancing qualities of the terrain now favor the enemy if the defense launches a counterattack. The defense may also find itself in a dangerous position if forced to withdraw under pressure. Clearly, this type of defense is best used against a heavy mechanized enemy with a deficiency in infantry.

The reverse-slope disposition is distinctive enough from other rear-edge dispositions to warrant a separate discussion. As with the related reverse-edge terrain features, the reverse-slope disposition provides the defense with cover from direct enemy fire. Depending on the degree of slope, this defense also degrades the effectiveness of indirect fire because of the difficulties associated with fire control corrections, the angle of trajectory, and the lay of the land. The goal of this defense is to gain surprise and destroy the attacking force in detail. Defeat of the enemy forces in detail can be accomplished in two ways: first, by ambushing the enemy while he is on the march (movement to contact), attacking a portion of his force as it moves into kill zones, and taking the advantage of the masking effects of the terrain from his follow-on forces; second, duping the enemy into deploying his forces early, allowing the effects of terrain and indirect fire to disrupt the synchronization of his attack so that portions of his force attack at staggered intervals.

The first method is likely to snare only the advance guard, while the main body retains the freedom of maneuver. Although the enemy will still encounter problems in attacking, the element of surprise is lost without significant loss to the attacker. The second method requires an extensive deception plan in order to cause the attacker to deploy the main body prematurely and to exhaust much of his munitions before encountering the main defense. Dummy positions and obstacles on the front slope are a prerequisite. Obvious dummy positions are interposed with well-camouflaged dummy positions. Dummy minefields and wire obstacles are useful in gaining the enemy's attention and perhaps causing him to expend some engineer assets in clearing them.

Enticing the enemy into deploying early enhances the effect of surprise on him. As the enemy assaults and sweeps through the objective, command and control are degraded. Assaulting forces tend to bunch up and lose cohesion as they advance. As the enemy line crests the ridge or hill, the skyline highlights the enemy soldiers, making them easy targets. It is important that the defensive positions along the reverse slope are located far enough down the slope (at least 100 meters) to allow a large portion of the enemy to advance into the kill zones before initiating fire. If the defense is fortunate to have a counter slope, positioned forces should fire on the advancing enemy to encourage him to deploy over the crest in haste. Establishing a trigger line for the main line of defense enforces fire discipline until the proper moment. Once the enemy hits the trigger point, the defense unleashes a storm of fire as in an ambush. Simultaneously, final protective fires along the military and topographical crest isolate the ensnared enemy. The effects are devastating, because the enemy is too close to call for indirect fire and is deprived of his heavier direct fire weapons. If the ensnared enemy breaks and runs, his casualties will steeply rise. Besides surprise, the real strength of this defense is that the enemy loses freedom of maneuver and suffers initial heavy losses without inflicting corresponding losses on the defender. Unless the enemy commander regains control of his unit, his follow-on forces can easily fall victim to other defensive positions as they seek to flank or rescue the ensnared element. Consequently, the enemy risks defeat in detail if he presses the attack.

As with other rear-edge dispositions, the reverse-slope disposition is not without risks. The defender loses any advantages in weapon ranges and the attacker hits the defense intact and in strength. Engagement is decisive and immediate. If the attacker determines before engagement that he is facing a reverse-slope defense, the defense is degraded. The attacker can measure his advance using the masking effects of the intervening terrain and can dispose his forces for the attack at a time and place of his choosing. Observed defensive indirect fire is practically nullified. If the attacker can penetrate the main line of defense, the whole defense could rapidly collapse, because the enemy can make excellent use of the ridge for supporting fires, compromising the defense by placing suppressive fires into the depths and breadth of the defense and suppressing the defenders. In this event, the defender has a difficult time shifting forces to the point of penetration or withdrawing. Because events unfold rapidly with reverse-slope defenses, a successful enemy penetration could buckle the entire battalion's defense before it has a chance to react. When a rear defense works, it is devastating; when it fails, it is catastrophic. In this sense, a reverse-slope defense possesses risks.

Even if the rear-edge defense is partially successful, the remaining enemy force and reinforcements have the opportunity to establish a foothold on the forward edge (woods, environs, slope) as a staging area for the assault or as a sector for positional warfare and attendant high attrition. To preclude this situation, a battalion counterattack is a crucial part of the defense. Without it, this choice of defense makes little sense and accepts risk without the decisive payoff. Since a battalion counterattack force is the prudent means to repulse the attacker, the company commander weighs the availability of such a force before deciding on this type of disposition.

Open-terrain dispositions. There is always the possibility that the company must defend a sector with little or no cover and concealment (bare slope and hills, open fields, desert). In fact, such sectors are not rare. In the twentieth century, several thousand engagements occurred in the open terrain of Flanders, North Africa, the Middle East, the steppes of Russia, and the Persian Gulf. The increasing lethality of weapons makes such terrain more dangerous than ever for the infantryman. As learned in World Wars I and II, no amount of fortification will guarantee protection.

Consequently, open-terrain dispositions require greater dispersion and depth, force protection, and an active defense.[8] The intent is to defend effectively the main line of defense with the minimum number of troops, allowing the remainder to focus on counterreconnaissance and ripostes to enemy attacks. The focus on counterreconnaissance during this type of fight must be to keep the dispositions of the defense from the attacker. Once the attacker has a good idea of the defensive dispositions, he can destroy these dispositions piecemeal.

Open terrain requires the establishment of an outpost sector to supplement the covering force area (CFA) because of the great mobility afforded the attacker and his ability to infiltrate reconnaissance elements into sector. As such, the company may be assigned to defend the outpost sector to the front of the battalion main line of defense. If the battalion elects not to establish an outpost sector, the company should establish one itself as part of the counterreconnaissance effort.

The outpost sector. The main task of the outpost sector is to deny to the enemy any information on defensive dispositions, presenting him with a vacant battlefield at the time of the attack. Located on a forward slope, the outpost sector is within one kilometer of the main line of defense and is assigned to a platoon, which establishes LP/OPs, counterreconnaissance patrols, assault teams, and one or two squad resistance nests. The LP/OPs and sensors detect enemy reconnaissance and alert the platoon CP. One or more assault teams (three to five soldiers with squad automatic weapons,

M203s, grenades, and AT weapon) intercept any detected intruders. Counterreconnaissance patrols routinely patrol the sector for enemy infiltrators.

The squad resistance nest (see figure 5-2) is used for the LP/OPs and counterreconnaissance patrols to rally on if attacked and is ideal for the assault squads to sally from when needed. Resistance nests consist of weapon positions at the apexes with a crew-served weapon bunker in the center, all connected by communications trenches.[9] For greater concealment, the nest should be located behind terrain (knoll or ground swell), and all trenches should be covered with a thin sheet of plywood and a thin layer of dirt. The garrison can push open the cover to fire from any point in the nest. In this sense, the resistance nest fields of fire are directed to the flanks and rear to prevent enemy support by fire position acquisition.[10]

The minefield should only have a single strand of wire to identify its boundaries from a short distance. Fighting positions maintain 50-meter intervals from each other as a measure against enemy indirect fire, and the protective minefield is 35 meters from the nest itself. The central, crew-served weapon (machine gun or AT missile) provides 360 degrees of coverage. If an armored platoon is attached to the company, the commander should attach a tank to the outpost sector platoon to destroy enemy mecha-

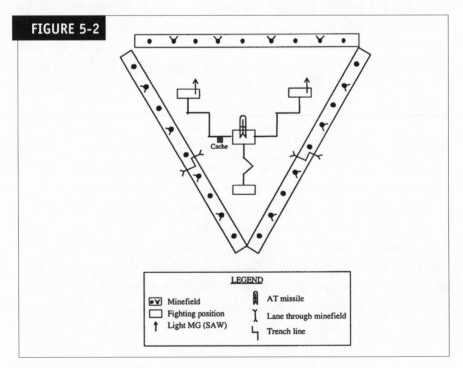

FIGURE 5-2

LEGEND

Minefield   AT missile

Fighting position   Lane through minefield

Light MG (SAW)   Trench line

Squad Resistance Nest

nized vehicles. Before an enemy attack, the outpost zone platoon retires to the main line of resistance. Its crew-served weapons withdraw into predesignated resistance nests, and its squads form assault teams and move to predesignated mobile fight sectors.

The main line of resistance. The main line of resistance (see "Company Open-Terrain Dispositions" figure 5-3) is located on a reverse slope if available and consists of squad resistance nests and independent assault teams. The reverse slope, no matter how slight (as in the case of an intervisibility line), is invaluable in preventing the enemy support-by-fire positions from supporting the attack. Intervals between nests may range from 300 to 600 meters, depending on the width of the defensive sector. Together with the cover and concealment, the dispersion prevents the enemy from massing artillery fires or isolating the point of penetration with smoke. Attempting to smother this defense with fires will prove ineffective and unsustainable. The resistance nests in the outpost sector are turned into dummy positions to draw enemy artillery fire. In this case, dummy weapons are placed in these nests for added realism. To separate the enemy armor and infantry, the forward nests and assault teams allow enemy tanks to pass through while engaging dismounted infantry and infantry fighting vehicles. The resistance nests in depth engage the tanks with flank or rear shots. The assault teams fight a mobile defense from favorable terrain or by sallying out of the nests.

Teams are assigned a sector to defend and conduct flank coordination with all neighboring nests and assault teams. They establish and coordinate withdraw routes through other sectors to mitigate fratricide. If not operating out of a nest, they establish bunkers within their sector. Use of existing bomb and artillery craters to excavate bunkers is effective, because enemy reconnaissance cannot easily identify which craters serve as bunkers. Nests may also incorporate craters as positions for the same reason. Assault teams ambush or attack targets of opportunity and are not tied to any one piece of terrain. They may also form hunter-killer teams for the sole purpose of destroying unsupported tanks and infantry fighting vehicles.

With this type of disposition and array, the enemy is forced to reduce each position during the attack. Each nest is defended at all costs and has a cache of ammunition, water, and food for prolonged resistance. This is a crucial factor in the defense, because it forces the enemy either to ignore the nests, suffering egregious loses, or to attempt to reduce each one, losing its momentum. Judicious use of obstacles within the depths of the sector is designed to help the resistance nests and assault teams stop enemy vehicles long enough to destroy them. Both the local assault team and squad resistance nest leader discuss the best locations for such engagements. Additionally, the forward observers record these microengagement areas in the fire plan.

FIGURE 5-3

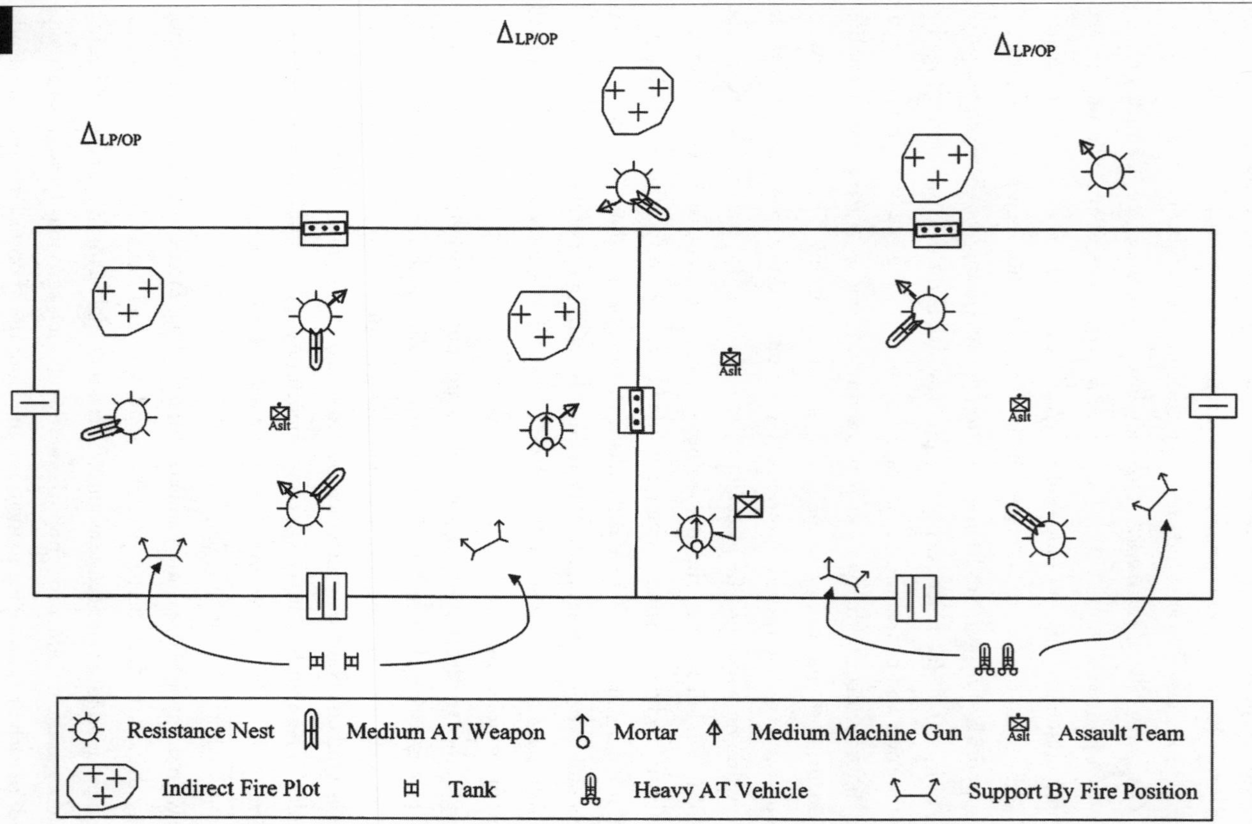

Company Open-Terrain Dispositions

> **TTP**
>
> Avoid isolated small groups of trees or a solitary house for positions, because the enemy forces will likely target these with artillery and air support. It is better to deploy dummy positions in these to induce the enemy forces to waste fire on them.[11]

The climax of the defense lies in the battalion or higher-level counterattack. The resistance of the company as well as the normal friction of the attack (loss of command and control, dispersal of effort, and chaos of battle) accelerates the attacker's culminating point. Counterattack forces come under the command and control of the company commander (regardless of differences in rank). Before the battle, the counterattacking force commander coordinates the route and possible support by fire positions in sector. Locating a liaison officer at the company CP is advisable to help the supporting force maintain contact with the company. Once the commander is alerted that the counterattack is coming into sector, he dispatches guides at the linkup points along the rear boundary to help guide the force into desired locations. Again, the company commander instills in his subordinates the need for each nest to fight on even when isolated and not to lose faith in the counterattack even if the local situation appears bleak. The stability of the resistance nests is the critical factor in this defense.[12]

Dispositions within the depths of restrictive terrain. Disposing forces within the depths of restrictive terrain (jungle, deep forest) offers many advantages, but also many risks. Engagements are characterized as close combat, meaning that both sides are decisively engaged upon contact.

The advantages. The attacker loses any advantage in firepower, because he cannot bring the full complement of heavy weapons within supporting range.

Depending on the depth of the forest, the attacker may not be able to bring the full complement of artillery forward because of the limited clear areas. Whatever indirect fire is available is less accurate, because observation of incoming rounds is difficult to spot for corrections. Close air support is virtually impossible to control, because the heavy foliage prevents identification of friend and foe. Foliage conceals frontline identification markers and dissipates smoke markers. The attacker's advance tends to follow roads and trails because of logistical realities. Even if the attacker maneuvers away from the roads and trails, he must eventually make contact with them again for resupply, reinforcements, and casualty evacuation. Worse, once off established trails, land navigation becomes infinitely more difficult, because the deep forest conceals landmarks.

A global positioning system (GPS) provides a solution to land navigation only for those elements that have it. At night, the ambient light is reduced to such an extent that night attacks are a foolhardy enterprise. Enemy elements that become separated have difficulty rejoining the parent unit. Higher command must divert forces from the front to control these arteries, or it risks having the defender sever them. Adding to this the extreme degradation of command and control, it is not hard to visualize how such a fight degenerates into chaos, resulting in little penny packet fights all along the front instead of a mailed fist.

Against this, the defenders can dig positions with overhead cover, using the dense foliage to conceal them from the attackers' observation even during contact. Fields of fire resemble tunnels no higher than a meter, making it very difficult for the attacker to recognize while walking. Prepositioned artillery and mortars can deliver devastating fires against the attacker, who is subjected to both shrapnel and the secondary effects of rounds splintering the trees, while the defender's overhead cover protects him. The few roads allow the defender to mass his heavy weapons, while forcing the attacking forces to diverge as they attempt to flank the defense. The defender can employ minefields rapidly, using the abundant camouflage of leaves and foliage. Company counter-reconnaissance can pick up the enemy rather early and attempt to force him to deploy prematurely.

Lastly, the defender can dispose a squad or platoon to sever the attacker's lines of communication if he fails to guard them. Severing the lines of communication need not be permanent, only long enough to halt the momentum of the attack and perhaps even cause the attacker to withdraw in despair.[13]

The risks. The company has no choice but to accept decisive engagement without the opportunity to engage the attacker at a distance. Close combat normally results in heavy casualties for both sides. Once he locates the main line of defense, the attacker can certainly infiltrate into the company rear because gaps are difficult to cover. To counter-act the disadvantages associated with restrictive terrain, a perimeter or horseshoe defense provides the most secure defense. The key in defeating infiltrations is time. Infiltrating units must obtain success quickly or be forced to withdraw for lack of resupply. In deep forest or jungles, a clearing can be an effective place in which to set up a strongpoint, although it is easier for the enemy to target with air and indirect fire. Against a lightly armed adversary, a strongpoint in an open clearing is the strongest defense.

The main line of resistance should be within the depths of the forest or jungle, because the edge is a natural target for enemy artillery. A strong outpost line is used to disrupt the enemy advance, to force him to deploy prematurely, and to confuse him as to the location of the main line of resistance. Snipers are in their element in this terrain and can create significant confusion and delays on the attacker. Forward observers can establish positions in trees to observe far-off clearings or avenues of approach. Use of mines and barbed wire interlaced among the trees and covered by fire greatly strengthens the defense. Lastly, a fire plan is essential in this type of fight with artillery targets planned along the flanks and rear also. Since visitors and new personnel are likely to become lost in such terrain, locating company identification signs along the rear boundary is helpful.[14]

Combat in mountains is similar to heavy forest and jungle combat, but ultimately, the advantage is to the specially trained mountain troop. The terrain is so unforgiving of mistakes and the weather so unpredictable that operating in mountainous terrain is hazardous even without enemy contact. Most of the operations will occur for control of the valleys. Valleys are where the logistics must flow. Only the lightly armed mountain infantry can take advantage of the numerous draws and ravines to infiltrate and maneuver around the defenders' positions. Because the attacker cannot sustain a prolonged engagement, it must be able to seize the key objective quickly or be forced to retire. Artillery, tanks, and even mortars are generally out of supporting range. Even though the attacking soldiers can drag smaller-caliber mortars with them, the number of rounds they bring along will be the limiting factor. Ideally, air power should be effective, because the defender is not as well concealed. The defender cannot dig in and must use rock structures as protection, which tend to be very dangerous to the defenders because of the secondary effects of rounds impacting against rocks. In reality, airpower is neutralized by the ability of the defender to cover air avenues of approach with well-hidden air defense missiles and guns. For the defender, the main line of defense should be along the valley walls in depth, because this affords the best protection. Likewise, the attack must progress along the valley sides, because the middle of the valley will be the kill zones.

Combination dispositions. Naturally, the commander can use a combination of dispositions in an attempt to exploit the strengths of terrain while minimizing the risks. To be able to strike the right balance is where the true art of tactics comes to the fore. The successful commanders in history were the ones who could recognize good places for disposition, which exploited the combat power of their forces while mitigating the enemy's.

The Array of the Defense—the Geometry of Positions.
Linear. A linear defense focuses all available firepower to the front and is very powerful when the enemy forces must attack frontally and the frontage has open, flat terrain. It

can be a straight line with platoons on line, a combination of two platoons forward and one back, or vice versa. Squads and positions within the array can be geometrically on line or can form a lazy W to add depth and disperse the pattern. The chief advantage of the linear defense is that it will inflict heavy losses on the attacker should he conduct a frontal attack. The disadvantage is that the linear defense is weak against attacks from the rear or flank. Psychologically, units arrayed in a linear defense are more likely to panic and break when confronted by an attack in the rear. The commander can counter the risks associated with the linear defense by arraying the supplemental and alternate positions in semi or full perimeters depending on the terrain. However, this precaution is only effective if the defense has time to react to any threat from the flank and rear and can withdraw to these positions. If the commander feels that such a threat is likely, he could man some or all of the supplemental positions at night and deploy into the primary positions during stand-to. If he has a squad reserve, this can be placed in a resistance nest to secure the company rear.

Perimeter. The perimeter is practical in circumstances (broken terrain or insufficient manpower) where gaps exist in the battalion defense. The battle position is actually a perimeter defense. The company can form one large perimeter or have its platoons form perimeters within the scheme of the defense. In turn, the platoons can even have their squads form perimeters—this is the basic strength of the seamless web defense. Perimeters are the most secure array inasmuch as they present the enemy with no soft underbelly. Yet, perimeter defenses do dilute the amount of firepower the defensive positions can mass on the attacker—by as much as two-thirds in pure geometric terms. The defender must be alert to enemy efforts to penetrate the perimeter, because this tactic represents the greatest danger to the integrity of the defense. Strongpoints and resistance nests are fortified perimeters. A strongpoint substantially strengthens the company defense and should be located on decisive terrain within the company sector. Before electing to establish resistance nests or a strongpoint, the commander must coordinate engineer assets and construction material, and determine whether there is enough time to construct these fortifications.

Use of Surprise in the Defense. Realistic dummy positions, well disposed and arrayed will often induce the enemy to deploy prematurely and to attack them with heavy weapons—leading to a waste of his munitions and exhaustion of his forces before the actual engagement begins. Such deceptions indirectly strengthen the defense and should be an integral part of every defense.

A surprise use of a weapon or equipment can also disrupt the attackers' rhythm. An effective technique is to use machine guns and M203s in the indirect fire mode, such as from reverse-slope or counter-slope positions. The use of spotlights (or vehicle lights), attached to trees and activated remotely, can disrupt the enemy attack or expose him when he is most vulnerable. All sorts of things are available in a war zone. Don't let the limitations of peacetime training stultify your ingenuity in combat.

The decision to use a hide position can also reap results far greater than the force committed. A small force emerging from and striking hard into the rear of the enemy attack can cause him no small amount of discomfort. If the timing is right, it could cause him to rout.

TTP

Thinking of new ways to employ surprise even in training provides a great lift to esprit de corps and helps establish a reputation for craftiness. Americans admire craftiness, and the commander should promote this type of thinking as long as it does not violate the law of land warfare.

THE ATTACK

Movement to Contact. Movement to contact (MTC) is the most prevalent type of attack, because it requires little preparation, allows swift execution, and provides great flexibility. Virtually every operation (except infiltration) involves a movement where contact is possible.

The purpose of an MTC is to make and maintain contact with the enemy. In such cases, the battlefield situation is fluid, with neither side dominating the intervening terrain. It is associated with offensive operations, but it is just as useful during counterreconnaissance, mop-up operations against enemy infiltration, or counterinsurgency operations. Whatever the mission, the company down through the platoon and squad levels organizes into an MTC, because this is the most flexible offensive action against enemy contact.

The corresponding control measure for an MTC is the direction of attack, normally with intermediate and main march objectives. This means that the company's main task is to clear a specific route for follow-on forces, eliminating obstacles and enemy detachments along the way. Ideally, an attached engineer squad or platoon assists in clearing the road. An implied task is to fix larger enemy forces for elimination by follow-on forces as necessary. Although it will result in control of lines of communication as it progresses, an MTC does not mean that the company walks down the middle of the road. There are tactical techniques that a company takes to clear a road without sacrificing tactical flexibility and security.

Upon receipt of an MTC mission, the commander meets with his company scout leader after his warning order and conducts a map recon. From this, they determine the most likely points where the enemy forces will establish OP, ambush, and delay positions. Company reconnaissance verifies these suspected sites, pinpoints other enemy positions, and reports findings immediately to the commander. The company scouts accompany the XO (or tasked leader) when he coordinates the passage of lines. The scouts should depart immediately after the OPORD, because the company time schedule may be advanced with little warning. If the commander feels comfortable with the scout section's readiness, he should consider letting them go before the OPORD so as to give it as much time as possible to perform a thorough recon. The scouts move along both sides of the assigned route, far enough away to escape enemy observation along the route. At each suspected site, the scouts reconnoiter from the perspective of the enemy's rear, using a selected recon technique (clover leaf, box, and so forth).

The scouts avoid enemy contact no matter how small, because their job is to collect information only. If time permits, the scouts reconnoiter the march objectives and a little beyond (unless limited by a control measure). The scout leader establishes a three-man surveillance team for the final march objective (and the intermediate if occupied) and then moves back toward the line of departure to establish communications with the commander for a recon report. Ideally, the scouts maintain radio communications with the company throughout the recon, but terrain may not support this. If the scout leader cannot gain radio contact, he dispatches one or two teams to report back with size, activity, location, uniform, time, and equipment (SALUTE) reports and sketches to the company in its attack position. Lastly, the scout leader establishes an ambush site along the route and activates it once (not before) the company crosses the line of departure (LD).

The company departs the LD, using the appropriate movement technique and formation. Note that the advance guard and main body do not travel down the road but move parallel to it and close enough to ensure that it is clear (between 10 and 300 meters, depending on the terrain). The commander must still be wary of enemy ambushes and should take the precaution of clearing any incidental ridge, even if it slows the advance. A prudent measure is for the commander to incorporate bounding overwatch in the movement. Even though it slows the rate of advance, overwatch positions provide immediate fire support if contact is made. The trail platoon (or attached engineers) clears the road proper, clearing mines and other obstacles. Lastly, a test vehicle (roller tank or a 2.5-ton truck heavily sandbagged) follows at a distance to test the road for pressure mines.

Enemy locations discovered by the company scouts are dealt with by the company as the commander sees fit—artillery and mortar fires, hasty or deliberate attack, or close air support. Indirect fire is preferred since it provides a good incentive for enemy soldiers to quit their positions before contact with the main body is made.

Actions on enemy contact. (See figure 5-4.) The commander must not allow his recon effort to lull him into a false sense of security. He must assume that the recon will miss some enemy positions or that the enemy may move in after the recon clears an area. Upon contact (with an enemy ground element or sniper), the element in contact immediately responds with fire in the general direction of the enemy fire. Paradoxically, soldiers are not to withhold fire until they pinpoint the enemy position.[14] Their task is to suppress or disrupt the enemy's aimed fire. Even if their fire completely misses the enemy by a relative long shot, he normally will flinch. (It is difficult to gauge the proximity of incoming fire.) Leaders urge everyone to aim low, because the tendency is to aim high while under fire.[16] The worst thing for soldiers to do is to seek cover and attempt to identify the source of fire or seek a definite target. In modern combat, the only indicator of the enemy's proximity is incoming small arms fire. Moreover, a disciplined enemy will take this opportunity to bring indirect and direct fire on the huddled masses, picking them off until what is left of the unit withdraws. Leaders must prompt their soldiers forward before enemy indirect fire starts impacting.[17]

Establishing a high volume of fire immediately gives the soldier a great psychological boost, buoying his spirits and providing a sense of morale superiority, while sapping the morale of the enemy. In this case, the enemy on the defensive, particularly conducting a delay, realizes that the fortunes of war are not in his favor at this moment. The enemy detachment may be exhausted from the weather conditions, hunger, fatigue, and thirst. A heavy fire response may just be the sort of inducement he needs to surrender or run. The leader cannot count on this response, but neither should he discount it, because it reaps such large benefits. A prompt base of fire buoys the spirit and gives the soldier a sense of moral superiority.[18] The act of doing something makes him feel in control and acts as a catalyst for the next action—forward movement.

During this action, leaders shout commands and demand information. Subordinates relay information back from the front and vice versa. Leaders direct activity or request information by unit designators (1st squad or 2d platoon) rather than by indi-

FIGURE 5-4

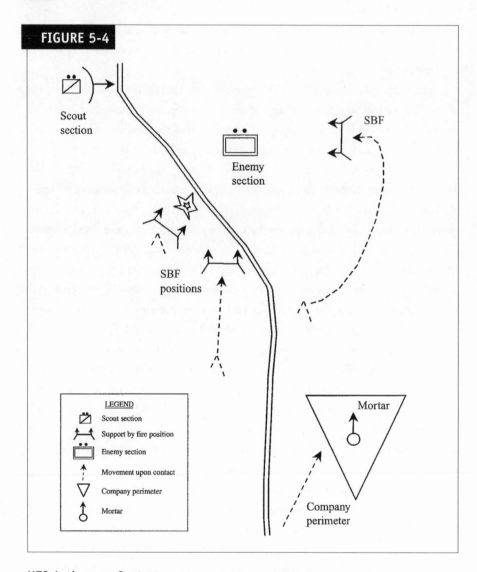

MTC Actions on Contact

vidual names (Lieutenant Huffacker or Staff Sergeant Torres). They may be a casualty, or they may not be within earshot. But normally, there is a soldier from the unit in question nearby who can respond to the question or directive. Subordinates must understand that shouting commands and relaying information does not cause the enemy to home in on the source. The enemy knows there is someone in the area, hence his reason for firing. Generally, the immediate area is a cacophony of sound, so the enemy will concentrate on observable targets and not sound sources. Squad leaders can help direct fires toward a particular target using tracer rounds or a laser device if the noise is too great (this can be part of the SOP) but should exercise some caution, because the enemy may pick up on this control measure. Shouting also buoys the spirits of the soldier. Once the firing begins and soldiers go to ground, the feeling of isolation sets in. Sounds of familiar voices and the requirement to respond to inquiries are reassuring. Experience reveals that talking and shouting not only counteract the feeling of isolation but also strengthen unit cohesion.[19] Hence, the active firing of weapons and personal intercommunication are excellent methods for energizing soldiers in the engagement.

The contact also results in a number of simultaneous actions from those elements not in direct contact. The commander informs the battalion tactical operations center (TOC) or TAC with "Contact, Wait Out," and then goes forward to investigate. The XO or 1SG ensures that the battalion TOC is apprised of the situation in case the commander cannot establish radio communications immediately. If the company scouts are nearby, they establish an ambush or blocking position along the enemy's most likely route of withdrawal. The platoons not in contact establish a hasty perimeter on the best defensible terrain nearby. The mortar section automatically lays in and provides immediate fire support from within the hasty perimeter. The scouts contact battalion mortars to identify targets behind the enemy position to isolate it. The platoon leaders go forward and link up with the commander. Additionally, the trail platoon leader performs a map check for a possible maneuver to the enemy rear or flank, particularly if the scouts are not nearby. If the commander wants this done immediately without consultation, he can direct this over the radio or have it relayed back, using a code word from the OPORD. The commander and platoon leaders decide on a course of action if the lead platoon requires assistance. This may take several minutes before it becomes clear. Needlessly deploying the main body into the fight leads to confusion, may precipitate casualties, and absolutely results in a delay, because the whole unit must consolidate and reorganize. The commander should exercise tactical patience before deploying the main body. Once the commander has sorted out the situation and issued orders, he gives the battalion TOC an abbreviated SALUTE, friendly situation, selected course of action, and whether support is needed at the moment.

If the commander chooses to envelop the enemy position, he takes control of the platoon in contact (the base platoon) and controls its fires as the enveloping force moves (squad or platoon). As he tracks the enveloping force's progress, he directs the base platoon to lift or shift fires. The enveloping force does not move through the enemy position. It establishes a base of fire from a good position until the enemy reacts to it (shifts forces to form new front). Once this occurs, the base platoon in contact moves forward and through the enemy position while the enveloping force suppresses or overwatches.[20] If the enemy does not react to the enveloping force, the base platoon continues to suppress until the enveloping force overruns the enemy. The commander can use his third platoon to reinforce the platoon in contact, the enveloping force, or to secure the perimeter as a reserve force.

If the enemy defense is too strong for the company to dispatch, the commander either reinforces the platoon in contact with the main body platoons, ensuring that the squads are oriented on the flanks and rear, or he pulls the platoon back to the hasty perimeter of the main body. In both cases, the commander apprises the battalion commander of the situation and makes recommendations. If artillery support is requested, the company perimeter should be at least 35 meters from the target. The commander also apprises the scouts of the artillery mission and shifts them laterally from the artillery gun-target line.[21]

Once the enemy is overcome, the company establishes a hasty perimeter triangle, using the assault platoon as the base leg (from 10 to 2 o'clock in the direction of the assault) with the other two platoons forming the other two legs. The commander avoids establishing a prearranged consolidation plan in the OPORD, where each platoon has a designated place. Normally, the engagement is chaotic and the platoons' final dispositions are rarely as planned. It is more effective for the company to form on the assault platoon no matter where it ends up and to conduct reorganization activities from there. Once consolidation and reorganization takes place, a new platoon takes the lead, with the original falling to the rear in the order of march. This policy distributes the stress of danger equitably and helps maintain morale. The commander does not wait long to resume the march (soldiers can conduct most reorganization activities on the march), because the inertia of long halts makes it harder to advance again.

During the MTC, the company trains remain at the attack position and displace forward at each enemy contact site or march objective once secured. The 1SG supervises medical evacuation and personnel accountability, while the supply sergeant exchanges preloaded magazines and filled canteens for empties. The supply sergeant also sweeps the site for discarded magazines, usable munitions and weapons, and equipment that can help sustain the company until it is resupplied by the battalion.

A variant of the MTC is the search and attack mission, normally associated with low-intensity conflict. In this case, company elements reconnoiter assigned areas independently until enemy contact is made or the enemy is discovered (covertly ideally). An effective technique for ensuring sufficient coverage of an area while maintaining a good security posture, particularly in heavy woods or jungle is the cloverleaf pattern. (See figure 5-5.) The platoon divides its area of responsibility into 100-to-400-meter-wide zones. Upon entering the zone, the platoon forms a hasty perimeter. From the perimeter, the platoon leader dispatches fire teams to the flanks and front 50 to 200 meters, depending on the terrain and density of vegetation. The fire teams look for signs of the enemy and report back with their findings. The platoon continues with this pattern until the entire area is covered.[22] The company element, which has made contact or has discovered the enemy without being compromised, contacts the company CP and discloses the details of the situation (a SALUTE format is sufficient). The commander informs higher headquarters, which either dispatches a force to attack the enemy (while the company continues with its mission) or directs the company to dispose of the enemy with its own forces. The company element in contact or providing surveillance dispatches guides to a linkup point for the reinforcements, which then meet with the element leader for a situation report (SITREP). At this point, the highest-ranking leader develops the situation and resolves it in a similar fashion to the conventional MTC.

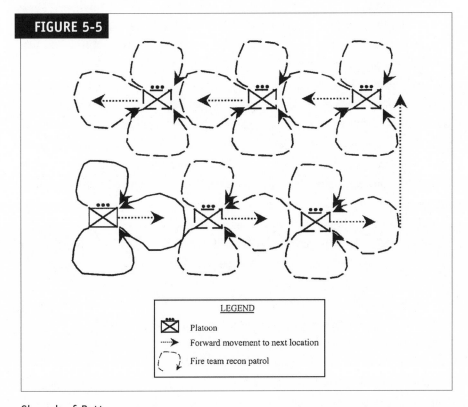

FIGURE 5-5

LEGEND

⊠ Platoon

┄→ Forward movement to next location

Fire team recon patrol

Cloverleaf Pattern

Infiltration. Infiltration is a type of maneuver. It is so difficult and so fraught with danger that the commander must include it as part of the tentative plan rather than delegating it totally as he would the land navigation plan. There are two assumptions required before an infiltration is considered. First, the enemy defense must have identifiable gaps; second, the infiltrating force avoids all enemy contact until it reaches the objective area. Normal objectives are soft targets, such as command posts, logistic units, or key terrain in the enemy rear, although they could culminate with an attack on an artillery position or an enemy defensive position from the rear. Infiltrations are very effective but require great confidence in junior leaders and an acceptance of decentralization. In World War II, the Germans were able to control defensive sectors with fewer soldiers by operating in two- or three-man teams, which infiltrated through Allied lines and kept their forces off balance.[23]

Locating gaps requires extensive reconnaissance and time. Such a mission is beyond the battalion scout platoon capabilities, because it has neither the manpower nor the interest in locating numerous infiltration lanes for each company. It is foolhardy to assign infiltration lanes to companies based on weak or incomplete reconnaissance. It is far better for the battalion to assign a wide zone to the company and allow the company to recon its own infiltration lanes. The company commander, in turn, provides the battalion with its lanes as part of the brief-back.

In this endeavor, the company requires at least two days to pinpoint gaps and ascertain the size of the infiltration force per lane. The larger the infiltration force is, the greater the probability of compromise. Yet, the smaller the infiltration force, the more time required for movement, linkup, and consolidation of the entire company at the ORP. If infiltrating by squads, the commander should figure on a two-hour consolidation period. Ultimately, the company commander determines the number of lanes and the size of each infiltration element based on the alertness of the enemy, the size of the assigned zone, the time available, and the terrain and weather.

Each infiltration element provides the commander with the land navigation plan to determine the time required for movement. Longer terrain-distance lanes require more time for execution.

During movement, the infiltration element avoids all enemy contact. If contact is made, the element attempts to break contact and continues the mission. If contact is decisive, the leader decides to eliminate the enemy element or breaks contact, carrying his casualties with him.

This leads to the acute problem of medical evacuation. Medevac helicopters and vehicles cannot operate behind enemy lines with impunity. Such activity is likely to alert the enemy that the contact was no mere patrol. The very real possibility of the enemy forces capturing or destroying a Medevac helicopter or vehicle during evacuation nearly precludes its use. Higher headquarters balks at using these assets before a major operation, because they will be needed for possible mass casualty evacuation. This problem requires the affected element to exfiltrate urgent casualties to a predesignated casualty collection point near the line of departure or contact or main supply route (MSR) and to continue with the mission.

For such contingencies, each squad needs a portable canvas stretcher that can be rolled up tightly and fit into a rucksack. Routine or non-life-threatening cases can exfiltrate to the casualty collection point with an escort, while the main body continues with the mission. Except for ambulatory cases, a casualty will require two to four bearers. With this manpower burden, it will not take many casualties before the company must abort the mission. It is counterproductive to train otherwise, for in combat, soldiers will not leave their critically wounded comrades in place. This fact alone highlights why it is so critical to avoid enemy contact.

Infiltration missions (except for raids) require immediate success with the objective, or they risk defeat in detail. Ammunition, water, and other critical supplies are not easily or readily replenished. If the friendly line of advance does not penetrate into the enemy rear and link up with the infiltration force in sufficient time (within a day), the enemy has the opportunity to eliminate it with overwhelming force.

For all these reasons, higher command needs to consider the risk-benefit factor before deciding on an infiltration maneuver. If it does not appear that higher command has thought this out, the company commander needs to remind his superiors of the risks.

Movement in Deep Forests/Jungle. As already discussed, movement in deep woods or jungles is very difficult during daylight and virtually impossible at night. Unless artillery can establish a position in a clearing within supporting range, the company will need to rely on the battalion and its own mortars. Even if fire support is established, forward observers will have a hard time adjusting fire, because the foliage makes it difficult to pinpoint where the rounds are landing. Since a proficient enemy will make it difficult for the commander to locate the main line of resistance, placing effective indirect fire on his positions will be even more difficult. Moreover, because of the heights of the trees, artillery will need to fire at a high angle or risk a higher degree of fratricide.[24] Nevertheless, the commander can take steps to overcome or mitigate the disadvantages to secure success.

Since command and control is very difficult to maintain, the commander remains closer to the front. Deploying into a skirmish line (on line) is ill advised, because the terrain splinters the company as it advances and causes company elements to become inter-mingled, making it difficult to control the unit once enemy contact is made. Advancing in company column (ranger file) exposes the unit to ambushes from enemy reconnais-sance patrols and security forces, causing significant delay and casualties, which may compromise the integrity of the attack. A solution is to have platoons and their squads in column. (See figure 5-6.) This formation helps the commander control the advance and maintain sufficient control despite the few trails. Each platoon has an advance guard for the purpose of discovering the enemy before it discovers the main body.

Additionally, a special hunter-killer team (HKT) comprising a squad automatic weapon (SAW) gunner, M203 grenadier, AT gunner, each with a supply of grenades and a demolition charge, should accompany each advanced guard in order to surprise and eliminate enemy armor and bunkers in the enemy outpost line or covering force area.

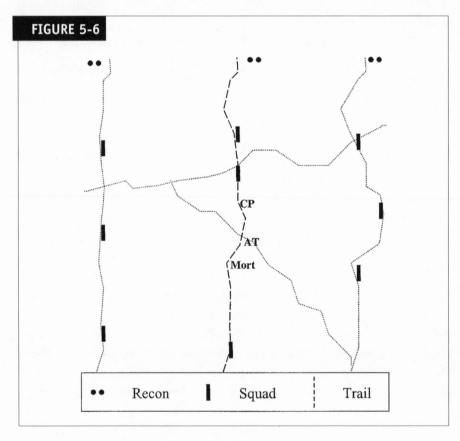

FIGURE 5-6

••	Recon	▮	Squad	┊	Trail

Movement Formation in Forest/Jungle

With such a disposition, the enemy cannot ambush a significant portion of the company. The company maintains the capability to respond to the threat quickly and with a greater chance of inflicting significant losses on the enemy. The commander needs to designate many phase lines on clearly defined linear features (roads, streams, ridges) in order to maintain the integrity of the advance. Experience shows that the enemy will likely deploy machine guns along roads. The best way to cross these danger areas is for each squad to cross them on line. That way, the squad has a better chance of safely crossing a road before an inattentive enemy can respond. If the company is forced to stop for a significant period (nightfall), it should always form a perimeter defense, even if it takes extensive time. During temporary halts, squads should deploy into a cigar-shaped perimeter. Because of the significant advantages the terrain offers to the defenders, everyone should always be vigilant against an attack from any direction.[25]

Each leader reminds his soldiers of the psychological impact of combat in such terrain. Noise is greatly amplified, making the size of the enemy unit seem larger. Soldiers can use this effect to their advantage by their own fires and also by shouting. Soldiers must know that machine gun fire has poor penetration, resulting in close combat with little fire support.[26]

Actions on the Objective: Deliberate Attacks. Actions on the objective begin at the objective rally point. It is here that the objective recon leader and the company leaders tailor the plan to the realities on the ground. The recon leader backbriefs with a detailed sketch of the objective area portraying enemy dispositions and array as well as the best fire support, assault, and security positions. During this time, the company forms into its task organization for the attack. The company does not task organize before this because it needs the flexibility to react to chance enemy contact during movement.

Task organization. The commander task organizes the company so as to accomplish the mission most effectively. Platoon and section integrity are not sacrosanct. The commander task organizes to exploit the leadership, manpower, and weapons capabilities. Nevertheless, the commander attempts to maintain as much integrity as possible, especially squad integrity, if tactically feasible. The generic task organization is as follows.

Security and fire support elements. The security and fire support elements comprise one platoon, the mortar section, and the antitank section. The tasked platoon leader assists the commander with the security and fire support plans, and task organizes both elements in accordance with the tentative plan.

Security element. The primary task of the security element is to isolate the objective area. (See figure 5-7.) During the assault, it engages enemy forces withdrawing from the objective as well as enemy counterattack or reinforcements. To be effective, the security element must have sufficient firepower. Since the security element cannot completely isolate the objective area, security positions cover the prominent avenues of approach. The enemy higher command places its reserve where it can react to threats throughout the sector and not just in our objective area. The reserve is likely to use the fastest route into the objective area because reaction speed is crucial to restoring the defense. The security element comprises one or two squads plus attachments. The senior squad leader is the security leader in charge and maintains radio contact with the platoon leader.

Each security position comprises at least a fire team and has a designator (that is, S-1) for quick identification. The platoon leader may consider placing a squad augmented with an AT team and a forward observer along the most likely enemy avenue of approach. Because of its limited strength, each security team seeks a local choke point along the avenue of approach and prepares an ambush.

The platoon leader provides depth to the security plan by designating alternate security positions (that is, A-1) as part of the security reaction plan and prepares the

FIGURE 5-7

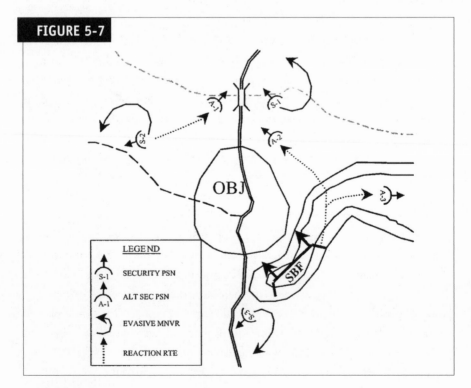

Security Plan

reaction plan—reaction force, alternate positions, routes, and the reaction trigger. The reaction force is composed of selected components of the fire support element and is directed by the platoon leader from security teams not in contact. Upon enemy contact, the squad leader alerts the platoon leader, "Contact, [position designation], out." The platoon leader issues a warning order to the selected reaction force. Once the squad leader develops the situation, he reports back within three minutes. If the platoon leader does not hear from him within that interval, he automatically dispatches the reaction force, which clarifies the situation and reports its findings. If the security element must withdraw from its position because of enemy pressure, it performs an evasive maneuver. It withdraws laterally and away from the objective area, attempting to draw the enemy with it. Failing this, the team maneuvers to harass the enemy from the flank or rear.

The security element also engages enemy units withdrawing from the objective area in order to destroy or induce surrender. This engagement could become the coup de grâce of a successful assault, totally transforming an enemy retreat into a rout. This opens the opportunity for a full-scale pursuit by higher command.

Lastly, the security element provides early warning during consolidation and reorganization. The security element squad leader also deploys three-person recon patrols to scout out the terrain and enemy activity. The commander decides whether the security element performs evasive maneuvers or withdraws to the company perimeter.

Fire support element. The commander and fire support element (FSE) platoon leader consider the best locations that isolate the breaching point and suppress the objective until the assault element penetrates into the depths of the defense. Rarely can one position accomplish both tasks. Moreover, dividing key weapons increases their survivability and flexibility in deployment. With such a division, the platoon leader controls one position while the platoon sergeant controls the other.

The organization of the FSE is much more than a line of machine guns suppressing the objective for 30 seconds, though this is often the case. Controlling the majority of the company's firepower, the fire support platoon leader focuses on destroying key enemy weapon systems rather than mere suppression. The distinction is subtle, for the goal of destruction automatically leads to the suppression of the enemy. Mere suppression of an area is only effective for the duration of fires.

The FSE leader selects alternate positions for the element to shift to in order to provide continuous support to the assault element as it advances through the objective. In such cases, half the element shifts while the other continues firing, creating a leapfrog effect.

> **TTP**
>
> To demoralize and confuse the enemy as to the true direction of the assault, the soldiers in the fire support element should shout out battle cries.[27]

The hunter-killer team is key to the fire support element. The platoon leader determines the number and composition of HKTs needed to accomplish the task. By inculcating a hunter psychology in the HKTs, key enemy positions are targeted, resulting in a degradation of the enemy's resistance. The destruction of key positions lowers the enemy's morale and causes individual soldiers to withhold fire for fear of being targeted. An effective weapon composition is a SAW, M203, and AT launcher with a fire team leader in charge. The HKT operates semiautonomously by maneuvering within the fire support position. The HKT does not open fire with the rest of the fire support element; rather, it seeks out targets of opportunity (machine gun positions, armored vehicles, key leaders) and engages them. Once the HKT destroys one target, it automatically deploys to another position so as to avoid enemy counterfires. HKTs may also be employed as the company feint, the security reaction force, local security to the fire support element, or the company reserve as determined by the commander.

In certain situations, a good fire support position may not be available, or the company objective must be divided into separate platoon-sized objectives. In such cases, the commander may opt to delegate fire support to each platoon or the main effort. The commander may decide to have the AT section responsible for security with perhaps a squad augmentation (if there is no armor threat) and place the mortar section under his direct control for possible use in the direct fire mode.

The commander should always seek to have at least one tank attached to supplement the fire support element, even if this means a linkup after the attack has begun because of the use of surprise. The linkup point and guide from the FSE is identified in the OPORD. Upon linkup, the FSE guide leads the tank to a forward position and informs the FSE leader of its arrival. One tank can completely destroy enemy bunkers and pillboxes very quickly, but it needs the infantrymen to protect and guide it into a good, protected position—ideally hull or turret defilade. Howitzers can be used in a similar manner with devastating results, if protected. The commander should never dismiss the chance to employ heavier firepower, if available.

Assault element. While the fire support and security elements set the conditions, ultimate success depends on the assault element. The manner in which the company assaults the objective depends on the enemy disposition and array. Guided by the principle of divide and conquer, assault tactics focus on defeating the enemy in detail. The intent is to defeat bite-size portions of the enemy defense and create a snowball effect as the momentum of the assault overwhelms the entire defense. The commander must accept the fact that the engagement is chaotic, making command and control an illusion. The piecemeal destruction of the enemy's constituent parts devolves to squads and fire teams. To ensure that the effort remains focused, the commander establishes control measures to prevent overextension of the assault element, making it vulnerable to enemy countermeasures.

If the enemy has pillboxes or fortified bunkers, the assault element does not attack these head-on. The FSE and local support-by-fire position suppress these strongpoints, focusing on the apertures, while the assault element focuses on destroying the local enemy positions on each side or the trenches to the rear. Once the bunker or pillbox is bypassed, the assault element destroys or breaches the bunker from the rear with demolitions and followed immediately by hand grenades. Enemy soldiers who emerge are engaged or taken prisoner.[28]

An added impetus to the power of the assault is the use of fixed bayonets. The display of bayonets is not intended to kill the enemy, but to create fear, to cause him to hesitate before reacting, and to induce him to surrender or flee. The commander wants to use every device possible to cause the enemy to give up, and it is a mistake not to use the bayonet simply because its psychological impact is intangible.[29] Moreover, the assault element should sound off with a battle cry once it has penetrated the enemy positions. This not only strikes fear into the enemy but also encourages the attacker. To reiterate, the use of the bayonet and war cry is put in use after the assault element has penetrated into the defense proper and not before. Soldiers do not fight to the death; they make a decision singly, and then collectively to quit the fight once they perceive that the battle is lost.[30] The bayonet and battle cry help them make that decision.

Disposition of the attack: Where to conduct the assault.

The rear assault. The rear of the enemy defense is the desired point of assault. Even if the enemy has established a perimeter defense, the focus of firepower, obstacles, and entrenchments are oriented to his front because that is where the greatest danger lies. If the assault successfully penetrates the enemy defense before he can reorient, the initial blow is often fatal, because the attacker overruns the defender's trains, command post, and mortar position, as well as his alternate and supplementary positions, depriving him of the ability to conduct a controlled and sustained fight. His position soon becomes untenable, forcing him to fight at a disadvantage or yield the position.

The commander has the choice of establishing the fire support position along the enemy's flank (exploiting the effects of enfilade fire) or from the rear on the flank of the rear assault (a fire support position along the enemy's front is generally not worth the risk of fratricide). As it advances, the assault element is likely to mask the fires of the rear fire support position. The FSE remedies this probability by shifting fire or by displacing by echelons as the assault element advances.

The psychological impact on the enemy is greater than a frontal attack even if the rear attack is weaker in firepower. Whenever the defender's lines of communication are severed or threatened, he suffers a loss of morale. With disciplined troops, this plight can be weathered, but troops of a lower caliber are more apt to break and flee.

This maneuver is not without risk, though. The enemy disposition may not have gaps that allow us to maneuver to his rear; time may not be available to travel the additional distance for a rear attack. The surrounding terrain may not support a rear attack. A fire support position may not be available on the enemy's flank or rear to support such a maneuver, though this is not a dominant factor for deciding on a rear assault. A cover or concealed approach to the assault position (line of deployment) may be lacking. If the enemy has established a defense in depth, we may not be able to attack one position without being targeted by defensive mutually supporting fire from another position (that is, the counter slope). If the defense is not quickly overwhelmed, the attacker may find himself attacked from the rear by enemy reinforcements. Casualty evacuation, resupply, and prisoner evacuation are also more difficult to manage. Lastly, the use of indirect fire is limited (depending on the artillery target line) because of the possibility of long rounds impacting into the assault element.

The flank assault. A flank attack seeks to roll up the enemy defense. There is more opportunity to employ an effective fire support position either from the enemy's front, rear, oblique, or flank. A flank assault will quickly mask the fires from a flanking fire support position, but shifting fires along the assault element's flank increases its security. The front or rear fire support position allows continuous suppression for the assault element by shifting fires and displacing fire support teams laterally as the assault element advances.

Generally, this maneuver incurs few risks. Enemy positions in depth could place enfilade fire on the assault element's flank. The defense may employ flank protective measures, such as obstacles or refusing each flank, providing the opportunity to reorient fires on the assault element. Even so, the flank assault provides the attacker with enough flexibility to counter such measures without incurring prohibitive casualties.

The frontal assault. The attacker tries to avoid the frontal assault, because this requires an assault against the strongest part of defense, thereby increasing casualties. Such an assault may be unavoidable, because the enemy defense has no gaps (perimeter, fortified zone, or bridgehead), or time does not permit an envelopment (movement to contact, synchronization of battalion's scheme of maneuver). This does not mean that frontal assaults are impossible to win. Many of the major operations of World War II were frontal assaults—the Normandy invasion, expansion, and breakout; the Pacific Island assaults; and Italy. The defenses were formidable and attacks appeared suicidal, but they were successful as a result of good planning and better execution. Because of the magnitude of difficulty, the frontal assault requires a phased execution.

Breach phase. The greatest danger during breach operations is the unnecessary massing of troops directly in front of the breach point. (See figure 5-8.) This exposes those not directly involved in the breach to direct and indirect fire. Generally, a squad is the largest entity needed to make each breach. The rest of the assault element is kept in the assault position: rearward, laterally away, and under cover.

The assault element NCO platoon sergeant (PSG) manages the forward deployment of the assault element through the breach (or replacement of the breach squad if it becomes combat ineffective). The assault element leader (PLT LDR) is forward positioned, where he can observe the progress of the breach, give orders to the assault element NCO, and relay intelligence to the commander and fire support leader.

FIGURE 5-8

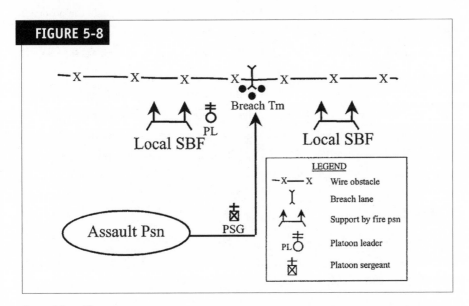

Breaching Phase

During initial planning, the commander decides whether to attempt a silent breach or not. Each has advantages that the commander must consider whenever planning any attack.

The silent breach seeks surprise and has the added benefit of conserving ammunition. The breach squad attempts to clear a lane as deep as possible before the defense brings it under fire. If successful, the assault element can penetrate into the depths of the defense and begin the attack at a tremendous advantage. The attacker cannot expect the defender to remain unaware of a silent breach for long and must inculcate in the fire support and assault elements the need for immediate fire support and assault through the breach once the enemy reacts. Once the breach becomes active, the assault element must be positioned to suppress local enemy positions, obscure the breach point with smoke, and secure the breach point with a bridgehead.[31]

> **TTP**
>
> To dampen the sound of cutting wire, each strand is wrapped with cloth or thick tape at the point of incision. The wire must also be secured by cloth strips or rope along both sides of the incision to prevent the wire (under tension) from snapping back.

The active breach allows the breach squad to work faster but may compromise surprise. Harassing and preparatory fires from artillery and mortars degrade the obstacles by cutting through wire and detonating mines. They also help to mask the work of the breaching team but alert the defenders that an attack is imminent. Such fires tend to suppress the defense but endanger the breach squad in the process. Because of their proximity, ease of control, and accuracy, the company mortars are ideal for masking and suppression near the breach point. Higher-caliber artillery or mortar fire is safer on the more distance sections of the defense. The attacker can still maintain the element of surprise if he initiates harassing fires early (24–48 hrs), thereby lulling the defenders into a routine.

> **TTP**
>
> The commander can request aircraft and artillery to place preparatory fires in and around the objective area for the purpose of creating craters, which provide opportunities of cover for the assault force.

Using the fire support element to suppress the defenders and simultaneously pro-tect the breaching team is generally not effective. The proximity and orientation of the FSE tends to draw enemy attention toward the locality of the breach, and the fire support position is normally too far away to provide effective suppression of positions closest to the breach point. Unless the FSE has sufficient ammunition, it will exhaust its supply before the breach and penetration phases are complete—depriving the assault of suppressive fires during the critical exploitation phase.

To compensate, the breach squad needs local fire support to help it isolate the breach point and is therefore organized into a breach team and a local fire support team. The platoon leader is located with the fire support team for command and control. The squad leader decides whether to split the fire support team on both sides of the breach point or to keep it consolidated. The FSE comprises one fire team leader, two SAWs, and two grenadiers. These personnel are additionally armed with light antitank weapons (LAW, AT-4), fragmentation and smoke grenades, chemlights, engineer tape, demolition charges, extra mine markers, and night observation devices (NODs). The breach team consists of one fire team leader, the breacher, and two assistants. The as-sistants check the obstacle for booby traps or warning devices and secure the wire on either side of the incision (to prevent the wire from snapping back), while the breacher prepares the wire for cutting (cloth or tape to dampen the sound). The breach team car-ries wire cutters, bolt cutters (for heavy gauge cable), a mine probe, engineer tape stripes or chemlights to mark the breach, gloves, night vision goggles (NVGs), and mine mark-ers. In the meantime, the platoon sergeant supervises the collection and assembly of demolition charges or bangalore torpedoes at the assault position. The platoon leader calls for these if needed. Because the bangalore is heavy and unwieldy, a selected team brings an assembled one forward for priming by the breacher and helps push it through the obstacle. The platoon sergeant selects other teams to use demolition charges against specific obstacles when called for. If the breach team must conduct an active breach with demolition charges as a result of discovery or need for speed, the squad leader will feed these forward from the fire support team. Once through the wire, the breacher begins probing for and marking of mines. The assistants mark the breach with engineer tape or directional chemlights.

Once through the obstacle belt, the breacher signals the squad leader to pass through the assault element (flashlight or radio). The squad leader, in turn, relays the signal through the platoon leader to the platoon sergeant to move the assault element through. The platoon sergeant sends one squad at a time through the breach to pre-clude congestion at the breach point. The last squad in the assault element picks up the assembled demolition charges (but not the bangalore torpedoes) for use against obstacles in depth. Once the assault element clears the breach, the breach team and fire support team occupies the breach perimeter to maintain security of the breach point.

Penetration phase. Once called to move forward, the lead assault squad moves 50–100 meters past the breach point and establishes the breach perimeter for the entire assault element. (See figure 5-9.) Once the entire assault element occupies the perimeter, the initial squad advances into the enemy defense to locate and destroy the enemy command post and mortar position. If the assault element breaches the enemy defense without detection, the lead squad should move directly to seek the enemy CP, giving the second squad the task of establishing the perimeter. The establishment of a perimeter is a risk, particularly if the enemy has detected the breach. The assault element loses time and the effect of surprise while the enemy reinforces the threatened sector and targets the assault element with direct and indirect fire. Nevertheless, such a technique is prudent. It serves as a base to reorganize the assault element before exploitation, to engage local enemy counterattacks or reinforcements, and as a no-fire area (NFA) for friendly direct and indirect fires.

FIGURE 5-9

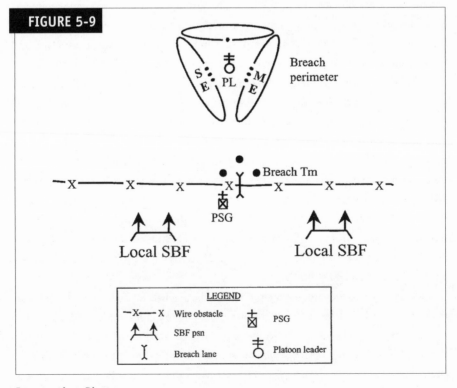

Penetration Phase

Exploitation phase. As the rest of the assault element closes with and destroys the enemy positions, the breach team and part of the local SBF occupy the breach perimeter. (See figure 5-10.) Against a perimeter defense, the assault element has the choice of rolling up the perimeter from one or both flanks at the breach point or of attacking throughout the perimeter simultaneously like an exploding bomb. The first technique is more methodical, entails less risk, and exploits the use of the fire support element; the second relies on shock and speed, and requires well-trained and experienced soldiers, because the probability of fratricide increases, and precludes the use of an FSE. Against a linear defense, it is best to roll up the flank in the direction of the ultimate objective,

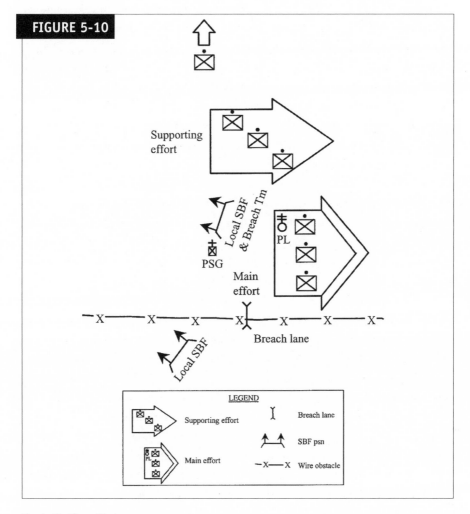

FIGURE 5-10

Exploitation Phase

using the breach team and local SBF to protect the assault element's rear. Until the objective is secure, the flow of reinforcements, casualty evacuation, evacuation of enemy prisoners of war, and supplies pass through the breach point.

Assault Formation (Arraying the Attack). How the assault element attacks is as important as where it attacks. Yet, the attack formation receives little attention in thought and practice. Adoption of a specific assault formation is largely dependent on the level of the enemy's defense and his determination to fight.

Linear assault (on line). This is the most prevalent formation and the least effective. It envisions soldiers organized on line with three- to five-meter intervals, sweeping the objective in one fell swoop. Generally, this formation works against weak resistance and few or no obstacles. It grants the commander the greatest degree of control but is very inflexible. Once a portion of the assault is delayed, the remainder must stop until the delay is overcome or disorganization of the formation is risked. The formation, not designed to overcome unexpected resistance or obstacles, quickly degenerates into a melee once enemy contact is made, results in dissolution of effort, and results in the loss of control of the overall attack by the commander. Nevertheless, if the objective area is small and enemy resistance is light, this is a perfectly viable formation.

Zone assault (nonlinear). Akin to World War I German storm troop tactics, the commander assigns zones of responsibility (main and supporting) for each platoon or squad, if applicable. This envisions that squads or fire teams advance as separate entities but generally on line with the main body until contact is made. Responsibility devolves to the squad or team leader in contact. The squad or leader determines whether additional support is required from platoon or higher and generally the degree of support. The platoon leader and company commander prioritize and determine the amount of support for each element or whether the pertinent squad should fix the point of resistance in order for subsequent squads to bypass it and thrust to the rear. At the company level, the objective is normally small enough to allow the commander and platoon leaders this amount of control. For the commander, his experience and position assure that the squad in need receives the swiftest and most effective support available. Each squad or team advances and eliminates enemy resistance within its power. The commander can provide extra impetus to the main assault element by holding a large reserve and committing it when needed. The commander accepts that the engagement will be chaotic and that control is tenuous at best. With this understanding, it is best to let the subordinate units fight the battle and provide leadership and fire support throughout. The commander can best retain control by establishing a limit of advance for each objective in order to preclude subelements from out-pacing the main body and to allow for consolidation and reorganization (even briefly). This keeps the assault focused and allows the fire support element to displace forward to a new position.

Use of Surprise in the Attack. The enemy may expect an attack but cannot determine the time and location. Attacks one hour before dawn are so commonplace that the defense is more alert then than at any other time. What if the company attacked during breakfast or dinner or even in the afternoon? Identifying the enemy's routine allows leaders to fix those times of the day when he is most and least alert. The commander needs to weigh the benefit of surprise with the risk of a daylight attack. If this is the case, an early evening attack may be correct.

As already discussed, an attack on the enemy rear or flank certainly causes alarm among the enemy defenders. The degree of surprise depends on how prepared the enemy is for such a contingency. Regardless, the enemy will devote the majority of his firepower and effort to his front.

A feint or demonstration magnifies the surprise of an attack elsewhere. A feint serves to catch the attention of the enemy, perhaps even causing him to reinforce the threatened area. The enemy may be distracted enough by the noise associated with the feint that the assault element may penetrate the defense unawares. The enemy may even be slow in reacting to the real threat as a result of uncertainty. The feint can keep the defenders transfixed by withdrawing and attacking from alternate positions successively away from the main point of attack.

SELECTING A COURSE OF ACTION

Time does not often permit the commander to develop two to three courses of action. Generally, he has a basic plan already formulated in his head or dictated by battalion headquarters. The process of developing three courses of action is legitimate at brigade and higher levels of command, because operations become more complex as a result of the involvement of more battlefield operating systems and the existence of a staff; but such a process is impractical at the company level. It is better for the commander to develop a tentative plan and modify it as more information comes in and in reaction to feedback from back-briefs, war-gaming, and rehearsals. War-gaming the course of action using the sand table (or at least a decisive point sketch) is the most practical, allowing the commander to visualize how the operation will unfold and what resources will need to be synchronized at each critical event. By having the XO portray a thinking enemy, the commander can focus on how to defeat the enemy rather than trying to implement a plan that may no longer be proper. This is what soldiers mean when they say, "Fight the enemy and not the plan." One of the most important aspects of the war game involves understanding the amount of force needed to defeat the enemy. If the XO makes a move that thwarts the plan, the commander must weigh the probability of the enemy making such a move and what resources the company will need to defeat it. The commander should provide this feedback to the battalion commander if additional resources are needed. At the very least, the battalion commander is now aware of the company commander's issue for consideration.

TTP

With a sand table, each side can represent their task unit composition with unit and weapon symbols on paper or cardboard. Cardboard pieces cut from index cards and glued to toothpicks are perfect for unit identification, maneuvering of the pieces, and stability from the wind. They are made durable and are weatherproofed by being coated with transparent tape, acetate, or transparent glue.

As a final part of the plan, the commander includes actions taken should the plan fail. Generally, this entails withdrawing to the ORP (in the attack) or the overrun rally point (in the defense) for consolidation and reorganization. There is nothing defeatist in planning for such contingencies. On the contrary, subordinates appreciate that the commander has the forethought to plan for such contingencies and will not be as demoralized once these plans are executed because it means that the company is regrouping for further action.

Sand Table Preparation. Normally a sand table of the decisive point is all that is needed. A sand table of the route is not worth the effort unless the unit is performing a movement to contact. The company RTOs begin preparation once the commander identifies the location of the OPORD. The commander gives the RTOs the suspense for completion, bearing in mind that he will need it for war-gaming. Preparation is as follows:

- Prepare the sand table platform. Excavate a 2 X 2 meter area, 12–18 inches deep. Move excavated soil to one side for use as prominent terrain.
- Make a reference sketch by placing a sheet of clear acetate (or clear plastic) over the decisive point on the map and tracing the grid lines and prominentterrain features (streams, roads, ridges/hills, and so forth). To portray the proper proportions of ridges and hills, trace the major (dark brown) contour lines (every 100 meter). This gives a quick and accurate reference.

- Affix grid lines (black yarn) along the edge of the sand table, using stakes. Place grid reference designators beside each stake.
- Build ridges and hills, using the grid lines as reference. To portray correct height, average the difference between the high and low points in the area of operation (AO). Everything above this height protrudes above the plain of the ground; everything lower than this interval is below the plain.
- Add the other features as well as control measures from the sand table kit.

Sand Table Kit

- Blue yarn for water features
- Red yarn for roads
- Black yarn for grid lines
- Green yarn for woods
- White yarn for control measures (LD, ORP, OBJ, and so forth). Cardboard markers for ID information (grid numbers, control measures), enemy and friendly unit symbols, and key weapon symbols. Gluing each marker to a staff (toothpick) allows you to secure the marker so that it is readable and wind resistant. Waterproofing the markers with rubber glue, scotch tape, or polyurethane makes it durable.
- Scissors
- Tape
- Extra 3x5 cards
- Matchboxes for urban terrain

START MOVEMENT

The commander's decision to move the unit forward is dependent on time and guidance by battalion headquarters. Movement forward may take place before or after the OPORD. For the attack, the advantages to moving after the OPORD are numerous. All on-site preparations (sand table, overlays) and coordination are accomplished faster and with minimum disruption. Other preparations such as zeroing weapons and NODs, inspections, and rehearsals take place in a relatively secure area.

The advantages to moving forward before the OPORD are no less important when time is a factor. Time may be so critical, the terrain and weather so difficult to navigate, and the distance to the line of departure so great, that the commander has no choice but to position the company forward. In such a case, the commander moves forward with his planning staff to the attack position to begin planning. The designated leader brings the company forward to the attack position as outlined in the warning order after the company has completed critical preparatory tasks (time driven). Other leaders performing coordination meet at the attack position in time for the OPORD.

For the defense, the commander should move the company near its defensive sector as quickly as practical, because time is normally of the essence. Giving the OPORD on or near the area to be defended assists in mission analysis, because it gives everyone a better feel for the terrain.

ISSUE OPORD

The commander needs to keep the OPORD as brief and visual as possible. Use of overlays and sketches, essential information printed on cards (signal information, task organization, timetables), and a sand table reduces time and ambiguity. The commander strives to avoid redundancy. If information is depicted on the overlay or sand table, he has no need to cover it in detail during the briefing. Each tasked subordinate back-briefs his task following the OPORD format. The XO determines the order of briefing. Each OPORD or annex briefer distributes paper copies or cards of detailed information such as pertinent grid coordinates (if not already provided on the overlay or sand table) and signal information (call signs, frequencies) for each leader. If this is not feasible, then each briefer places the information on a placard for all to see and copy at their leisure, eliminating the need to dictate routine information. Remember, the most important parts are the plan and its execution. The OPORD should be limited to an hour or less.

The XO briefs paragraph 1, Situation, as it applies to the company mission. He briefs only those enemy forces pertinent to the objective area. The commander briefs the task organization, mission, and portions of the execution: commander's intent, tentative plan, movement instructions (not the route); subunit instructions; reorganization and consolidation, coordinating instructions, and the command portion of command and signal. The platoon leader tasked with route preparation briefs the route plan. The FIST officer briefs the fire support plan. Tasked leaders brief their respective annexes. The 1SG or supply sergeant briefs service support. The commo NCO briefs the signal portion of command and signal. The majority of command and signal is SOP. Code words and signals need not change from mission to mission. The probability of the enemy intercepting and exploiting company signals is remote. Primary instructions are given by code word via radio or orally, if practical. Signals (pyrotechnics or weapons fire) are the alternate method for relaying instructions.

The participants of the company OPORD comprise platoon leaders, platoon sergeants, section leaders, all RTOs, attached leaders, and recon element leaders.

CONDUCT RECONNAISSANCE

The process of reconnaissance begins with the commander conducting a map recon during development of the tentative plan. Reconnaissance is further refined with the development of the sand table. However, there is no substitute for physical reconnaissance by patrols. Only the company reconnaissance can provide the specifics on the enemy in the company area of operations quickly and accurately. No other unit or technological wonder will have the required interest or the capability to provide the unit with the intelligence it needs to succeed.[32] The objective and route reconnaissance elements assist in the completion of the plan and the quick movement of the unit to the objective.

THE OBJECTIVE RECON ELEMENT (OFFENSE)

Both the OBJ and route recon elements depart after the OPORD. They take no overlays or OPORD notes with them. They make marks on their map as a memory device, but nothing obvious. The OBJ recon has specific tasks by priority:

- Reconnoiter as small teams completely around the objective area. Determine trace of main line of defense, location of obstacles by type, and location of crew-served weapons, command posts, and LP/OPs. Recon draws a sketch of the objective in the same fashion as a sector sketch and notes enemy patterns of behavior—if and when he dispatches patrols, how and when he conducts field feeding, his level of discipline and alertness, reactions to contact, and his level of morale.
- Recon assault position or line of deployment, fire support, and security positions.
- Recon and secure ORP and linkup point. Mark linkup point, ensuring that it is no closer than one kilometer from OBJ. Make physical contact with route recon element at linkup point in order to pass a sketch of the OBJ for immediate delivery to the commander.

If the enemy habitually conducts counterreconnaissance, the commander may strengthen the OBJ recon element to a recon-in-force and give it the additional task of eliminating or degrading the enemy counterreconnaissance effort. The effect is more psychological than physical. If the recon-in-force can force the enemy to withdraw his counterreconnaissance into the safety of the defenses, it sets favorable conditions for the attack. By yielding his outpost sector to the attacker, the defender forfeits the ability to see and a degree of moral superiority. The recon-in-force acts aggressively when it encounters the enemy. Realistically though, even a recon-in-force cannot afford to become decisively engaged if the defender is determined to fight it out, possessing the advantage of proximity to his base. The recon leader must be able to read the enemy forces astutely. But he must act aggressively. This willingness to engage may cause a larger enemy force to break contact and retire. If the recon-in-force acts timidly, the enemy will react more boldly. The recon leader just has to learn how to read a situation.

The OBJ recon destroys or forces the withdrawal of any LP/OP it encounters. The recon then withdraws to observe the defender's reaction. Failure of the defender to dispatch a reaction force may indicate a lack of resolve or organization. Before the attack, the commander may direct the recon to probe the enemy defenses with recon by fire in order to pinpoint crew-served weapons. The recon is careful not to become decisively engaged by conducting the probe as a recon by fire. If the enemy lines are within friendly mortar range, the recon leader plots target reference points on his withdraw route in order to help it break contact if the enemy tries to pursue it.[33]

Once the OBJ recon accomplishes its tasks, the recon leader leaves a surveillance team to monitor the objective while the main body moves to the ORP. After securing the ORP, the leader identifies guides to lead the assault, fire support, and security elements into position. Next, he prepares the ORP for occupation, using a triangular configuration for night occupations and marking the apexes as visual guidance. Lastly, with a security team, he departs for the linkup point to make contact with the company and guide it to the ORP. Once the company occupies the ORP, the recon leader updates the commander and subordinate leaders, referring to the OBJ sketch. The commander takes this opportunity to make final adjustments to the plan.

Higher command may be reluctant to allow each company to dispatch recon elements for fear of alerting the enemy. The fear that surprise will be compromised does not negate the need to uncover the enemy dispositions. The use of reconnaissance is no surprise to the defender. He knows that an attack is coming, but not the exact time and place. Regardless, it is better to lose a little surprise if the alternative means stumbling around the battlefield in a panic because the company cannot find the objective. If the operation is a battalion-level attack and the battalion TO wants the battalion scouts to conduct the recon, the company requests to have its recon element accompany them, because it can recon the company positions and disseminate battlefield intelligence quickly to the company.

Recon elements do not have the benefit of preparation and rehearsal that the rest of the company has. To compensate, the recon element comprises the security element and perhaps a portion of the fire support element. With this in mind, the recon element has the opportunity to recon its security positions in detail and is in a position to occupy them swiftly once the company arrives at the ORP.

There is also the danger of the defender destroying the recon element or inflicting heavy casualties. Rather than avoiding reconnaissance, the company, and by extension the battalion, plans for this contingency, augmenting the company with battalion scouts or infiltrating a medical team forward.

THE ROUTE RECON ELEMENT (OFFENSE)

The route recon element has a three-fold task: identify enemy units along the route for bypass (if conducting an infiltration mission) or for attack (movement to contact missions); investigate the route for trafficability and alternate routes if enemy forces or obstacles block passage; and guide the company to the linkup point. The route recon resists departing from the planned route, because a deviation takes the unit from the planned route and into the unknown. Shortcuts at night are not worth the risk of getting lost. Upon completion of the route recon, the route recon leader leaves two-man teams at each major checkpoint and the linkup point. Ideally, the remainder of the route recon links up with the company at the attack position or a checkpoint near the line of departure. The recon leader also positions teams at points near discovered enemy positions and terrain where the terrain is difficult to negotiate. Guides are a supplement to land navigation, not a replacement. The company still verifies each leg.

The recon leader may elect to mark the route as an aid to navigation. The leader carefully weighs the use or placement of markers; although they help verify the route, they may also compromise the company if discovered by the enemy. The danger of compromise increases as marker placement nears the objective area. Because of the danger, neither recon element places markers within one kilometer of the objective. Day markers are any devices that verify the route and indicate the proper azimuth. Example markers are as follows: a pile of rocks with a pointer; stones arranged in an arrow; markings on trees; and chalk marks on trees. Effective night markers are infrared chemlights, directional chemlights, or flashlights. Night vision devices are able to detect these at a distance.

Without guides or markers, the linkup at the linkup point is extremely difficult and time-consuming at night, even if the point is on prominent terrain. If link up between the company and the recon team has not been effected and radio contact exists, the recon team marks the linkup point (infrared chemlight hoisted up a tree) upon notification. If no radio contact is made, the recon team automatically marks the linkup point at the planned time of occupation (IAW OPORD). If the company cannot link up within 30 minutes, it continues toward the ORP.

Direct radio communications may not be possible because of terrain, weather, or distance. If battalion scouts are operating in the area, company RTOs may use them as a relay through the S2-S3. The route recon may take a regular or field expedient

OE-254 antenna and set up a relay station in accordance with the commo sergeant's line-of-sight profile plan.

Generally, *defensive area reconnaissance* is not needed, because of the proximity of the defensive sector and time constraints. If the commander feels that it is needed, he dispatches the recon element after the warning order. In such cases, the commander and selected personnel may decide to accompany the recon element to the defensive sector for preparation of the OPORD. The objective recon element then serves as both a security and a reconnaissance element, providing immediate defense of the sector from enemy reconnaissance and terrain information to the commander and the RTOs preparing the sand table.

CONDUCT INSPECTIONS

Division of labor ensures that selected leaders inspect critical equipment for functionality, serviceability, and accountability. Formalizing this in the SOP ensures that no preparation task or supporting task is overlooked and that the responsible leader is held accountable when it is. Leaders begin inspections between the company warning order and rehearsals. It is best to conduct inspections early to identify discrepancies and rectify them in time. Leaders also use this time to ask subordinates questions regarding the mission. This process reinforces the dissemination of essential information; clarifies task, purpose, and intent; clarifies misunderstandings; and engages the soldier in the corporate effort.

CONDUCT REHEARSALS

Most rehearsals are talk-through discussions because of time constraints. The most important rehearsal is actions on the objective. If practical, the commander conducts this in terrain similar to the objective area. Otherwise, he refers to the sand table. This rehearsal helps everyone see the execution of the plan in action. The commander's role is as coach and narrator, talking through each phase and explaining how each task fits into the overall plan. He may also explain where he has accepted risk during critical phases. Platoon, squad, and section rehearsals take place before the company rehearsal to ensure that subordinate leaders are well familiar with the mission and also to discuss problem areas discovered with the plan during their rehearsal. During the rehearsal, the commander makes modifications to the plan as he deems necessary as a result of the discussion. Note that the rehearsal is not the forum for subordinates to voice major changes to the plan, but rather is the venue to refine and synchronize the plan. The commander will take his observations from his rehearsal to the battalion rehearsal as well.

At each level of the rehearsal, leaders discuss the chain of command in preparation for the loss of leaders during the operation, to ensure that the momentum is not stalled by such a loss. During lulls between rehearsals, soldiers should cross train on key weapons to ensure that they are kept in action should the primary weapon holder become a casualty. Lastly, a review of enemy aircraft and vehicles also lowers the chance of fratricide.[34] Leaders can discuss these subjects at any time, as long as it is done.

COMPLETE THE PLAN

The plan is not complete at the OPORD. The commander refines the plan as a result of feedback from back-briefs, rehearsals, and new intelligence. The commander will likely need to show the changes on a sketch in the defensive assembly area, the attack position, or even the ORP. In the attack, the most likely changes would be the location of the assault position or the fire support position. In the defense, viewing the actual terrain probably will require changes in dispositions. Discovery of enemy fortifications, tanks, or special weapons may require attachments or tanks, engineers, or more artillery. The commander should not hesitate in requesting them. Oftentimes battalion and higher staffs do not address these new developments, because they are too busy with future plans or coordination. The commander helps remind them of the immediate task at hand.

TALK TO SOLDIERS

The final step in any operation is for the commander to speak with the soldiers before departure of the assembly area or attack position. It is unrealistic to talk to the company as a whole, so he should talk to the platoons whenever they have five minutes of lull time. This talk is not a motivation speech. Soldiers do not want or seek such inspirations, particularly if it is a long conflict. Soldiers want information that will keep them alive and help them defeat the enemy.

He wants to emphasize the key to success is to approach with caution and then execute on-the-spot decisions with dispatch. Everyone must learn the latest methods for reducing casualties.[35] Veterans do not need this advice. It is aimed at the new soldiers, who mistake rashness with heroics. He wants to safeguard new soldiers until they have learned how to perform in combat without becoming needless casualties.

The commander uses this time to remind everyone of new enemy tactics and weapons. He then reminds them of new company TTPs that will negate the enemy tactics and weapons. This helps reduce soldier apprehension and reminds them that they have the superior edge over the enemy.[36]

The commander also takes time to remind soldiers of the effects terrain and weather will have on weapons, munitions, noise levels, physical health, and the mental psyche. For example, deep snow or mud dampen the effects of indirect fire, and heavy woods decrease the effective range of small arms, accentuate noise, and increase the soldier's feeling of isolation. Dense woods and urban terrain amplify noise, making it appear as though the enemy force is stronger than it actually is. Vast expanses of unbroken terrain seem to have a depressing effect on the soldier's psyche over time. Soldiers will develop trench foot within 24 hours if they fail to dry their feet and change socks.

The commander focuses on actions on the objective and enemy contact. He reminds everyone that prompt use of organic weapons (those in the company's normal complement) on contact will reduce casualties because it suppresses the enemy. He strongly reminds them not to bunch up during the fight, because this makes them a more inviting target to the enemy machine guns and indirect fire. He reminds them to maintain good light and noise discipline. He reminds everyone of the importance of keeping key weapons and leadership positions filled once casualties mount. Most important, the commander must be positive and confident in his manner and speech. A leader's countenance and speech are infectious. Dejection and uncertainty become a self-fulfilling prophesy of defeat even before the mission begins. This does not mean that the commander sugarcoats tactical realities. He lays out the risks and explains what actions he has taken to ensure success. The commander does this for every mission, because it serves the soldier when things go wrong.[37]

Lastly, the commander tells everyone to expect friction on the battlefield— the plan may go awry and everything may seem to have gone wrong. But the enemy has a host of problems and mistakes to contend with also. As is the case in so many operations, it is the initiative and aggressiveness of the individual soldier that will figure out how to fix the problem and get the momentum of the attack going again. Brilliant plans do not win battles, soldiers do.

In this role, the commander becomes a coach, reminding everyone of their training, new TTPs, and expectant psychological pressures; he is a counselor, assuaging fears, cracking a few jokes, and giving them confidence; and he is a father figure, reminding them of their duty, reiterating how this operation is linked to ending the conflict and that he cares for their welfare. These talks are an integral part of every operation, and commanders should regard them as important as issuing the OPORD.

CONCLUSION

The whole idea of preparing for a mission must be considered in its entirety. Time is the most crucial factor in the process, and the company must take every opportunity to create the conditions for success. Because of instruction in military schools, leaders try to follow the decision-making process and troop-leading procedures too closely. At company level, the leaders must swiftly glean the essential of the intelligence preparation of the battlefield (IPB) and move on to the tentative plan. Too often, planners become fixated on the particulars of OCOKA and METT-T, which have too many unknowns to validate the time expended. The tentative plan should also receive only the bare minimum of time, using a sketch of the area of operations as the planning vehicle. The details and changes to the final plan will result from back-briefs, rehearsals, intelligence from the company scouts, and leaders' recon. Using this pragmatic process, the commander develops a flexible plan, which addresses the realities of the situation. A plan is not really complete until the final review in the objective rally point. The commander that binds himself to a plan that was developed in the dark will likely wreck his company during the execution of the mission.

Lastly, the commander must cultivate and maintain the soldier's morale. It is as important as the maintenance of equipment. The nebulous region of morale is often ignored in peacetime because it does not become a real issue then. The injection of fear, stress, and fatigue make morale an essential part of the process. If mail and hot food were the only factors that determined good morale, no army would be successful in war, because both are often lacking at the front line. The commander must use all his powers of persuasion, force of will, and persistence in ensuring that his subordinates maintain focus on the mission and not become distracted by the school process, boredom, and lethargy. He must infuse in everyone that all obstacles can be overcome by a team effort. Within the company, there is always someone who sees the solution to a problem; the commander must learn to recognize such a person no matter what his rank. The true spirit of mission-type orders is to empower subordinates with responsibility to accomplish an operation within the mission statement and commander's intent. If the company can establish an environment honoring the personal initiative of the local leader to take action in pursuit of mission success, even if acting as small as a two-man group, it will amaze higher command and confound the enemy.[38]

TABLE 5.1 PREPARE FOR MISSION	
Task	**Responsibility**
Receive Warning Order(s)	CO RTO
1. Receive battalion mission, S2 enemy SITEMP, company AO and type operation.	CDR
2. Initiate intelligence preparation of the battlefield (IPB)—MCOO and refined enemy SITEMP (refer to Make Tentative Plan below).	CDR
Receive the Mission	CDR/FIST OFF
1. Copy overlays, update SOI and encryption fills.	CO RTO

TABLE 5.1 *(continued)*

2. Focus on understanding the purpose of the company's mission and how it contributes to the attainment of the battalion mission.	CDR
3. Receive battalion fire support overlay, fire support priority of fires, and the number of target reference points allotted to the company.	FIST OFF
Conduct Mission Analysis	CDR
1. Develop restated mission (task and purpose—Who, What, When, Where, Why):	
a. Understand higher command's mission (one level up) and commander's intent (two levels up).	
b. Analyze AO and higher headquarters' concept of the operation statement.	
c. Identify specified tasks: battalion concept of the operation statement, specified tasks to the company, and coordinating instructions.	
d. Identify implied tasks: unspecified tasks needed to accomplish specified tasks (passage of lines, breaching operations, etc.).	
e. Extract mission essential tasks to produce the restated mission.	
2. Identify the enemy's center of gravity—the one thing that, if attacked, will cause the enemy's attack or defense to crumble.	
3. Identify the decisive point in the fight—the point (geographically or correlation of forces in the fight) where the company starts winning and the enemy starts losing.	
4. Determine critical facts and assumptions:	XO
a. Facts—known data. • Enemy—identified locations, composition, etc. • Friendly—review assets available: attachments/ detachments, **Unit Manning Roster** (chapter 10), readiness, and CL III/V/water status. • Terrain and weather data.	
b. Assumptions: data that replaces missing or unknown facts—must be valid (likely to occur) and necessary (essential for plans development).	

TABLE 5.1 *(continued)*	
• Enemy—SITEMP, EVENT template, and likely rate of attrition before engagement. • Friendly—projected repairs on weapons/ equipment, incoming personnel, and resupply. • Weather effects on enemy and friendly forces.	XO
5. Determine limitations:	XO
a. Constraints (tasks the company must perform— detachments, retaining and seizing specific terrain during operation, performing certain tasks in support of higher command activities).	
b. Restraints (activities the company cannot do—radio listening silence, limit of advance, direction of attack, no recon and counterreconnaissance forward of FLOT).	
6. Calculate preliminary analysis of time (1/3–2/3 rule) for troop leading procedures.	CDR
Issue Warning Order (within 30 minutes of return)	CDR
1. Check attendance of key leaders.	1SG
2. Ensure all know that this is the warning order.	CDR
3. Give enemy and friendly situation as it pertains to company mission.	XO
4. Give restated mission.	CDR
5. Give preliminary analysis of time for preparation (backward-time planning):	CDR
a. Depart assembly area/patrol base.	
b. Conduct final inspections.	
c. Eat ration meal.	
d. Conduct rehearsals (SQD/PLT/CO).	
e. Conduct mission brief-backs.	
f. Test fire/zero weapons; prepare demolition/ special equipment.	
g. Conduct initial inspections.	
h. OPORD time and location.	
i. Coordination and OPORD planning preparation back-brief.	

(continued)

TABLE 5.1 *(continued)*

6. Issue special instructions/delegation of tasks:	CDR
a. **OPORD** preparation and briefing. (chapters 6 and 18)	CDR
1) Task organization	CDR
2) Enemy and friendly situation.	XO
3) Terrain and weather.	CDR
4) Mission.	CDR
5) Commander's intent and basic plan.	CDR
6) Movement order of march, technique, and formation.	CDR
7) **Land Navigation Plan**—from attack position to objective.	PLT LDR
8) Tasks to subordinate units.	CDR
9) Fire support plan.	FIST OFF/MORT SL
10) Reorganization and consolidation plan.	CDR
11) Coordinating instructions.	XO
12) Service support: logistical support.	SUPPLY SGT
13) Service support: medical support and casualty evacuation plan.	1SG
14) Command.	CDR
15) Signal.	COMMO NCO
b. **Tactical Foot March Annex**—from assembly area to attack position, pickup zone, or defense assembly area. (chapter 7)	PLT LDR
c. **Air Assault Plan.** (chapter 13)	PLT LDR
d. Motor transportation movement plan.	PLT LDR
e. **Passage of Lines** coordination and briefing. (chapter 12)	PLT LDR
f. **Relief in Place** coordination and briefing. (chapter 21)	PLT LDR
g. Offense reconnaissance—conduct route and objective reconnaissance; focus on named areas of interest (NAIs)	CO SCOUT LDR

TABLE 5.1 *(continued)*

and targeted areas of interest (TAIs); prepare linkup site and ORP; recon support, assault, and security positions; provide guides for each position.	
h. Defense reconnaissance and security—Conduct route recon to AO; conduct screen of defensive positions until occupation. Select positions that allow observation of NAIs and TAIs.	CO SCOUT LDR
i. Inspect attachments for mission readiness; brief them on applicable company TAC SOP tasks; if attachments not present for OPORD, brief them and link them up with the applicable company element IAW the plan.	1SG
j. Prepare terrain model (sand table) or sketch of area of operation.	CO RTOs
k. Provide SOI extracts and radio encryption fills for duration of mission. Provide company scouts with extracts and fills immediately.	COMMO NCO
l. Conduct line-of-sight profile of terrain from the attack position to the objective (offense) or from the sector to the battalion TOC and TAC (defense). Identify radio relay locations to allow the company to maintain communications with the battalion TOC and TAC.	COMMO NCO
m. Special MOUT preparations: concussion grenades, grappling hooks with ropes, flashlights secured to rifle barrels, laser pointers secured to rifle barrels, demolition breach charges, chalk, ladder, smoke grenades, ammo package affixed to rucksack frame, gloves.	ASSAULT PSG
n. Deliberate breach plan.	CDR
o. Identify equipment and demolition required for mission: obstacle breaching (demolition kit, bolt cutters, bangalore torpedo kit, shape charges, grappling hooks) or obstacle preparation (shape crater charges, concertina wire, mines, chain saws).	PLT LDR or ENG SL
p. Prepare markers for wire breach, mines, mine field lane, cleared bunkers, cleared rooms and buildings.	PSG
q. Medical evacuation plan coordination with S1, IAW **Consolidate and Reorganize.** (chapter 17)	1SG

(continued)

TABLE 5.1 *(continued)*	
r. Identify and request mission essential Class V (ammo, pyrotechnics, demolition, mines), Class II and IV (sandbags, 4 X 4 inch supports, and barrier material) and CL I water. Do not include normal sustainment supply requests.	1SG/PSG/SEC LDRs/SUPPLY SGT
s. **Resupply Plan** and transportation of excess equipment and personal gear (IAW soldier's load guidance). (chapter 23)	SUPPLY SGT
t. Prepare target overlay for route and objective.	FIST OFF
u. Identify number and type of rounds and distribution requirements for mortars.	MORT SEC LDR
7. Synchronize watches.	CDR
Make Tentative Plan (METT-T)—OFFENSE	CDR
1. Develop course of action:	
a. Identify area of operations (AO) and area of interest (AI).	
b. Analyze terrain (extracted from map recon and updated from intelligence updates, recon reports) of objective area in terms of OCOKA:	
1) Distance available for **O**bservation and fields of fire: best place for fire support position(s). Best place for an assault position, which avoids enemy observation and primary fields of fire.	
2) Available **C**over and **C**oncealment from ORP to assault and fire support positions.	
3) Location of natural (swamp, cliff, streams, thickets, etc.) and existing manmade **O**bstacles (minefield, wire, terrain modifications, etc.).	
4) Identification of **K**ey and decisive terrain whose control provides marked advantage—control over the AO, a nodal point (intersection, choke point), the linchpin of enemy defense (dominating hill or ridge), and support positions (ridge or hill).	
5) Identify **A**venues of approach into objective area in order to isolate objective area from enemy	

TABLE 5.1 *(continued)*	
reinforcements, and friendly access to fire support and assault positions.	
c. Weather (effects on friendly and enemy weapons and maneuver) in terms of:	CDR/XO
1) Mobility and speed of movement: mounted and dismounted movement. If weather prevents off-road movement, early control of avenues of approach becomes critical.	
2) Visibility in meters.	
3) Effects on NBC use.	
4) Effects on smoke.	
d. Enemy situation (from S2 IPB templates—refine array two levels down):	XO
1) Organization for the defense—doctrinal template.	
2) Composition: type unit, weapons, and equipment.	
3) Disposition: conformation to terrain—situation template.	
4) Projected strength during company attack:	
a] Of forces in objective area.	
b] Degree of preparation—hasty/deliberate/ strongpoint positions.	
c] Ability of adjacent enemy units to influence the defense.	
d] Reinforcements or counterattack force; size, composition, location, time of reaction, and probable routes—event template.	
e] All fire support within range.	
f] NBC capability.	
5) Peculiarities: Strengths and weaknesses from doctrine, weapons, morale, logistics, mobility, command and control, aggressiveness, discipline.	
6) Select most likely and most dangerous course of action.	
e. Troops available (array two levels down):	CDR

(continued)

TABLE 5.1 *(continued)*	
1) Strength: **Unit Manning Roster.** (chapter 10)	CDR
2) Attachments and combat support available.	
3) Consider **U.S. Weapon Capabilities** in plan. (chapter 11)	
f. Develop course of action (one only) against the enemy course of action most dangerous to company success. COA must be suitable (accomplishes mission within commander's intent), feasible (within company capabilities), and acceptable (reasonable costs).	CDR
1) Assign each subordinate element a task and purpose.	
2) Select security positions to isolate the objective area.	
3) Select fire support position(s), which suppresses and isolates the objective area.	
4) Select assault position with covered and concealed approach and a local fire support position to isolate the initial point of penetration and local suppression.	
5) Consider use of surprise and deception: • Time of attack • Location of assault • New use of key weapon • Feint/demonstration • New assault technique.	
g. Critical event-time planning (backward-time planning):	CDR
1) Achieve decisive point.	
2) Begin assault.	
3) Conduct deliberate breach.	
4) Occupy assault position.	
5) Occupy support position(s).	
6) Occupy security positions.	
7) Depart ORP.	
8) Conduct leaders' recon of objective.	

TABLE 5.1 *(continued)*	
9) Occupy ORP.	CDR
10) Occupy linkup point.	
11) Cross line of departure (LD) or passage of lines.	
12) Depart attack position.	
2. War game COA to synchronize the attack: using map, sketch, or sand table, war game from decisive point, backward to ORP and then forward to consolidation and reorganization. Identify NAI for reconnaissance to verify enemy dispositions, TAI for reconnaissance to verify high payoff target locations or engagement areas against the enemy counterattack force, and decision points to trigger execution of a TAI or other friendly actions.	CDR/XO
3. Develop into tentative plan and task organization.	CDR
Make Tentative Plan (METT-T)—DEFENSE	CDR
1. Develop course of action:	
a. Identify AO and AI.	
b. Analyze terrain (OCOKA):	CDR/XO
1) Distance for **O**bservation and fields of fire: open and flat terrain requires more antitank (AT) weapons and fewer small arms weapons to provide sufficient coverage; close and broken terrain requires fewer AT weapons and more small arms weapon density for sufficient coverage.	
2) **C**over and concealment available for defensive positions and routes to supplementary and alternate positions; availability for enemy approaches into sector.	
3) Identify Natural **O**bstacles to vehicles: • Slow-go terrain: % slope = 30–45% uphill; stream—flow < 5 ft/sec, depth < 4 ft; trees—diameter = 2 in, density > 20 ft. • No-go terrain: % slope $\geq$ 45 % uphill; streams—flow $\geq$ 5 ft/sec, depth $\geq$ 4 ft, bank > 4 ft; trees—diameter = 6–8 in, density $\leq$ 20 ft. Urban area > 500 meters. Ice thickness 31.5 in for tanks. • Identify natural choke points: use for engagement kill zones and obstacle	

(continued)

TABLE 5.1 *(continued)*	
emplacement—prioritize placement by type (wire, mines, ditch, etc.).	
4) Identify **K**ey terrain: determine features that dominate the AO, a key nodal point, a decisive feature to serve as the keystone of the defense, key features to serve as supporting redoubts. ID potential fire support positions for the enemy.	CDR/XO
5) Identify and prioritize mounted **A**venues of approach for relative force ratios and key weapons' allocation. Determine the size mobility corridor into the sector (Co—500 m, Bn—1.5 km, Rgt—3 km) in order to calculate how much firepower the enemy can bring to bear on the defense. Consider relative force ratios against dismounted AA.	
c. Weather (effects on enemy and friendly weapons and maneuver):	
1) Mobility: potential for enemy off-road movement and access to key fire support positions. Effect on counterattack force.	
2) Visibility in meters: consider countermeasures (thermal sights/NODs for key weapons, battlefield illumination, flares, illumination from fires).	
3) Use of NBC and smoke: Dependent on wind direction and speed, and inversion conditions.	
d. Enemy situation (from S2 IPB templates—refine array two levels down):	XO
1) Organization for the attack: Doctrinal template—attack from march or attack from a position in direct contact with the battalion.	
2) Composition: Type unit—typical weapons and equipment.	
3) Disposition: Situation template—attack formation conformed to existing terrain and weather. For attack from a position identify likely ORP, fire support, assault, and security positions.	
4) Projected strength upon contact:	

TABLE 5.1 *(continued)*	
a] Projected strength after attrition through CFA.	XO
b] Projected arrival of reconnaissance, first, and second echelons. Event template.	
c] Organic and likely direct support fires (arty, mortars, etc.).	
d] Air and NBC capabilities.	
e] Most likely method of breaching obstacles.	
5) Peculiarities: Strengths and weaknesses from doctrine, weapons, morale, logistics, mobility; C2, aggressiveness, and discipline.	
6) Select most likely and most dangerous enemy COA.	
e. Troops available (array two levels down):	CDR
1) Strength: **Unit Manning Roster.** (chapter 10)	
2) Attachments and combat support available	
3) Consider **U.S. Weapons Capabilities** in plan. (chapter 11)	
f. Develop course of action (one only) against the enemy COA most dangerous to company success. It must be suitable (accomplishes mission within commander's intent), feasible (within company capabilities), and acceptable (reasonable costs), distinguishable, and complete.	CDR
1) Assign each subordinate element a task and purpose.	
2) Align defense where the relative force ratios favor the defense—choose terrain that forces the enemy to commit its units piecemeal.	
3) Select disposition so that positions are mutually supporting and not easily isolated for defeat in detail. Select supplementary and alternate positions so that the defense maintains cohesion throughout the fight.	
4) Identify areas for risk: Places where economy of force can be used; hide positions from which key weapons, snipers, or company elements are to emerge or attack the enemy from the flank or rear.	
5) Consider use of surprise and deception: • Dummy positions	

(continued)

TABLE 5.1 *(continued)*	
• Use of hide positions • New use of weapon • Placement of obstacles	CDR
6) Consider which enemy action makes the defense untenable and requires the withdrawal of the defense. Consider best axis for higher-command counterattack force.	
g. Consider time available and whether defense can be completed with assets available (digging equipment, sandbags, etc.) or whether positions need to be dug by priority (key weapons first). Use backward-time planning:	CDR
1) Defend NLT time.	
2) **Priorities of Work in Defense** complete. (chapter 19)	
3) Begin priority of work.	
4) Occupy defensive positions.	
5) Depart assembly area.	
2. War game COA to synchronize the defense: using map, sketch, or sand table, war game from decisive point backward to initial contact with enemy and then forward to consolidation and reorganization. Identify decision points— points where enemy actions trigger a local counterattack or commitment of reserve or cause realignment of defense. Identify named areas of interest (NAI) for reconnaissance to pick up enemy advance into sector, targeted areas of interest (TAI) for reconnaissance to spot high payoff targets en route, and decision points to trigger execution of a TAI or other friendly actions.	CDR/XO
3. Develop into tentative plan and task organization.	CDR
Start Movement	CDR
1. In attack: Movement to attack position.	
2. In defense: Movement to defense assembly area.	
Issue OPORD	CDR/TASKED LDRS
1. Oral Briefing: OPORD format to provide organization and structure to dissemination. (chapters 6 and 18)	

TABLE 5.1 *(continued)*	
2. Rely on visual presentation: sand table and graphical sketches or overlays.	CDR/TASKED LDRS
3. Tasked subordinates are responsible for success of their tasking and not solely for planning and briefing.	
Conduct Reconnaissance	CDR
1. Map recon and company route and objective recon.	
2. Request for Information (RFI) and intelligence updates from Bn S2.	
Conduct Inspections	CDR/1SG
1. Weapons (minimum-bolt clean, *lightly* oiled, and function check):	PLT LDR
a. Crew-served machine guns. Check for sample range card.	PSG
b. Squad Automatic Weapon (SAW). Check for sample range card.	SL
c. M203 Grenade Launcher.	TM LDR
d. M16A2 rifle.	TM LDR
e. M202, LAW, AT4 (check for damage only).	PLT LDR
f. M47 Dragon (accountability and functioning of all components).	AT SEC LDR
g. Mortars (accountability and functioning of all components).	MORT SEC LDR
2. Ammunition (clean and properly carried to protect from filth and damage):	PSG
a. Machine gun: For movement, a 20-round belt loaded for instant response; assistant gunner clips remaining and succeeding belts onto loaded belt immediately during action (practice this during crew drills). Remaining belts are carried in 2-quart canteen pouch or butt pack with strap (never inside of rucksack; however, the pouches may be attached to the outside of the rucksack).	PSG
b. Rifle: accountability and distribution, and proper allotment of magazines, which are fully loaded.	SL

(continued)

TABLE 5.1 *(continued)*	
c. M203: accountability and distribution. *Ensure grenadier has his vest.*	SL
d. Grenades: Distribution, number, and type. Place excess in 2-quart canteen pouch or butt pack with straps.	TM LDR
e. Mortar rounds: Distribute to platoons IAW commander's soldier's load plan.	MORT SEC LDR
3. Night observation devices (NOD) (batteries properly installed and NODs functional):	SL
a. PVS-7: Soldier demonstrates donning of head harness and clipping of PVS-7 to harness (blindfolded).	SL
b. PVS-4: Inspect SAWs for bracket. Inspect for bracket fastener. Ensure soldier zeros PVS-4 during test fire.	TM LDR
c. Dragon night sight (AN/TAS-5): Batteries and coolant bottles properly charged/filled. Functions check; soldier demonstrates attaching AN/TAS-5 to sight (blindfolded).	AT SEC LDR
4. Test fire and zero weapons (only upon commander's approval).	PLT LDR/SEC LDR
5. Inspect antiobstacle and antibunker devices (demolition kit, bangalore torpedo, satchel charges): check completeness and distribution for soldier's load; test 6–12 inch section of time fuse or detonation cord; check igniters and blasting caps for damage; ensure explosives and fuse and primers are carried separately; prepare demolition for quick assembly.	PLT LDR
6. Inspect field expedient antitank (AT) devices (Molotov cocktail, eagle fireball, eagle cocktail, towed charge, pole charge, and satchel charge).	PSG
7. Soldiers: Equipment layout for accountability and soldier's load; check soldier's uniform and noise discipline. Check first aid pouch for bandage (for seal), DA Form 1156 (casualty feeder report), and DA Form 1155 (witness statement). Ensure form headings (name, SSN, rank, unit) are complete.	SL
8. Medic's aid bag (for basic load and additional concerns of the commander such as extra IV bags).	PSG/CO MEDIC

TABLE 5.1 *(continued)*	
9. M17A2 protective masks (for seals and filters); M256 detection kits, M8 alarms for designated NBC Teams If NBC threat.	NBC NCO/TM LDR
10. Rope bridge and rappelling equipment: accountability and serviceability of 120-foot ropes (for fraying, dirt, wetness, mildew), 12-inch sling ropes and snap links.	PSG/SL
11. Soldier brief-back on mission details.	All LDRs
12. Final commo check with secure devices and inspection of SOI cheat sheet for duration of mission.	COMMO NCO/ RTO
13. Engineer attachment's equipment (accountability and serviceability).	PLT LDR
14. COMSEC and OPSEC violations check: only CO CP has CEOI during mission, subordinate units carry cheat sheets; mission graphics on maps erased and mission notes destroyed.	PSG/SL
15. Assembly area/attack position sterilized in preparation for departure.	1SG/PSG

Conduct Rehearsals			CDR
Rehearsal	*Level*	*Type*	
Actions on the objective to include **Consolidation and Reorganization.** (chapter 17)	Platoon	Full rehearsal	PLT LDR/SEC LDRs
Breaching operations and marking of breach.	Platoon	Full rehearsal	PLT LDR
Clear and mark bunker/ trench line/ building/ room.	Platoon	Full rehearsal	PLT LDR
Crew Drills (machine gun/ AT/mortar).	Section	Full rehearsal	PSG/SEC LDRs
Loading and off-loading of aircraft.	Chalk	Full rehearsal	CHALK LDR
Actions on contact.	Squad	Talk through	SL
Linkup operations.	Squad	Talk through	SL
Actions at rally points.	Squad	Talk through	SL
Passage of Lines. (chapter 12)	Squad	Talk through	SL
Fire Support Plan.	CO CP	Talk through	CDR/FIST/ MORT SEC LDR

(continued)

TABLE 5.1 *(continued)*			
Actions on the objective: breaching operations through **Consolidation and Reorganization.**	CO	Talk through and full rehearsal	CDR
Actions at pickup zone and loading zone.	CO	Full rehearsal	PZ CONTROL OIC/NCOIC
Actions on contact.	CO	Full rehearsal	CDR
Complete the Plan			CDR
1. Issues resulting from leader back-brief.			
2. Changes resulting from rehearsal.			
3. Issue FRAGOs			
Talk to Soldiers			CDR
1. Brief reminders of new enemy weapons and tactics to include vulnerabilities.			
2. Brief reminders of terrain and weather effects on weapons, munitions, noise levels, physical health, and mental stress.			
3. Brief reminders of new TTPs, the need to keep key weapons manned, and leadership positions filled once casualties occur.			
4. Remind everyone that friction is always a factor, but that the enemy has a host of problems of his own to contend with too. Success depends on the initiative of everyone to keep the attack going forward or the defense intact.			

Source: U.S. Department of the Army, *Mission Training Plan for the Infantry Rifle Company,* Army Training and Evaluation Program No. 7-10-MTP (Washington D.C.: 1988), 5-217–5-226. "Prepare for Combat" (7-2-1046) formed the foundation document for this portion of the SOP. U.S. Department of the Army, *Intelligence Preparation of the Battlefield,* Field Manual No.34-130. (Washington D.C.: U.S. Government Printing Office, 8 July 1994), B-8–B-38.

NOTES

1. Jay Luvass, trans. and ed. *Frederick the Great on the Art of War* (New York: The Free Press, 1966), 142–43, 365. Frederick the Great called this *Coup d'oeil*—"The gift or instinct of a general that allows him to distinguish at a glance the strong and weak points of the terrain." It is divided into two parts: "the ability of judging how many troops a given position can contain. . . . The other and by far the most superior talent is to know how to distinguish at first sight all the advantages that can be drawn from the terrain."

2. Dwight D. Eisenhower, *Crusade in Europe* (Garden City, N.Y.: Doubleday and Company, Inc., 1948), 240. Eisenhower's insight into this problem has often gone ignored. The U.S. service schools still teach that weather is neutral despite the plethora of practical experience. German experiences in Russia during the fall and winter of 1941 and in the Ardennes in 1944, as well as the Allied experiences in Italy 1943–44 and in France during the fall of 1944, illustrate how weather has hindered offensive operations when enemy resistance was nil.

3. U.S. Department of the Army, *Tank Platoon,* Field Manual No. 17-15 (Washington, D.C.: U.S. Government Printing Office, April 1996), Figure 1-8. This endnote refers to the tank diagram only.

4. James Lucas, *War on the Eastern Front: The German Soldier in Russia 1941–1945* (Novato, CA: Presidio Press, 1979), 125–28; Wolfgang Fleischer, *Die deutschen Sturmgeschütze 1935–1945* (Wölfersheim-Berstadt: Podzun-Pallas GmbH, 1996), 36.

5. U.S. Department of the Army, *Small Unit Actions during the German Campaign in Russia,* Department of the Army Pamphlet No. 20-269 (Washington, D.C.: U.S. Government Printing Office, 1953), 112, 114. This cooperation between infantry and tanks was already common practice in the German Army; Doubler, 31–62. U.S. infantry-tank cooperation was forged during the Normandy campaign and effectively turned the tables on the Germans until the end of the war. These teams destroyed German bunkers, pillboxes, strongpoints, and field fortifications wherever established, to include towns and cities. During the fight for the hedgerows, the 29th Infantry Division organized combat teams consisting of one tank, an engineer squad, and an infantry squad with a light machine gun and a 60 mm mortar. Moving together, the infantry suppressed, while the engineers reduced obstacles. The tank occupied a good position, identified by the infantry, to destroy enemy points of resistance. Forward observers mounted the back of the tanks to help them observe impacting rounds, pp. 49 and 51. Such teamwork is crucial during fighting in deep forests. The ability of the defender to mass his AT weapons on trails makes it imperative for infantry to suppress these, while the tanks eliminate bunkers. This is slow work and cannot be hurried without risking heavy casualties, p. 194.

6. U.S. Department of the Army, *Military Improvisations during the Russian Campaign,* Department of the Army Pamphlet No. 20-201 (Washington, D.C.: U.S. Government Printing Office, 29 August 1951), 22. The Germans adopted this method of mutual protection while fighting in the heavy forests in the north of Russia.

7. George S. Patton Jr., *War as I Knew It* (New York: Bantam Books, August 1981), 335.

8. Lupfer, 30; Martin Samuels, *Doctrine and Dogma: German and British Tactics in the First World War* (Westport, Conn.: Greenwood Press, 1992), 78. The German Army leadership of World War I reached these conclusions after two years of fighting and produced one of the most successful defensive doctrines in history. Unfortunately, its success was overshad-

owed by the events of 1918 and overcome by events in World War II. This doctrine needs greater examination and adaption for modern combat.

9. Lupfer, 15. The Germans used resistance nests (*Widerstandnester*) in both the outpost line and main line of resistance to defeat enemy reconnaissance, provide early warning, and contain the enemy attack.

10. Irwin Rommel, *The Rommel Papers,* ed. B. H. Liddell-Hart, trans. Paul Findlay (New York: Da Capo Press, 1953), 128–29. Rommel was impressed by Tobruk's fortified positions and made special note of the effectiveness of this simple camouflage. Heinz Werner Schmidt, *With Rommel in the Desert* (New York: Ballantine Books, 1951), 81–82. The design of the resistance nest is adapted from Rommel's use of strongpoints on the Solumn front, North Africa, 1941. Paddy Griffith, *Forward into Battle* (Novato, Calif.: Presidio Press, 1991), 109–13. Griffith cites Schmidt and observes that this concept was derived from World War I defensive doctrine. He also illustrates how the British used similar tactics at El Alamein.

11. Patton, 327.

12. The genesis of open-terrain dispositions comes from the following sources on German defensive doctrine in World War I: Graeme Wynne, *If Germany Attacks,* Reprint 1940 (Westport, Conn.: Greenwood Press, 1976); Lupfer, 11–12; Samuels; War College, *German and Austrian Tactical Studies. Translations of Captured German and Austrian Documents and Information Obtained from German and Austrian Prisoners—From British, French, and Italian staffs,* compiled and edited at the Army War College (Washington, D.C.: U.S. Government Printing Office, 1918), 121–23. The Germans often made use of craters for bunkers and converted resistance nests in World War I. References to use of open-terrain disposition during the North African campaign and during the Normandy campaign are as follows: Schmidt, 81; John Keegan, *Six Armies at Normandy* (New York: Penguin Books, 1984), 192–93, 200–19; Hans von Luck, *Panzer Commander* (New York: Dell Publishing, 1989), 189, 192–200.

13. Doubler, 172–197. The Battle for the Huertgen Forest lasted about three months, involved six divisions, and resulted in 33,000 casualties, virtually decimating all units involved, 193; John English, *On Infantry* (New York: Praeger Publishers, 1981), 97.

14. Erhard Rauss, Hans von Greiffenburg, and Waldemar Erfurth, *Fighting in Hell: The German Ordeal on the Eastern Front,* ed. Peter G. Tsouras (Mechanicsburg, Pa.: Stackpole Books, 1995), 277–79. This book is actually a compilation of U.S. Department of the Army Pamphlet 20-230, *Russian Combat Methods in World War II,* U.S. Department of the Army Pamphlet 20-291, *Effects of Climate on Combat in European Russia,* U.S. Department of the Army Pamphlet 20-292, *Warfare in the Far North,* and U.S. Department of the Army Pamphlet 20-231, *Combat in Russian Forests and Swamps.*

15. Doubler, 14. An American lesson during World War II was not establishing a heavy volume of fire upon contact. Because of prewar training, soldiers did not engage enemy forces they could not see for fear of wasting ammunition.

16. Rommel, 7. Rommel drew the same conclusions regarding the immediate use of firepower upon contact during the 1940 campaign in France. "I have found again and again that in encounter actions, the day goes to the side that is the first to plaster its opponent with fire. The man who lies low and awaits developments usually comes off second best . . . [soldiers at the point must] open fire the instant an enemy shot is heard . . . even when the exact position of the enemy is unknown in which case the fire must simply be sprayed over enemy-

held territory. Observation of this rule, in my experience, substantially reduces one's own casualties." Patton, 327–28. Patton reached the same conclusion, noting "that it is much better to waste ammunition than lives." English, 49.

17. Doubler, 13–14. Another costly lesson for the Americans. Upon contact, units learned that the best way to keep casualties low was to close with the enemy rapidly. Patton, 319–20. Patton reflected that "hitting the dirt" caused high casualties, because the enemy already had the avenue of approach pretargeted with direct and indirect fire. The Germans would fire a machine gun, often in the air, just to cause the Americans to seek cover. Upon immobilization of the unit, the fire would saturate the area.

18. Patton, 321.

19. English, 145.

20. Patton, 322, 326, 330. Patton reminds the reader that the purpose of movement is to gain a good position to direct fire on the enemy. He recommends that the unit fix the enemy with a third of the unit and envelop him with a larger force. Once the enemy reacts to the envelopment, the fixing force moves forward to finish the enemy.

21. Daniel P. Bolger, *The Battle for Hunger Hill* (Novato, Calif.: Presidio Press, 1997), 286. Compare this tactic with that used by Lieutenant Colonel Bolger's platoons during his Joint Readiness Training Center (JRTC) rotation.

22. Albert N. Garland, *Infantry in Vietnam: Small Unit Actions in the Early Days: 1965–66* (New York: Jove Books, 1985), 124–25.

23. English, 144.

24. Rauss, 262–63, 271.

25. Ibid., 272–75.

26. English, 97–98.

27. DA Pam 20-269, 44. The Germans used this TTP on occasion with good effect.

28. Doubler, 121–23. American units learned costly lessons on reducing fortified lines. Once they focused first on eliminating the supporting positions, isolating and destroying bunkers and pillboxes became fast work. Patton, 325, 329. Patton thought that surprise attacks against pillboxes were the most effective, provided the enemy is in the pillboxes and not the trenches behind them. If the enemy is occupying the trenches, the unit should place indirect fire onto the trench to force the enemy into the pillboxes, since an enemy in the pillboxes is less dangerous to the attack. He recommended that units must task organize into pillbox-busting teams consisting of two machine gun teams, an AT missile team, a light machine gun, two to four riflemen, and two demolition men. Reconnaissance should determine if the pillboxes at the breach point are mutually supporting. If so, they must be attacked simultaneously. Lastly, this type of operation should be conducted at night.

29. David Chandler, *The Campaigns of Napoleon* (London: Weidenfeld and Nicolson Ltd., 1966), 342, 344. The French use of the bayonet charge illustrates the psychological impact on the enemy. Few soldiers were actually wounded by bayonets; rather, units broke and ran before French infantry could close on them because the prospect of being bayoneted was too terrifying to withstand. Griffith, 20–22, 26–28, 36. Looking at the British perspective, Griffith concludes that the bayonet charge decided most battles. The British tactic was to fire a volley then charge. The combination of both acts often caused the enemy to break and flee the battlefield. He also observes that few wounds were the result of the bayonet. Further on in his book, Griffith makes the argument that, in practice, the use of the bayonet charge

up to World War I was still the decisive tactic in battle. He makes a case that the effectiveness of firepower was overrated and that casualties during World War I can be attributed to a loss of momentum resulting from dispersed formations and the attackers stopping to fire their weapons in the attack rather than an obsolete bayonet tactic. Once the attack stalled, soldiers were exposed to enemy fire longer. John Ellis, *Eye-Deep in Hell: Trench Warfare in World War I* (Baltimore, Md.: The Johns Hopkins University Press, 1991), 78–79. British casualty figures from World War I reveal only .032 percent of wounds were the result of bayonets. U.S. figures reveal only .024 percent. These figures are probably low since most bayonet wounds were fatal and hence would not be recorded. When confronted by a bayonet attack at close range, most defenders either ran away or surrendered.

30. English, 223. English reiterates what most veterans know—loss of hope ultimately decides engagements.
31. Ibid., 79. The German method for breaching obstacles was to suppress local defenses, obscure the breach point, close on the other side of the breach, and defeat the defending unit.
32. Doubler, 26. Commanders during World War II learned to rely on their own reconnaissance patrols in order to get the intelligence they needed.
33. English, 128.
34. Doubler, 23, 26. Deficits in these subjects remained in most units throughout World War II.
35. Rommel, 133. "In these small-scale infantry tactics, in particular, what is wanted is a maximum of caution, combined with supreme dash at the right moment." Rommel was a big advocate of teaching soldiers "tactical tricks" to reduce casualties.
36. Doubler, 256–57. Commanders discovered that soldiers were less fearful of enemy weapons and tactics if they were familiar with them. Once familiar, they could develop countermeasures.
37. Doubler, 250, 254–55. Effective leaders talked to their men frequently in such a manner. The greatest deficiencies for the U.S. infantrymen were bunching together on contact, not using their weapons automatically, and poor noise and light discipline.
38. English, 221. English believes that the spirit of mission-type orders was a great advantage for the German Army in World War II.

Chapter 6

Offense Operations Order

The operations order (OPORD) is the commander's tool for communicating and disseminating the mission plan to his subordinates. The commander can have the most brilliant plan ever conceived in military history; but if he cannot communicate it, it is practically worthless. Since it is the commander's tool, he has the leeway to organize it any way he wishes, as long as the critical information is passed.

The OPORD should be as visual as possible, meaning that the use of sketches, overlays, sand tables, and charts are used to communicate the plan clearly and quickly. This method results in a concise, intelligible plan that exploits memory retention. The strength of the company OPORD is that it is an oral presentation with as little written material as possible. It streamlines the whole process and can be used even when little time is available between preparing the plan and briefing it.

The worksheet serves two purposes. First, it provides structure for the briefing. The briefer treats it as a checklist as he prepares his portion of the OPORD. It allows him to collect his thoughts, to ensure critical information is passed, and to manage the briefing time. Adherence to the worksheet helps the briefer maintain focus. As a guideline, it is particularly useful when the planner is tired and time is limited. The structure also provides a pattern the recipients can anticipate for note taking. Second, it sets the standards for subordinates to emulate. It assists in their intellectual and military development by providing them with a model for planning and briefing. Subordinates have the leeway to create their own OPORD format as long as the decision is based on logic. Ultimately, they judge the value of the format as a conveyer of information.

> **TTP**
>
> Make separate charts of the task organization, the weather data, critical events, and subunit instructions. Laminate them and use grease pencil to fill in the blanks. Use these charts with the sand table.

Task Organization Matrix. Use of a matrix helps the planner quickly allocate his resources. It also informs subordinates unambiguously and immediately of the composition of these resources. The commander may also use it to allocate crew-served weapons or specialty teams as he plans and war games. Displaying the task organization matrix during the OPORD allows subordinates to see how the commander has allocated the company resources in support of the plan. The commander should not use it to allocate task and purpose to the subordinate units. To do so clutters the matrix and can lead to confusion. It is best to use the subunit instruction chart for the allocation of task and purpose. A platoon is normally the base unit for a specific element. An example Task Organization Matrix is as follows:

TABLE 6.1 SAMPLE TASK ORGANIZATION MATRIX

Company Element	Assault Element: 1st Plt (+)	Fire Spt Element: 3d Plt (—)	Security Element: AT Section (+)	Other:
1st Platoon	1st Plt (ME)			
2d Platoon	2d Plt (—)(SE)	1 MG		Feint: 1/2; 1 MG
3d Platoon		1/3, 2/3; 2 MG	3/3	Co Recon: 3/3
AT Section		AT1	AT 2, AT 3	Co Recon: AT sec
Mortar Section		1 & 2 sections		
Attach: Eng Sqd	Eng Sqd (breach)			
Attach: Stinger Tm		Stinger Tm		

The commander has his radio-telephone operator (RTO) complete the task organization matrix chart and display it at the sand table for subordinates to study and copy before the OPORD.

Situation. Briefers provide only that information on enemy and friendly forces that is pertinent to the mission and that was not covered in the warning order. The battalion commander's intent is passed verbatim. The commander can provide his interpretation of the battalion commander's intent in his intent statement. The missions of adjacent units are addressed by task and purpose. The briefer does not need to address the details of adjacent units (grid coordinates and routes) unless they directly affect the company mission. Subordinates are not going to remember the details anyway.

Terrain and weather are addressed as they affect the plan. They are used to help explain specific planning decisions as well as warnings.

Restated Mission. The mission is always conveyed twice. The commander reads it slowly for subordinates to copy. Nothing is more irritating than to have the commander rapidly read the mission statement as though his subordinates are experts in shorthand. The commander is responsible for ensuring subordinates understand the mission. The commander may also have an RTO place the mission statement on a piece of cardboard and display it at the sand table.

Execution. While referring to the sketch or sand table, the commander explains the plan. FM 100-5 defines the commander's intent as, "a concise expression of the purpose of an operation, a description of the desired end state, and the way in which the posture of that goal facilitates transition to future operations."[1] The purpose explains how the company mission is linked to higher HQ plans. The desired end state translates to seizing key terrain that exposes the enemy to destructive fires and ruptures his defenses. Seizure of key terrain that secures the avenues of approach for follow-on forces is also a corollary to the attack. The subsequent posture describes how company success expedites the plans and opportunities of higher headquarters. The intent is concise and straightforward.

As the commander briefs the concept of the operation, he wants to introduce the plan in general terms. The basic plan complements the commander's intent by revealing how the company is going to approach the problem in general terms. Without reference to specific subordinate units yet, the commander conveys the type of maneuver, disposition, and array of the attack. At this point, subordinates make a mental check to ensure that the mission, intent, and graphics are not in contradiction. If any is out of line, the subordinate should bring it up at the end of the briefing. The commander wants the plan to be absolutely clear in everyone's mind. For this reason alone, simple plans are more powerful than complex plans.

At this point, the commander addresses movement and the timetable. He may also elect to have the land navigation plan and route briefed at this point, because this is a logical place to discuss it. Remember, if the company does not make it to the objective on time, it jeopardizes the plans of higher headquarters.

The commander addresses the actions on the objective in detail by critical events and subunit, using task and purpose. The company RTOs display the actions on the objective matrix at the sand table next to the task organization matrix. The commander explains the cadence and tempo of the attack from the departure of the objective rally point (ORP) through to the seizure of the objective or accomplishment of the mission. While going through each event, he integrates how the fire support plan and deception plan support the assault plan. The commander shows how all the pieces come together at each critical event to support the progression of the engagement. Although the basic plan will probably remain, the commander will need to make modifications to the final plan as the recon element and battalion S2 provide new information.

The commander addresses what actions to take if the attack stalls (regroup at the ORP). This is by no means fatalistic or defeatist. It is a professional approach to military operations. If some argue that such talk will lower morale, then the unit probably suffers from morale problems anyway. Regardless, the purpose of rallying the unit is to consolidate and reorganize and to seek a continuation of the attack.

Coordinating instructions are normally those items that are common to all missions but that are still essential. This is an excellent place for the briefing of annexes. Although the commander reminds everyone to review the SOP for consolidation and reorganization, he must still have a basic plan for the hasty defense. He should identify where the general defensive positions will be during consolidation and reorganization (C&R). This should not be too detailed, because it will probably change during the course of the assault; however, by addressing C&R as an event, the company will automatically perform its tasks when alerted. This eliminates the "smoking and joking" attitude during training and the frequent paralysis of company elements in combat after the fight is over.

The commander takes this opportunity to highlight priority information requirements (PIR) for the reconnaissance element to verify the initial assumptions of the tentative plan or to fill in the intelligence gaps. Generally, the commander uses PIR to verify assumptions and alert battalion HQ of a problem that endangers the mission.

Rules of engagement (ROE) are normally used in operations other than war (OOTW). The problem with ROE is that they are typically long, legalistic, and nebulous. The use of the acronym RAMP[2] gives the individual soldier enough guidance to act in all situations without the need of a personal attorney.

Return fire with well-aimed fire

Anticipate attack

Measure the amount of force used

Protect with deadly force only human life and property designated by commander

Service Support. The majority of this information is valuable to the first sergeant (1SG), executive officer (XO), supply sergeant, and company medics. The briefer portrays most of the information graphically and passes the rest rapidly. After the OPORD, the 1SG conducts a fast meeting with these personnel and platoon sergeants concerning the details of service support.

Command and Signal. Most of this information is SOP. In fact, action code word or signals and method of markings should be standardized so that soldiers can recognize them and react immediately. For example, the signal for initiating fires could be machine gun firing from the support element. The code word for switching to the alternate frequency could be "Wolfhound." If possible, the company limits the use of pyrotechnics for signaling, because the enemy by coincidence or design could use the same signal. The enemy does not need to know company pyrotechnic signals to create confusion; he could simply begin randomly firing off all the pyrotechnic munitions he has.

Standardization of code words and signals is not a gross violation of communications security (COMSEC), because the enemy is not likely to focus counterintelligence efforts against a company level—the payoff does not match the effort. It is highly improbable that the enemy would be able to break the company signals or code words, track them to the objective area, and employ countermeasures at the time of action. At the company level, the benefits of standardization far outweigh the risks of compromise.

Using company HQ, platoon, and section standardized code names are also useful:

HQ	Mongoose
1st	Redskin
2d	Bushmaster
3d	Headhunter
AT	Cobra
MORT	Scorpion

Sometimes RTOs do not respond when radioed because they have forgotten their call sign or are not paying close attention. The code name is the best way to get their attention. Code names are also useful on a telephone or radio when discussing other subunits, because using a signal operation instruction (SOI) call sign could be confusing. The use of frequency-hopping radios makes the use of code names nonproblematic, because the enemy cannot intercept the traffic. The thrust of the use of code names and signals is to give the company a command and control edge over the enemy. If the company can act or react faster than the enemy can (inside the decision cycle), it has a better chance of maintaining or seizing the initiative.

NOTES

1. U.S. Department of the Army, *Operations,* Field Manual No. 100-5 (Washington, D.C.: U.S. Government Printing Office, 1993), Glossary-2. One problem with U.S. doctrine is the rapidity in which terms, concepts, and procedures change. For stability sake, the commander updates the SOP, at the most, annually, or after a company level ARTEP. This methodology makes most sense, because the after-action reviews address needed changes and come as no surprise when the SOP is revised.
2. Bolger, 98–100. Lieutenant Colonel Bolger gives an in-depth explanation of the use of RAMP and its successful use at JRTC and Haiti.

TABLE 6.2 OFFENSE OPORD WORKSHEET

| Company Element | Task | | Organization | |
	Assault Element	Fire Support Element	Security Element	Other
1st Platoon				
2d Platoon				
3d Platoon				
AT Section				
Mortar Section				
Attachment				
Attachment				
Attachment				

I. Situation (from battalion OPORD):

  a. Enemy forces:

    1) Disposition/array/composition/strength/forces on objective/ reinforcements/fire support:

    2) Peculiarities/habits: mealtime, alert, and security procedures, light/noise discipline, camouflage, and deception measures:

  b. Friendly forces:

    1) Brigade and battalion mission and commander's intent:

    2) Mission of unit to the left:

    3) Mission of unit to the right:

    4) Mission of unit to the front:

    5) Mission of unit to the rear (reserve, or follow-on units):

    6) Fire support (FA, CAS, AT, tanks, ADA):

  c. Detachments: gaining unit POC, time and location of linkup.

(continued)

TABLE 6.2 *(continued)*

d. Terrain and weather (refer to map, sketch, sand table when briefing):

1) Terrain effects on operations for friendly and enemy forces:

a) Distance available for **O**bservation and fields of fire.

b) Available **C**over and **C**oncealment.

c) Location of natural and man-made **O**bstacles.

d) Identification of **K**ey Terrain.

e) Identification of **A**venues of approach.

2) Current weather data available (from BN OPORD or S-2):

EENT	Sunset	Moonrise	Temp high/low	Wind speed/direction
BMNT	Sunrise	Moon set	% illumination	% precipitation

3) Weather effects on operations for friendly and enemy forces:

a) Mobility:

b) Visibility in meters:

c) Effects on use of NBC and smoke:

e. Probable enemy course of action:

II. Restated mission (from mission analysis) [Read twice]:

III. Execution:

a. Concept of the operation (refer to map, sketch, or sand table):

1) Commander's intent and basic plan:

a) Purpose of the operation:

b) End state signaling success:

c) How success facilitates objectives of battalion mission:

2) Movement:

a) Order of march:

TABLE 6.2 *(continued)*

b) Technique (traveling, traveling overwatch, bounding overwatch) for critical events:

c) Formation (wedge, column, vee, echelon left or right) for critical events:

d) Planning matrix for critical events (refer to map, sketch, or sand table):

Critical Events	NLT Time
1. Depart assembly area to attack position:	
a. Start point (SP)	
b. Release point (RP)	
2. Depart attack position (SP):	
3. Perform passage of lines or begin movement:	
a. SP	
b. Point of departure (PD)/line of departure (LD)	
c. RP	
4. Pass by checkpoints/phase lines	
5. Occupy linkup point	
6. Occupy ORP	
7. Occupy security positions	
8. Occupy support position(s)	
9. Occupy asslt psn (day)/prob line of deployment (PLD) (night)	
10. Assault objective(s)	

(continued)

TABLE 6.2 *(continued)*

b. Actions on the objective—subunit instructions:

Combat Event	Asslt Element (T/P)	FS Element (T/P)	Sec Element (T/P)	Other (T/P)

1. Actions on the objective key:

a) Typical events: occupy positions, preparatory fires, breach, penetration, exploitation, seizure, consolidate and reorganize.

b) Typical main effort tasks: breach, penetrate, destroy key positions, seize main objective.

c) Typical supporting effort tasks: cover ME, attack by fire targets of opportunity, suppress/isolate area of penetration, seize/neutralize peripheral positions, seize secondary objective.

d) Typical fire support element tasks: suppress objective, destroy key weapons, fix enemy units, feint/demonstrate, cover advance of ME, displace to new position.

e) Typical security element tasks: occupy positions/isolate objective area, fix/delay reinforcements, interdict retreating enemy, occupy LP/OPs, reconnaissance.

f) Typical control measures: objectives, phase line, assault position, direction of attack/axis of advance, final coordination line (lift/shift fires), limit of advance.

TABLE 6.2 *(continued)*

2. Reconnaissance tasks (focus on the intelligence gaps):

3. Deception plan (feint, demonstration):

4. Actions to take if attack is repulsed or stalled:

c. Fire support plan:

1. Artillery support plan for route, objective, and consolidation and reorganization.

2. Priority of fire support and when to lift and shift fires.

3. Method of control (on call, signal, scheduled, etc.).

4. Priority alignment of high payoff targets with weapons available:

a) Bunker: LAW/AT-4, M203, machine gun, M224 60mm mortar in direct-fire mode.

b) Tank: M47 Dragon, AT4, field expedient AT device.

c) APC/AFV: LAW/AT4, M203, M-60 MG with armor-piercing rounds.

d) Trench section: Company mortars, M203, grenades, and then clear.

d. Reorganization and consolidation hasty defense guidance:

1. Positions located 100–200 meters from objective.

2. Key weapons cover each avenue of approach and choke points.

3. LP/OPs located on each avenue of approach.

4. Task security element as recon patrol if not designated as LP/OP.

5. Use TRPs (from artillery support plan) for orientation of fires.

6. Method of fire control against targets:

a) Sector or engagement area.

b) Cross fire.

c) Nearest half/far half.

d) Trigger point.

e) TRP.

f) Terrain feature.

g) Restricted fire line (RFL).

(continued)

TABLE 6.2 *(continued)*

7. Priority of targets: Command vehicles, engineer vehicles, tanks, AFV, dismounted troops.

8. EPW guard responsibility.

9. Remind subordinates to refer to **Consolidation and Reorganization.** (chapter 17)

e. Coordinating instructions:

1) NBC MOPP level:
MOPP 0: Mask carried.
MOPP 1: CPOG worn.
MOPP 2: CPOG, boots worn.
MOPP 3: CPOG, boots, mask worn.
MOPP 4: CPOG, boots, mask, gloves worn.

2) CCIR:
Priority intelligence reporting (from S2—Information on the enemy that identifies his intent):

Enemy friendly forces information (from S2—What the enemy is seeking from us to identify our intent):

Essential elements forces information (from commander/S3—friendly actions/resources needed that affect the success of the mission):

3) Constraints and restrictions to mission:

4) Brief-back times and rehearsal times (if not designated in warning order):

5) Exfiltration route (if applicable):

6) Annexes:

    a) Air assault/vehicular/water movement.

    b) Foot march.

    c) Passage of lines.

    d) Land navigation plan.

7) ADA alert status:
- Red: air attack is imminent or in progress.
- Yellow: air attack is probable.
- White: air attack is not probable.

TABLE 6.2 *(continued)*

• Weapons hold: do not fire except in self-defense. • Weapons tight: engage only positively identified hostile aircraft. • Weapons free: fire at any aircraft not positively identified as friendly.

8) Rules of engagement (ROE)—RAMP (OOTW only)
 • **R**eturn fire with well-aimed fire.
 • **A**nticipate attack (Hand SALUTE).
  Hand-what is in their hands?
  Size—how many?
  Activity—what are they doing?
  Location—are they within range?
  Uniform—are they in uniform?
  Time—how soon before they are upon you?
  Equipment—if armed, with what?

• **M**easure the amount of force used (VEWPRIK)
  Verbal warning.
  Exhibit weapon.
  Warning shot.
  Pepper spray.
  Rifle butt stroke.
  Injure with bayonet.
  Kill with fire.

• **P**rotect with deadly force only human life and property designated by commander
9) Preparation of ORP and linkup point:
IV. Service Support:
a. Location of combat trains:
b. Location of field trains:
c. Location of company trains during consolidation and reorganization:
d. Location of battalion LRP:
e. Resupply plan:
f. Location of aid station (if not located with combat trains):
g. Location of company aid post during consolidation and reorganization:
h. Casualty evacuation plan (from attack position to objective):
i. Location of battalion casualty pickup points:

(continued)

TABLE 6.2 *(continued)*

j. Location of company and battalion EPW point, guard and transportation plan:
k. Meal cycle and time of water resupply:
l. Trace of battalion MSR.
m. Tentative LZ/PZ locations for resupply and medevac:
V. Command and signal:
a. Command:
1) CP location during movement and assault:
2) Chain of command:
3) Second in command's location during movement and assault:
4) Battalion TOC and TAC location throughout mission:
b. Signal:
1) Priority of communication: Radio, wire, and messenger; visual, sound.
2) SOI duration and DTG of change:
3) Radio encryption secure fills: Time and location of fill update; duration of each fill and the corresponding positions for each duration:
4) Code words or signals used during mission (as applicable):
a) Initiate fire (signal/code word):
b) Initiate breach (signal/code word):
c) Initiate assault (signal/code word):
d) Lift fire (signal/code word):
e) Shift fire (signal/code word):
f) Switch to alternate frequency (code word):
g) Commit reserve (signal/code word):
h) Objective secured (signal/code word):
i) Consolidate and reorganize (signal/code word):
j) Consolidation and reorganization complete (signal/code word):
5. Challenge and passwords used throughout mission:

TABLE 6.2 *(continued)*

6. Running password:
7. Number combination password:
8. Method of marking breach lanes (wire and minefields):
9. Method of marking mines/booby traps:
10. Method of marking cleared bunker/room:
11. Method of marking entry point into building:
12. Method of marking cleared building:
13. Method of marking casualty collection point at night:

Chapter 7

Foot March Planning

Foot marches are an integral part of war for the infantryman. Even in the modern age, the availability of transportation is not assured, timely, or tactically sound. Infantry companies must conduct foot marches weekly to maintain stamina. If units do not sustain this training, nonbattle casualties will rise, movement will be desultory, and soldier stamina will decrease. In short, if the company cannot get to the battle, it fails the battalion.

Historically, the average progress of an advancing army is 12 miles a day. Rapid foot movements (forced marches) over a short period of time in past wars are a matter of record. Although impressive and laudable, such feats are also rare and do not come without cost to the soldier. Units resort to forced marches (20 to 30 miles over a 9 to 12 hour period) when necessary, but it must also be recognized that the debilitating effects on the readiness of the unit will soon become apparent. Certainly, such a pace rarely extends past three days without exhausting the troops. Commanders should concentrate on conducting foot marches as a matter of routine without making each one a forced march. Forced marches are excellent vehicles for building physical fitness and esprit, but in an extended conflict, the company must be able to march day after day without an appreciable loss in strength. The goal is to execute a foot march well, not to set rate and distance records.

Foot marches are conducted within friendly lines; otherwise such movement past the line of departure is characterized as a movement to contact. The distinction lies in the probability of enemy contact. A foot march moves more quickly and is not organized for enemy contact in the same manner as a movement to contact. If the probability of enemy contact with ground forces is high, a movement to contact is the better movement option.

The annex format is straightforward and succinct. The Enemy Forces paragraph pertains only to the enemy forces that will likely affect the foot march. The same applies to the Friendly Forces paragraph. There is no need to repeat information in the annex if it is covered in the OPORD unless the planner needs to highlight an effect on the foot march itself. For example, if a mechanized unit is scheduled to use the same route, the company will probably move along but off of the road, which will result in a slower rate of march. The most effective method for dissemination is the strip map or overlay. It is succinct and clear. The body of the annex deals with administrative matters, and once planned and executed a few times, a foot march can be planned and executed within 30 minutes. In similar fashion, a road march (vehicular movement) can use the same format with minor modifications as they pertain to the peculiarities of a road march.

For extended distances, the rate of march, water consumption, and the soldier's load assume greater import. The battalion issues the distance to be covered each day. The company calculates the rate accounting for the effects of weather and terrain over which it will traverse. The commander informs the battalion of the closure time and shares the rate calculations from the route overlay, if requested. Frequent rests are extremely invigorating.

Soldiers have their feet checked and cared for. It is amazing how effective drying, powdering, massaging, and wearing dry socks are to the stamina of the soldier. Soldiers rotate their socks. Two pairs should be hanging out to dry on the rucksack or the load-bearing equipment. In inclement weather, socks will dry if placed between the T-shirt and outer jacket (place the sock ends in the waist band). If water is plentiful, a simple wash of the socks extends their life and performance.

Enforcing water consumption prevents many problems with heat injuries. If a soldier falls victim to a heat injury, the commander must strongly inquire down the chain of command exactly how this came to be. Continual sipping of water by taking a mouthful and allowing it to trickle down the throat permits the body to absorb most of it before it becomes urine.

The commander demands that the downward chain of command manages the soldier's load closely. A comfortable load is 48 pounds for extended foot marches.[1] Gimmicks such as directing that heavier equipment be rotated frequently are usually ignored. A soldier carrying a machine gun will not demand that his comrades share the burden. It is a matter of pride that he maintains possession of it. Instinctively, soldiers feel that they are not doing their share if they transfer some of their load to another soldier. If the leadership cannot decide which items should be deleted from the load, the soldiers by default will. History rarely mentions the amount of equipment discarded in war by the soldier or the number that succumbed to carrying heavy loads, but rest assured these were frequent occurrences and need to be addressed. The first sergeant and supply sergeant must transport the company baggage (duffle bags and company equipment). The commander must exploit all means of transportation to lighten the load.

The bottom line is that once soldiers are inured to foot marching, they can perform forced marches and enter into an engagement in strength. But the simple foot march must be practiced frequently in order for the unit to benefit. A unit that performs a simple foot march weekly has better readiness than a unit that performs a 12-mile forced march monthly and a 20-mile forced march quarterly. Leaders need to get away from impressing superiors with amazing feats and stick to the business of soldiering.

TABLE 7.1 FOOT MARCH ANNEX

I. Situation: As it applies to the foot march.

a. Enemy forces:

1) Likely type of enemy contact. Ground—composition, probable strength, and type (partisans or regulars; delaying force, LP/OPs or patrols); Indirect fire—harassment and interdiction, bombardment.

2) Capabilities: Air, artillery, and NBC.

3) Most probable course of action upon contact.

b. Friendly forces:

1) Battalion mission and commander's intent (if given).

2) Unit on line of march to the left.

3) Unit on line of march to the right.

4) Unit to the front.

5) Unit to the rear.

6) Fire support (FA, ADA).

c. Weather and terrain:

1) Current data available (from Bn OPORD or S2):

BMNT:	EENT:	Sunrise:	Sunset:	Moonrise:	24-hr Forecast:
% illumination:	Start/end NVG:	High/low temp:	Moon set:	Wind speed and direction:	% precipitation:

2) Weather effects on foot march for friendly and enemy forces:

a. Soldier's load: no more than 72 lbs. Weigh the need to wear wet weather gear or cold weather gear, or the need to lighten the soldier's load as a result of heat and humidity. Refer to **Land Navigation Plan** for effect on the rate of march. (chapter 8)

b. Mobility (effect on rate of march):

(continued)

TABLE 7.1 *(continued)*

c. Visibility in meters: because of inclement weather or percent illumination.
• The need for traffic control points or guides if critical navigation points are obscured.
• Effect on likely places enemy LP/OPs along the route.

d. Effects on use of NBC and smoke: wind and weather conditions affecting enemy and friendly use.

3) Terrain:

a. Available cover and concealment on route. Identify danger points and plan countermeasures.

b. Effects of man-made and natural obstacles on the rate of march.

c. Effects of choke points on the rate of march.

II. Mission: Who, What (distance of foot march), When (start and end time), Where (destination), and Why.

III. Execution:

a. Commander's intent: The essential factors that will make the foot march a success (speed, security, timeliness, unit integrity).

b. Plan: Strip map or overlay.

1) Start point (SP): Location and time.

2) Route: Phase lines, checkpoints.

3) Traffic control points (TCP).

3) Release point (RP): Location and time.

4) Assembly area/attack position: Location and closure time.

c. Movement:

1) Order of march:

2) Formation: Traveling or traveling over watch.

3) Technique: Column, double column along road side, wedge, vee, etc.

4) Determine total distance to be traveled:
• Use pipe cleaner to measure distance along road (margin of error—add 10% to distance covered).
• Make tic marks on map along the route to show distance covered.

5) Determine effects of elevation on rate of march (**Land Navigation Plan**—10:1 rule). (chapter 8)

TABLE 7.1 *(continued)*

6) Guideline speed: 4.8 kph (3 mph).

7) Rate of march (speed plus 10-minute rest break per hour) [Guideline]:

	Day	Night
Road	4 kmph (2.5 mph)	3.2 kmph (2 mph)
Cross Country	2.4 kmph (1.5 mph)	1.6 kmph (1 mph)
Trail	3 kmph (1.9 mph)	2.2 kmph (1.4 mph)

8) Interval between platoons [Guideline]: 25 meters—night; 50 meters—day.

9) Interval between men [Guideline]: 5 meters.

10) Calculate length of column (Lgthcolm) = number of soldiers x table factor + total column intervals between units.
Table Factor: Single column (5m/man) = 5.4; column of twos (5m/man) = 2.7

11) Calculate pass time (PST) = length of column ÷ rate of march

12) Calculate duration of march to include complete closure: T = D/R + Total rest halt time + PST

13) Determine effects of soldiers load on rate of advance: Subtract 2 km off the total distance covered in six hours for every 10 lbs above 40 lbs.

14) Use **Land Navigation Worksheet and Sketch** for briefing. (chapter 8)

d. Time schedule:

1) Initial inspection:

2) Final inspection:

3) Initial rest halt: After initial 45 minutes of march, rest for 15 minutes. Readjust equipment and load distribution; inspect feet.

4) Follow-on rest halts: Every 50 minutes of march, rest for 10 minutes.

e. Subunit instructions:

1) Lead element: Designate pace man to maintain designated speed. Designate advance guard for early warning.

2) Rear element: Designate rear security to guard against enemy efforts to track the unit.

3) XO:

    a) Establish water resupply point along the route.

    b) Request medic and ambulance support.

(continued)

TABLE 7.1 *(continued)*

c) Establish traffic control points IAW plan.
d) Organize guides at RP to lead company elements into respective positions of the assembly area or attack position.
4) 1SG: TC company vehicle. Pick up stragglers.
5) Commo NCO: Complete commo checks 60 minutes before SP. Ensure commo sheet is current for each RTO for the duration of the march.
6) FIST OFF and MORT SEC LDR: Fire support plan for route. Keep Bn FSO informed of unit location and allow sufficient time for execution of smoke missions without causing the unit any delays.
f. Coordinating Instructions:
1) Squad leaders and team leaders conduct initial and final inspections of soldiers for proper uniform and load IAW packing list. Ensure that the soldiers load is distributed as evenly as practical and that no individual load surpasses 72 lbs.
2) Platoon leaders provide commander with personnel status (PERSTAT) of the number of soldiers participating in the foot march, one hour before SP and a PERSTAT update at each rest stop for strength monitoring.
3) PSG and section leaders recon route to SP before foot march.
4) Leaders check feet at least once every 8 miles. Leaders check one level down to ensure everyone is checked. Medics assist team leaders. Leaders ensure that soldiers consume at least two mouthfuls of water, thirsty or not.
5) Soldiers will take frequent sips of water throughout the march. Do not gulp water. A good method for maximizing the effects of water intake is as follows: take a mouthful of water, tilt head back, and allow the water to trickle slowly down your throat. This slakes thirst over the long term and allows the body to absorb more water.
6) Soldiers massage and powder their feet at the initial rest stop and every 8 miles thereafter. Alert medics to any hot spots or friction spots immediately.
7) Stragglers are to remain on the route for vehicle transportation to the RP.
8) During rest halts, the company forms a cigar-shaped perimeter (hedgehog) on both sides of the route. Each platoon/section moves into available cover and concealment along the route and designates security guards along the flanks.

TABLE 7.1 *(continued)*
9) Platoon leaders radio the commander upon crossing phase lines or passing a checkpoint.
10) At intersections without TCPs, the last two soldiers in the platoon/section halt and ensure that the next platoon/section visually sees and acknowledges the correct direction before rejoining their parent unit.
IV. Service support:
a. Water resupply point: plan for every 12 miles in normal weather, every 4 miles in hot weather.
b. Medics: maintain stock of foot powder, moleskin, and blister treatment devices.
c. Location of company aid post and battalion casualty collection points.
V. Command and signal:
a. Command:
1) Co CP location:
2) Chain of command:
b. Signal: Extracted from OPORD.

Source: FM 21-18, 3-10–3-16. The manual provides rates of march, length of column, and pass time. Steve Boga, *Orienteering* (Mechanicsburg, Pa.: Stackpole Books, 1997), 45–46. Pipe cleaners work exceptionally well for measuring road distances on a map.

NOTES

1.   FM 21-18, 5-4. Calculating the effects of the soldier's load on the rate of march provides the commander with a realistic appreciation of time and space considerations.

Chapter 8

Land Navigation Planning

Few tasks create more anxiety for the leader than land navigation. For an officer and noncommissioned officer (NCO), land navigation skills reflect directly on their competence. Land navigation kudos are neither cumulative nor perpetual. A leader can be half Crow Indian, half mountain man; been everywhere, done everything. But if he gets the unit lost just once, he will never hear the end of it. Whether by disdain or playful bantering, soldiers will regale each other with "The time *he* got us lost" war stories. Subliminal machismo is the main culprit. The soldier who possesses a knack for land navigation is well respected and at times revered. As a result, the pressure on the leader to be on course can be so great that he becomes obsessed—at the expense of good tactical sense. The Army is partly responsible for this mentality. One sure way of failing a patrol in Ranger School is for the leader not to know instantly the patrol's location whenever the ranger instructor asks. There is no leeway for consultation. He either knows it or doesn't. The ranger student is inculcated with the belief that he cannot delegate land navigation to subordinates, carries this experience to his unit, and perpetuates this micromanagement of land navigation.

During movement, the commander should focus his attention on the tactical situation and the mission rather than on keeping his nose pressed against the map and compass. His immediate thoughts should be on the unit's tactical disposition (use of cover and concealment, movement technique), reaction to enemy contact, and the time schedule. As with the captain of a ship, he checks the unit's location through his tasked navigator, stops the unit whenever it loses its track or navigation or whenever there are feedback conflicts, and takes action (map check, scouting party) to get the unit back on track. Good land navigation rests primarily on a good navigation plan.

Land Navigation Planning. Besides reducing the probability of getting lost, the land navigation plan provides the commander with the estimated duration of movement, giving him an excellent insight into time and space considerations. Forearmed with this information, the commander has an idea of how much lag time he has for leaders' recon or contingencies.

> **TTP**
>
> Use the Global Positioning System (GPS) as an aid to navigation, but not as a substitute for land navigation planning.

The commander does not plan the detailed route. He does not have the time. He delegates this task to a platoon leader (hereafter called the chief navigator) during the warning order and gives him general guidance on the route and time constraints. This frees him to work on the decision-making process and the operations order (OPORD). Using the **Land Navigation Planning** checklist and the **Land Navigation Worksheet** (tables 8-1 and 8-2), the chief navigator develops the detailed plan and back-briefs the commander when he is finished or discovers a problem. Once the commander approves the plan, all radio-telephone operators (RTOs) (section, platoon, company, and attachments) copy the land navigation overlay (figure 8-9) for their respective leaders. The first sergeant (1SG) provides quality control for this phase. The chief navigator briefs the plan during the OPORD.

Plan Development. The route consists of a series of smaller segments called legs. Unless the objective is close by, the route consists of numerous legs. A common mistake is to plan a route with only one or two long legs, hoping that one azimuth and straight line distance will make up for the difficulties in terrain. Unless the terrain is incredibly gentle, straight-line routes expend considerable time and exhaust soldiers. Once committed to this inflexible plan, leaders are reluctant to deviate from the azimuth, even if the terrain warrants it, for fear that deviations will lead to inaccuracies, thereby increasing the odds of getting lost. This forces the unit to move over compartmentalized terrain and numerous hills, wade through swamps and streams, and claw through scrub brush and wait-a-minute vines just to remain on course. The greatest fear becomes how enemy contact will disrupt the land navigation effort rather than how the unit will react to it. In effect, the commander becomes captive to an impractical and inflexible method of navigation. Sound familiar?

A good land navigation plan uses a route consisting of multiple legs. Each leg is demarcated between two distinct and recognizable start and end points (also called waypoints), using collecting features, such as a stream, road, wood line, or lake.[1] When traversing featureless terrain, legs are demarcated by time, rate, and distance. Each leg portrays the magnetic azimuth (not grid), distance, and estimated consumption of time. The planner attempts to limit legs to 1,000 meters or less, because pace count inaccuracies, particularly in rough terrain, increase over long distances. A new leg doesn't necessarily mean a new azimuth. It is simply a method for verifying the unit's location and starting a new pace count. The start and end points are means for the unit to pause and verify the progress along the route. Most important, this progress is disseminated throughout the unit for everyone's knowledge. That way, if elements become separated from the company, the new leader has an idea as to his present location and can proceed from there.

TTP

The planner's land navigation kit should include the following:

- Hand calculator
- Magnifying glass
- Pipe cleaners
- Military protractor or clear ruler
- Laminated map scale index
- Laminated copies of field expedient direction determination techniques.[2]

Terrain handrails also assist in route development. A handrail is a terrain feature parallel to the direction of movement such as a ridge, stream, road, railroad embankment, power lines, or wood line. As the name implies, a handrail serves to guide the unit along its route with assurance.[3]

If the collecting feature at the end of a leg is a distinct point, such as a road intersection, the navigator should deliberately offset 10 degrees to the left or right of the point. Each degree of offset results in a deviation of 18 meters to the left or right of a point at 1,000 meters. Having deliberately offset from the point, the navigator can move straight to the point (left or right) at the end of the leg in order to maintain course accuracy.[4] Since intersections are likely to draw the enemy's attention, it is better to send a recon team to verify the location of the point and tell the navigator how far the unit is from the point.

Another navigational aid is a catching feature. This is a clearly identified terrain feature that alerts that the unit has moved too far on a leg. It becomes critical if there no good collecting features to define a leg clearly.[5]

FIGURE 8-1

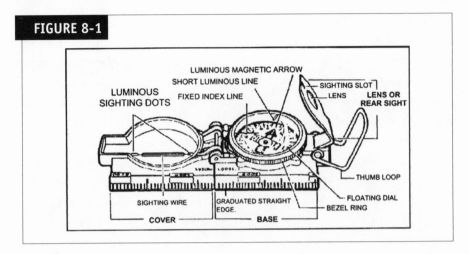

Lensatic Compass

Selection of terrain features to delineate each leg is particularly important for night navigation. The navigator cannot rely on the recognition of distant terrain features (ridge lines, mountains, hills, woods, and so forth), because his depth perception is degraded at night, causing these features to blend together or lose their distinctness. At night a patch of woods could look like a hill or a ridgeline. Trees which line a stream can completely disguise its existence. In this sense, triangulation at night is rarely possible. Verification requires someone to physically check out the feature.

Determination of Azimuth between Two Points. To determine the azimuth between two points,[6] the planner can use a protractor or a Lensatic (military) compass (see figure 8-1)[7] with a map to get the same results. The method with a compass is as follows (see figures 8-2 and 8-3):

- Place a ruler or draw a line between the two points directly on the map.
- Place the graduated straight edge of the compass parallel to the line with the front of the compass pointing toward the direction of travel.
- Determine the grid azimuth. Rotate the bezel ring so that the short luminous line is parallel to the north-south grid lines on the map and is pointing toward grid north. Rotate the compass until the luminous line is aligned with the north-seeking arrow. The reading on the compass (fixed black index line) is the grid azimuth. For civilian compasses, align the bezel ring interior lines parallel to the north-south grid lines of the map with the arrow marker (not the needle) pointing toward grid north.

FIGURE 8-2

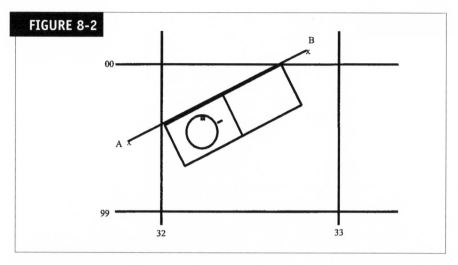

Northern Azimuth

- Determine the magnetic azimuth (also called "accounting for declination"). At the bottom of each map is a declination symbol, normally displaying three arrows. We are concerned only with two. Locate the line with either a star symbol or the letter *G* at the tip. This is grid north. If in doubt, this line is parallel to the north-south grid lines. Next, locate the magnetic north line. It is a solid arrow cut in half with the letter *M* at the tip. This line is either left of (west) or right of (east) the grid north line. The angle between the grid and magnetic lines is the declination. Keeping the compass on the line of travel, rotate the bezel ring toward magnetic north (left is counterclockwise, right is clockwise) the number of degrees between the grid north and the magnetic north lines. (See figures 8-4, 8-5.) Align the short luminous line with the north-seeking arrow and read the azimuth from the fixed black index line—this reading is the magnetic azimuth.[8]

- For precise declination calculations, draw the grid north and magnetic lines on a scratch pad referring to the declination diagram on the map. Next, draw a reference line to the right of the grid north line and at a right angle. Now, draw an arc between the grid and magnetic north lines and insert the number of degrees separating the two (as portrayed on the map); this is the declination angle. Continue by drawing an arc between the grid and reference lines and insert the grid azimuth; this is the grid azimuth angle. The angle between the magnetic and reference lines is the magnetic azimuth. To achieve this, add or subtract the declination angle from the grid azimuth angle. (See figures 8-6,

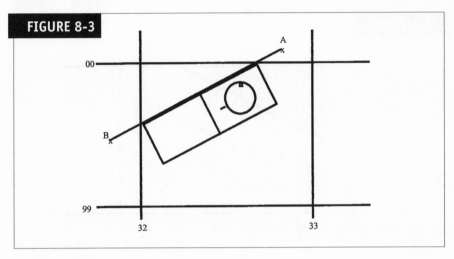

FIGURE 8-3

Southern Azimuth

8-7.) Example: For an eastern declination (the magnetic north line is to the right of the grid north line), the magnetic azimuth is the grid azimuth angle minus the declination angle. For a western declination, the magnetic azimuth is the grid azimuth angle plus the declination angle. This method is used to convert from magnetic to grid azimuths too.[9]

TTP

Each click on the bezel ring represents 3 degrees.[10] This is useful to know when changing azimuths at night.

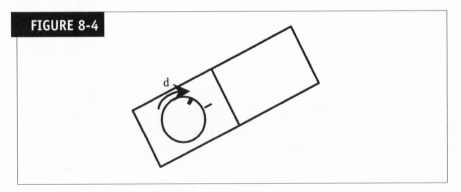

FIGURE 8-4

Eastern Declination

FIGURE 8-5

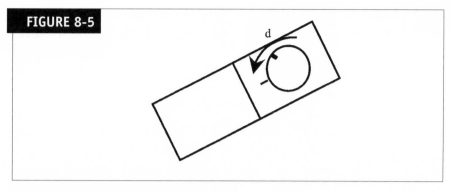

Western Declination

Time-Distance Calculations. Coming to grips with the factors that affect the movement time schedule is no simple matter. Accurate calculations are a product of experience and experimentation. Four factors influence the rate of movement of dismounted infantry: the soldier's load, conditions (weather and climate, percent illumination), vegetation, and terrain. The soldier's load should be below 72 pounds, with 35 pounds being the ideal load. Obviously, the lighter the load, the easier to maintain a steady rate of advance. For planning, subtract two kilometers off the total distance covered in six hours for every 10 pounds carried above 40 pounds.[11] Weather and climate conditions are very difficult to quantify and vary so much that their effect must be recorded over time. It may suffice to understand that in deep snow, thick mud, or heavy rain resulting in slick ground, the company will get there when it gets there. The amount of illumination also affects the rate of advance. The darker it is, the slower the rate of movement. But with enough night training, soldiers can maintain a sufficient

FIGURE 8-6

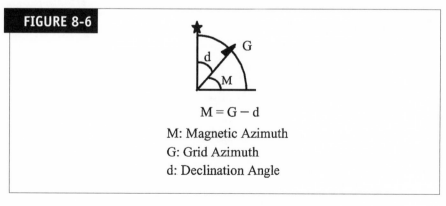

$$M = G - d$$

M: Magnetic Azimuth
G: Grid Azimuth
d: Declination Angle

Eastern Declination Angle Equation

rate in line with the night movement planning rate. Like weather and climate, determining the effect of vegetation on the rate of movement is a virtual unknown until the unit traverses through it. Generally, the unit can avoid heavy vegetation, which may increase the total distance traveled, but not the rate of movement generally. Changes in elevation are no easy matter and are the most crucial factor in route planning.

Calculation of horizontal distance. For flat terrain, simply measure the distance along a leg, using a protractor or laminated map scale. Add 20 percent to the straight-line distance to account for slight course deviations and minor changes in elevation. To determine distances along a road, trail, or stream, use a pipe cleaner with annotated map scale tick marks. Trim the pipe cleaner to an even number on the map scale (for example, 10 kilometers). Bend the pipe cleaner to conform to the road and place a tick mark on the map to indicate the distance covered. To account for error, add 10 percent to the total distance measured.[12] Use a pipe cleaner for movements along the same contour interval (for example, moving around hills or along ridges).

Calculation of elevation conversion. When the route traverses elevated terrain, the planner must account for the effect the slope will have on the rate of march. First, verify the contour interval from the map information! Using a magnifying glass, count the number of contour lines the leg crosses. Subtract the lowest contour crossed from the highest contour level crossed—this reflects the change in elevation or vertical distance.[13] Using the general rule of 10:1, convert the change in elevation to the horizontal equivalent. This means that each meter of rise or fall is equivalent to 10 meters of flat distance for purposes of determining the rate of march.[14] This figure is called the elevation con-

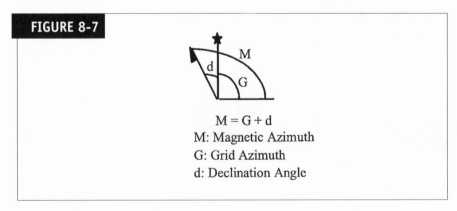

FIGURE 8-7

$$M = G + d$$
M: Magnetic Azimuth
G: Grid Azimuth
d: Declination Angle

Western Declination Angle Equation

version. Because the 10:1 rule already accounts for course deviations, the planner does not add the 20 percent deviation factor. Note that this is not the actual distance the unit must traverse for pace count purposes. For the pace count, the planner should use the Pythagorean Theorem ($a2 + b2 + c2$), where a is the horizontal distance, b is the vertical distance, and c as the slope distance.

Calculation of combined distance. The elevation conversion is added to the horizontal distance to get the combined distance. For example, the horizontal distance between the bottom and top of a hill is 1,000 meters. The change in elevation is 200 meters. The elevation conversion is 2,000 meters. The combined distance for time is 1,000 m + 2,000 m + 3,000 meters.

Calculation of time expenditure. The general planning rate of march on a road is 4 kilometers per hour (kph) during the day and 3.2 kph at night. For flat, open terrain, it is 2.4 kph (day) and 1.6 (night). For daylight movements in other terrain, refer to the following rates:[15]

- Tropical rain forest: 1 kph
- Deciduous forest: 0.5 kph
- Secondary jungle: 0.1–0.5 kph
- Tall grass: 0.5 kph
- Swamps: 0.1–0.3 kph
- Rice paddies (wet): 0.8 kph
- Rice paddies (dry): 2.0 kph
- Plantations: 2.0 kph
- Trails: 3.0 kph

The rate of march factors in a 10-minute break for every hour on the march. Even if such rest halts are not planned, the planner should leave it in to account for unscheduled halts (navigation check, security halt, and so forth). Using the timerate formula, calculate Time = Combined Distance ÷ Rate of March. Continuing with the above example, the night cross country time expenditure over a distance of 1,000 meters with a rise of 200 meters is 3 kilometers ÷ 1.6 kph = 1.9 hours.[16] The land navigation planner can use this calculation to determine whether it is faster to negotiate elevated terrain or bypass it. He must bear in mind that the physical burden on soldiers favors bypassing such obstacles whenever possible. Historical experience reveals that night movements in jungle or deep forests are too slow to be reliable for calculating time and space considerations. Even with emerging technologies for night vision, such movements may still be impossible to execute.

A factor often overlooked is the amount of time it takes for the unit to completely close on the objective rally point (ORP). Most movements are conducted in one column formation, especially at night. The commander must tell the planner how many soldiers will be involved in the movement in order for him to calculate the length of the column. The table factor accounts for the interval between each soldier and the space he fills. The total of column gaps accounts for the gaps (usually 50 meters at night) between the advance guard and between the distinct company elements. Usually, because of the darkness and difficulty of the terrain, the only gap will be between the advance guard and the main body.[17]

These planning figures are a starting point for planners. As part of the after-action review (AAR) process, the commander tasks the land navigation planners with tracking and recording actual expenditure of time for each leg. This results in a planning archive for march rates on legs over various terrain, in differing conditions, and carrying different loads. Tedious and time-consuming? Yes. But it provides the company with precise planning figures for estimating the amount of time required to cover distances. The commander needs this time estimate for determining the correlation of time and space. It also provides junior leaders with an appreciation of such matters, particularly when they become staff officers or NCOs. The commander provides this feedback to the battalion commander, the executive officer (XO), and the operations officer (S3) immediately, because it affects the synchronization of the plan (time and space). Armed with such figures, the commander can bring the S3 back to reality when he plans for the company to cross the line of departure (LD) at 2100, cover 15 kilometers in mountainous terrain, and attack an objective at 0600.

To digress a bit, the same discipline applies to mechanized and motorized movement. The disadvantage in a mechanized movement is that tracked and wheeled vehicles normally cannot move along a straight azimuth. They must contour around obstacles from point A to point B of each leg, making the use of pipe cleaners to determine distances more important. In comparison with other mechanical measuring devices, the pipe cleaner is probably more accurate and easier to manipulate. The good news is that elevations and vehicle loads are hardly limiting factors. But weather affects mechanized and motorized movements more so than dismounted movements. Thus, the mechanized unit needs to maintain a log of actual movements too.

Time Schedule. The time schedule tracks the progress of the unit along its route. Starting with the LD time, each leg's time expenditure is applied to real time. In this manner, the commander can track the progress, alerting him early if the unit falls behind schedule and giving him time to make adjustments. Lastly, the schedule accounts for occupation time of the ORP. The unit should occupy the ORP at least two hours from the attack time to account for the leaders' recon completion of the plan, and the movement into final positions. If the schedule reveals that the unit cannot complete its movement on schedule, the chief navigator back-briefs the commander immediately, with recommendations.

Route Overlay. Once the commander approves the route, the chief navigator makes the master overlay. The overlay portrays each leg with the magnetic azimuth, actual distance, and time expenditure. It also depicts the terrain delimitations of each leg, rails, and limit of advances. Each RTO produces a copy of the master overlay for his leader in time for the OPORD.

Movement. Everyone is responsible for land navigation during movement. The commander emphasizes this at the OPORD, and each subordinate leader echoes this philosophy during his OPORD. Although each soldier does not have a map and compass, leaders need to disseminate the route plan to each soldier and keep him informed of the unit's location throughout the movement. Armed with such information, he is better prepared to carry on with the mission should he become separated from the main body or should his leaders become casualties.

Pace Count. Determining pace count is best calculated over a long course with varying terrain. Once a soldier has determined his pace, he should memorize it or write it on a laminated card and place it in his cap or helmet. In elevated and rough terrain, a normal pace count becomes inaccurate, requiring a modified pace count. Repeated training and experience are the best methods for determining an accurate pace count in all types of terrain. Each soldier should note how the various terrain, weather, and light conditions affect his pace count and record it on his place count card. As with time expenditure calculations, making an archive of modified pace counts helps the unit move with more confidence. Leaders must identify those soldiers with a knack for accurate pace counts and use them.

During movement the lead platoon reports the start and end point of each leg, referring to the Global Positioning System (GPS), if available. Each leader verifies the navigational assessment and concurs or dissents. If the dissension appears valid, the platoon leaders report to the company command post (CP) for a map check. The commander makes the final decision regarding navigational questions as he factors in the influence of time and the tactical situation. To reduce the frequency of navigational dissents, the lead platoon leader focuses on reading the route. He is located next to his main compass man and pace man for continuous consultation. He keeps his thumb on the route marked on the map to help him track progress rapidly. The compass man sights in on landmarks, called steering marks, up ahead to keep the unit moving on the correct azimuth.[18] The platoon leader reads the map on the move, checking it frequently in order to maintain terrain association. At night, this means that he must use his filtered flashlight frequently, even if it detracts from light discipline. If he keeps the flashlight close to the map (penlights are the most effective), the amount of light escaping is very small. He focuses on remembering as much of the map details as possible and on reading the terrain ahead looking for terrain features that will verify the location. He also maintains a broad field of vision, constantly looking to the sides and rear in order to gain an appreciation of the terrain around him.[19] The platoon leader uses GPS only as the final verification of his location. It is better for him to gain an appreciation of the terrain and his skills as a land navigator, than to become mesmerized by a readout console. Since land navigational skills are perishable, the unit must not rely solely on a piece of equipment for navigation, for when it is not functioning, or the batteries die, the unit must be able to continue the mission. In this capacity, the lead platoon leader is not paying much attention to the tactical side of the operation. The platoon sergeant assumes this role and advises the platoon leader when the tactical situation does not merit moving through certain terrain. If the two disagree, the matter is brought to the commander for resolution.

For obstacles blocking the route, the lead platoon leader reports the type of obstacle (swamp, water, clear cut) to the commander and bypasses it. The accepted method is to use the box method: add or subtract 90 degrees from the azimuth, move on this new azimuth until the obstacle is no longer in the path, move another distance along the original azimuth, then add or subtract 90 degrees to get back on the original course. (See figure 8-8.) For minor obstacles, the unit can use the zigzag method. Similar to the box method, the company adds or subtracts 60 degrees, moves a certain distance, then adds or subtracts 60 degrees and moves the same distance to get back on the original route (forming equilateral triangles along the route). Such a technique is useful when negotiating steep terrain too.[20]

The company always navigates to an attack point within one kilometer of the ORP.[21] It is located on or near an easily recognizable and unique terrain feature. The company can use this as a linkup point with its scouts, which can occupy and mark it to ease the linkup. If it is not used for a linkup, it makes an excellent point from which a quartering party can easily locate and set up the ORP for occupation. In this manner, the occupation of the ORP is performed quickly.

Getting Back on Track. Despite all this planning, the unit may get off track as a result of enemy contact, bypassing an obstacle, unrecognizable terrain features, or navigational errors. The worse decision is to continue wandering about in the hope that the unit will stumble on a recognizable terrain feature. The commander halts the company and calls his platoon and section leaders and their navigators to the CP for consultation. Together, they attempt to pinpoint their position, referring to the last identified terrain feature the unit passed. If light conditions are favorable and some clearly identifiable terrain features are in view, the leaders can use the intersection or resection method to determine the unit's location.[22] After considering the possible points where the unit might have strayed, the commander dispatches a couple of patrols to verify the unit location. The commander informs the patrols of the unit's possible location and what terrain features (or depending on the detail of the map, benchmarks, religious shrines or distinctive trees) to look for. He sends back one patrol to the last verified position to attempt to see where the company got off track. He also gives them a return time (15–30 minutes) to preclude them from trekking about for hours.

If unsuccessful, the commander can request an artillery or mortar spotting or illumination round on or over a checkpoint, target reference point, or the objective.[23] Another technique is to have the company recon element (at the objective area) move to a known point away from the objective and pop a star cluster flare. A more dangerous method is to have higher headquarters contact an electronic warfare unit triangulate the company position from its radio transmissions.

Natural Lines of Drift. Natural lines of drift features are ridgelines, spurs, streams, valleys, and so forth—anything that makes movement easier. The enemy also

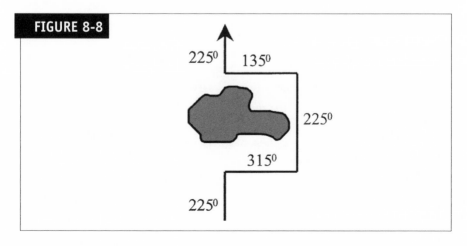

FIGURE 8-8

225^0 135^0

225^0

315^0

225^0

Box Method

recognizes this fact and is likely to focus his reconnaissance effort along them, particularly around the objective area. If following a ridge, the actual route should be between the crest and the valley floor. Use of animal trails are fast tracks and generally too numerous for the enemy to cover. The enemy's assets are limited, particularly on the defense, so he is forced to focus his efforts on choke points and lines of drift. If the company is forced to use an avenue along a line of drift, the commander should determine how far along to use it, dispatch an advance guard to clear the way, and follow behind it at a distance where the main body is not engaged if the advance guard makes enemy contact.

CONCLUSION

Delegating land navigation to a subordinate ensures that this crucial aspect of the operation receives the proper attention it deserves even when time is compressed, allowing the commander to concentrate on the tactical plan. Preparing a land navigation plan gives subordinates a better appreciation of the terrain and the time and space factors associated with it. This will reduce the amount of friction associated with movements and also make them better leaders and staff officers/NCOs later on in their careers. Lastly, the commander can focus his attention on the tactical situation during movement, knowing that subordinate leaders are tracking land navigation properly.

TABLE 8.1 LAND NAVIGATION PLANNING
1. Receive initial guidance from commander during warning order regarding the route:
a. General primary and alternate routes with cover and concealment.
b. Times for SP, occupation of ORP, and time of attack.
c. Soldier's load—maximum weight ceiling.
2. Plan tentative primary and alternate routes using map recon:
a. Designate the legs of the route:
1) The start- and endpoints (waypoints) of each leg are demarcated by collecting features (clearly defined and recognizable terrain features as an aid to navigation).
2) Distinct changes in elevation (slopes) are designated as legs also.
3) Navigational handrails (terrain features which run parallel to the leg) such as streams, ridges, roads, etc. are used as a navigational aid whenever possible.

TABLE 8.1 *(continued)*

4) When handrails are not available, plot the leg endpoint to a collecting feature, such as a road or stream intersection, deliberately aim off (about 10 degrees) to the left or right of the point.

b. Determine the magnetic azimuth of each leg:

1) Using a protractor or a compass directly on the map, determine grid azimuth.

2) Convert grid azimuth to magnetic azimuth on each leg.

c. Determine the distance in meters for each leg:

1) Using the scale on the protractor or the map, measure the horizontal distance of each leg. Add 20% to distance measured in order to account for minor deviations and changes in elevation.

2) For road or trail movement, use pipe cleaner to measure distance traveled. Add 10% to distance to account for error.

3) Determine the change in elevation by subtracting the lowest contour line from the highest contour line of each leg.

4) Using 10:1 rule, convert change in elevation into horizontal distance equivalent for time calculation (e.g., 100 meters change in elevation equals 1 km of level ground), called elevation conversion. Do not add 20% to the horizontal distance.

d. Estimate travel time per leg:

1) Rate of March:	*Day*	*Night*
Road	4.0 kph	3.2 kph
Cross country	2.4 kph	1.6 kph
Trail	3.0 kph	2.2 kph
Deciduous forest	0.5 kph	
Tropical rain forest	1.0 kph	
Secondary jungle	0.1–0.5 kph	
Tall grass	0.5 kph	
Swamps	0.1–0.3kph	
Rice paddies (wet)	0.8 kph	
Rice paddies (dry)	2.0 kph	
Plantations	2.0 kph	

2) Time expenditure = (Horizontal + Elevation Conversion Distances) ÷ Rate of March.

(continued)

TABLE 8.1 *(continued)*

3) Soldier's load factor: Subtract 2 km for every 6 hours of movement for each 10 lbs. load over 40 lbs (i.e., a 70 lb. load results in a loss of 6 kms over 6 hours marched from the total planning distance).

e. Calculate closure time.

1) Calculate length of column (Lgthcolm) = number of soldiers x table factor + total column gaps between units. Table factor: Single column (5m/man) = 5.4; (2m/man) = 2.4.
2) Closure time = length of column ÷ rate of march.

f. Identify a catching feature (clearly defined and recognizable terrain feature) to alert the unit that it has gone too far on a leg (use if no catching feature exists for a leg).

g. Identify an attack point within 1 km of the ORP on or near recognizable terrain feature to serve as linkup point for company recon or as a point for precision navigation to ORP.

3. Back-brief commander:

a. Provide commander with the **Route Worksheet** and **Route Overlay.**

b. Upon approval of the route, add rally points (RP).

c. Coordinate with FIST for TRPs along the route to include RPs.

4. Prepare route briefing for OPORD:

a. Company, platoon, and section RTOs report to platoon CP to copy route overlay.

b. Platoon RTO (and assistants) prepares route sand table if required.

TABLE 8.2 LAND NAVIGATION WORKSHEET

Leg #	CP/Grid	CP Description	Azimuth	Horizontal Distance (+ 20%) *	Change in Elevation	Elevation Conversion (10:1)	Combined Distance HD + EC	Time Expenditure T = CD/Rate	Time Schedule SP: 1900	Remarks
1	m/123456	LD	47°	960 m	—	—	960 m	36 min	1936	Rate / 1.6 kph
2	k/123456	stream	47°	360 m	—	—	360 m	14 min	1950	
3	a/123456	ridge bottom	88°	1320 m	—	—	1320 m	50 min	2030	
4	s/123456	stream	354°	450 m	+ 40 m	400 m	850 m	32 min	2102	Contour Interval = 20 m
5	g/123456	ridge bottom	354°	400 m	+ 100 m	1,000 m	1400 m	53 min	2155	
6	b/123456	trail	354°	350 m	- 80 m	800 m	1150 m	43 min	2238	
7	z/123456	ridge bottom	354°	600 m	- 20 m	200 m	800 m	30 min	2308	
8	f/123456	Highway 1	336°	960 m	—	—	960 m	30 min	2338	
9	c/123456	hill	306°	1500 m	—	—	1500 m	56 min	0034	
10	e/123456	river bend	260°	420 m	—	—	420 m	16 min	0050	Linkup Pt
	ORP								0115	Closure Time: 25 min
	Total			7320 m				5.6 hrs	0500	Attack

*Note: Do not add 20% course deviation for legs traversing elevated terrain; the 10:1 rule already factors that in.
Add 10% to distance measured on a road movement to account for measurement error of pipe cleaner.

FIGURE 8-9

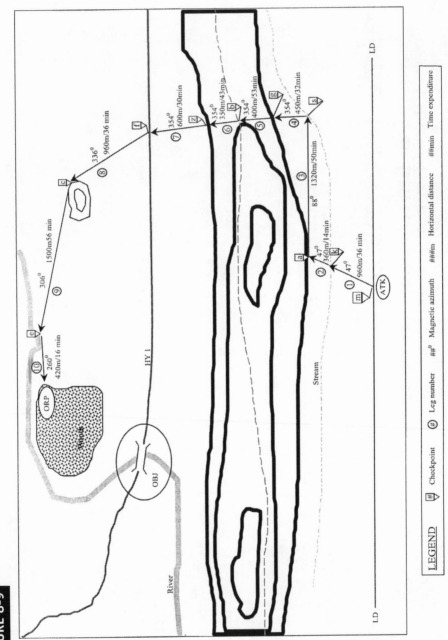

LEGEND   ⊞ Checkpoint   ⑂ # ▷ Leg number   ##⁰ Magnetic azimuth   ###m Horizontal distance   ##min Time expenditure

NOTES

1. Boga, 180. "[C]ollecting feature: a feature such as a reentrant or pond, that crosses your path . . . and is used to funnel you or direct you along the way." W. S. Kals, *The Land Navigation Handbook* (San Francisco, Calif.: Sierra Book Clubs, 1983). Kals uses collecting feature and catching feature interchangeably. FM 21-26 only uses the term catching feature.

2. U.S. Department of the Army, *Map Reading and Land Navigation,* Field Manual No. 21-26 (Washington, D.C.: U.S. Government Printing Office, 7 May 1993), 9-6–9-10.

3. Boga, 184. "[H]andrail: linear feature used to guide an orienteer along a route . . . such as trails, streams, fences, and power lines."; FM 21-26, 11-2.

4. FM 21-26, 9-6.

5. Boga, 180. "[C]atching feature: a feature such as a road, fence, or hillside, that is usually . . . perpendicular to your route. It catches the attention of the orienteer who has moved beyond [a leg] and tells [the unit that it] has gone too far."; FM 21-26, 11-2.

6. Kals, 106–108, 112–13; Boga, 45–46.

7. FM 21-26, 9-2.

8. FM 21-26, 11-1–11-3.

9. FM 21-26, pp. 6-6–6-8.

10. FM 21-26, 9-1, 9-4. Refer to FM 21-26 for a detailed explanation on using the bezel ring at night.

11. U.S. Department of the Army, *Foot Marches,* Field Manual No. 21-18 (Washington, D.C.: U.S. Government Printing Office, June 1990), 5-4.

12. Boga, 45–46.

13. FM 21-26, 10-1–10-3. FM 21-26 provides a detailed explanation of working with contour intervals.

14. Kals, 48, 207; Boga, 116–17. Boga uses a formula of 15 meters of rise equaling 100 meters of level ground.

15. FM 21-26, 13-6. No guidance is provided for night movements. In heavy forests or jungles, night movement is so difficult that the distance covered is not worth the exertion required.

16. FM 21-18, 3-11; Kals, 48–50. Kals cites three other formulas for march rates and time expenditure calculations over elevated terrain: *Wood Method*—Plan a rate of march of three kilometers per hour. For each 300-meter rise, add 60 minutes (that's one minute per 5 meters). For each 300-meter fall, add 30 minutes. This method assumes a light load of only 30 pounds per soldier. *Wolf Method*—Plan a rate of march of 3.7 kilometers per hour. For each 200-meter rise, add 60 minutes. *Karlschmidt Method*—Plan a rate of march of one kilometer per 10–15 minutes. For each 100 meter rise, add 15 minutes.

17. FM 21-18, 3-12–3-14.

18. FM 21-26, 11-10.

19. Boga, 51–52.

20. Kals, 210–15; FM 21-26, 9-5; Boga, 57.

21. FM 21-26, 11–12. The navigation attack point is actually a catching feature. At this point, the navigator relies on a strict azimuth without deviation.

22. FM 21-26, 6-9–6-12.

23. FM 21-26, 13-3.

Chapter 9

Soldier's Load

Like the weather, everyone talks about the soldier's load but does nothing about it. Generally, Army leadership gives the soldier's load lip service, because calculating it requires detailed planning and assumes risk. Junior leaders must jealously guard the amount of weight their soldiers must carry into combat. Proper monitoring of the soldier's load requires junior leaders to monitor individual loads, because the weight each soldier must carry depends on his duty position and the immediate mission. Calculations require attention to detail, because the details account for the load creep.

While in Afghanistan, I had heard a story in which the Taliban referred to the American soldiers as turtles, because as they trudged up the mountainous terrain with all their body armor, helmets, and rucksacks, they looked like ponderous turtles. While it may be true that body armor ameliorates the extent of injuries, it also reduces the soldier's agility and speed, keeping him in the kill zone longer. The extra gear also makes it much tougher to pursue fleeing insurgents in mountainous terrain.

Historical studies conclude that the optimum training load for soldiers is 30 percent of the soldier's body weight but not more than 45 percent of his body weight. Thus, the bigger the soldier (muscle not fat), the more he can carry. This does not mean, however, that the larger soldiers become the company pack mules. Accounting for an average body weight of 160 pounds, this translates to an ideal load of 35 pounds, a comfortable or sustainable load of 48 pounds, and a maximum load of 72 pounds. Hot weather and dehydration also exacerbate the burden of the load, causing fatigue, which in turn causes a loss of will power and courage. No amount of physical training is going to strengthen soldiers to surpass these loads.[1]

Obviously, the combat load should be as light as possible. Historical evidence reveals that a relationship exists between combat stress (danger and fear) and physical effort. Clausewitz regarded physical effort as another source of friction, because its limits could not be gauged, seeming to "chain the spirit and secretly wear away men's energies."[2] S. L. A. Marshall noted that fear zaps a soldier's strength. A heavy load not only depletes a soldier's strength, but it also makes him more susceptible to the effects of combat stress. A vicious cycle results, fatigue intensifies the influence of combat stress, which in turn exacerbates the effects of fatigue. The end products are exhaustion, paralysis, and panic, rendering the soldier ineffective on first contact. That this connection is not made in peacetime is because of the lack of danger.[3]

Getting soldiers and equipment to the battlefield is not enough—the soldier must be in a condition to fight. Since maneuver is key to tactical success, leaders are responsible for ensuring that soldiers carry the minimum weight necessary for unit accomplishment. The battalion staff has no business dictating what the soldier's load will be in this regard, since staff tendencies lean toward equipping the soldier for every contingency.[4] Responsibility therefore falls on the company commander, because he considers all factors that affect the mission. His goal is for the soldiers conducting the assault to bear only 32 pounds and for the soldiers in the fire support and security elements to bear only 48 pounds.[5] The soldier's load is always his calculated risk in this regard. To mitigate this risk, the resupply plan during consolidation and reorganization takes on critical emphasis. Before the unit departs the assembly area, soldiers place the equipment they will need upon consolidation and reorganization into their A-bags (duffel bag) or rucksacks (if not taken) and consolidate them. The supply sergeant loads these into trucks or configures pallets for sling loading forward during consolidation and reorganization. The supply sergeant supervises the work detail and ensures that unit integrity is maintained. The supply sergeant accompanies the baggage forward and supervises the distribution to company elements during consolidation and reorganization once the critical items of ammunition and water have been resupplied.

Because the mission dictates which equipment the soldier must carry, the commander and his leaders consider what equipment to add to the minimum fighting load to produce the approach march load. Table 9-1 reflects the minimum-approach march load; the true approach-march load will also include items in the combat mission and sustainment load. The approach-march load will exceed the maximum load of 72 pounds for specific soldiers if the leadership applies it blindly across the company, so the company leadership needs to ensure that soldiers carrying a heavy fighting load have a minimum approach-march load. To make heavier loads more bearable and to allow the company to maintain a good march rate, heavier equipment is rotated during the approach march. Soldiers with light loads are designated to carry a mortar round each to ensure the mortars have enough rounds to sustain the mission.

> **TTP**
> Soldiers rotate heavy equipment at the end of each land navigation leg. Team leaders supervise.

Table 9-1 provides a framework for soldier's load planning. Given the rapidity in which new equipment is fielded today, this book can only provide a snapshot of equipment weights. Leaders can obtain weight data from various websites, field manuals and technical manuals. Whenever discrepancies exist among manuals, leaders should select from the manual which is dedicated to the piece of equipment. The websites listed at the end of Table 9-1 provide only a small example of the resources available on-line. Lastly, leaders should verify the true weights using scales and test the weight themselves by use of a road march. Only by comparison can they judge proper burden.[6]

A review of the combat and existence loads within the planning weight data reveals that the goal weights are difficult to attain. This shows just how difficult it is to maintain the soldier's load to a reasonable level. Normally, the uniform common to all and the basic ammunition loads are dictated from above. Adding in the approach-march load causes the soldier's load to exceed the maximum load of 72 pounds—in some cases exorbitantly. The commander is obligated to brief the battalion commander of the consequences of the burden on the soldier and the mission. He should also make recommendations of where the cuts should be made. Some recommended trimming is as follows:

- Move protective mask to A-bag unless there is a chemical threat
- Move bayonet to rucksack
- Move one canteen of water to rucksack
- Cut the basic load of ammunition in half
- Replace Dragon AT weapons with AT4 and LAWs whenever a light armor threat is likely
- Move sleeping pad, poncho and liner, sleeping bag, sleeping shirt, two pair of socks, towel, and undershirt to the A-bag

TABLE 9.1 SOLDIER'S LOAD PLANNING*

Item	Quantity	Weight	Cumulative Weight
Uniform Common to All			
ACU	1 ea	2.95	
Boots, combat (TWCB)	1 pr	6.00	
Drawers, cotton	1 pr	0.10	
Undershirt, cotton	1 ea	0.30	
Socks, cushion	1 pr	0.20	
Helmet, Kevlar (Medium)	1 ea	3.70	
MOLLE Load Bearing Vest with belt & first aid pouch	1 ea	3.00	
Camelbak ® with 100 oz of water	1 ea	7.38	
Subtotal		23.63	**23.63**
Fighting Load (Goal: < 48 lbs)			
Leader			
M4	1 ea	5.65	
30 rds 5.56 w/magazine	7 ea	7.07	
Flashlight	1 ea	0.80	
Compass	1 ea	0.25	
Subtotal		13.77	**37.40**
Rifleman			
M4 Carbine	1 ea	5.65	
30 rds 5.56 w/magazine	7 ea	7.07	
Bayonet w/scabbard	1 ea	1.80	
Subtotal		14.52	**38.15**
Grenadier			
M203 (empty)	1 ea	3.00	
30 rds 5.56 w/magazine	7 ea	7.07	
40 mm rd	36 ea	18.00	
Vest	1 ea	0.40	
Subtotal		28.47	**52.10**

TABLE 9.1 *(continued)*

Machine Gunner			
M240B Medium MG	1 ea	26.70	
M9 pistol w/2 mags	1 ea	2.90	
7.62 Lnk (100 rds)	2 ea	14.00	
Subtotal		44.50	**68.13**
Machine Gun Assistant Gunner			
M4 Carbine	1 ea	5.65	
30 rds 5.56 w/magazine	7 ea	7.07	
7.62 Lnk (100 rds)	2 ea	14.00	
Subtotal		26.72	**50.35**
RTO			
M9 pistol w/2 mags	1 ea	2.90	
SINGARS (RT-1583E) w/battery	1 ea	9.00	
Subtotal		11.90	**35.53**
SAW Gunner			
M249 SAW	1 ea	16.41	
5.56 lnk (200 rds)	2 ea	13.84	
Subtotal		30.25	**53.88**
Mortar Gunner			
M225 mortar	1 ea	15.30	
M9 pistol w/2 mags	1 ea	2.90	
Subtotal		18.20	**41.83**
Mortar Ammo Bearer			
M8 Base plate (small)	1 ea	3.80	
M4 Carbine	1 ea	5.65	
30 rds 5.56 w/magazine	7 ea	7.07	
M170 Bipod	1 ea	15.40	
Subtotal		31.92	**55.55**

(continued)

TABLE 9.1 *(continued)*

Mortar Squad Leader			
M4 Carbine	1 ea	5.65	
30 rds 5.56 w/magazine	7 ea	7.07	
Mortar poles and bag	1 ea	2.00	
M67 sight	1 ea	2.25	
Subtotal		16.97	**40.60**
Mortar Section Leader			
M4 Carbine	1 ea	5.65	
30 rds 5.56 w/magazine	7 ea	7.07	
Ballistic computer (M30)	1 ea	8.00	
Mortar poles and bag	1 ea	2.00	
M67 sight	1 ea	2.25	
Subtotal		24.97	**48.60**
AT Gunner			
M4 Carbine	1 ea	5.65	
30 rds 5.56 w/magazine	7 ea	7.07	
Command Launch Unit	1 ea	14.08	
Javelin round	1 ea	34.98	
Subtotal		61.78	**85.41**
AT Assistant Gunner			
M4 Carbine	1 ea	5.65	
30 rds 5.56 w/magazine	7 ea	7.07	
Javelin round	1 ea	34.98	
Subtotal		47.70	**71.33**
AT Team Leader			
M4 Carbine	1 ea	5.65	
30 rds 5.56 w/magazine	7 ea	7.07	
Binoculars	1 ea	3.20	
Javelin round	1 ea	34.98	
Subtotal		31.92	**74.53**

TABLE 9.1 *(continued)*

AT Section Leader			
M4 Carbine	1 ea	5.65	
30 rds 5.56 w/magazine	7 ea	7.07	
Binoculars	1 ea	3.20	
Subtotal		15.92	**39.55**

MINIMUM APPROACH MARCH LOAD (GOAL: <72LBS)

Item	Quantity	Weight
Assault Pack, MOLLE	1 ea	3.00
E-tool with carrier	1 ea	2.50
MRE	2 ea	3.00
Parka, Gortex	1 ea	1.50
Socks	1 pr	0.30
Towel	1 ea	0.20
Undershirt, cotton	1 ea	0.30
Subtotal		**10.80**

MISSION CONTINGENCY/SUSTAINMENT LOAD ITEMS

Item	Quantity	Weight
Bag, duffel	1 ea	3.50
Bag, Sleeping, MOLLE	1 ea	16.80
Bag, waterproof	1 ea	0.80
Bayonet w/scabbard	1 ea	1.80
Binoculars	1 ea	3.20
Black gloves and snap link	1 pr	0.75
Body Armor, Interceptor (complete)	1 ea	16.40
Bore light, Laser (AN/PEM-1)	1 ea	0.28
Canteen, cup, water (1 qt)	2 ea	2.60

(continued)

TABLE 9.1 *(continued)*

Cap, BDU	1 ea	0.30
Cap, pile	1 ea	0.26
Combat ID for Dismounted Soldier (CIDDS)	1 ea	2.00
Flare, trip	1 ea	1.00
Flash (M202A1) multishot rocket launcher	1 ea	26.60
Grenade, fragmentary (M67)	1 ea	0.88
Grenade, smoke (AN-M8 HC)	1 ea	2.00
Grenade, colored smoke (M-18)	1 ea	1.19
Grenade, white smoke (M-83)	1 ea	1.00
Grenade, incendiary (AN-M14 TH3)	1 ea	2.00
Grenade, concussion (MK3A2)	1 ea	0.98
Grenade, stun (M-84)	1 ea	0.52
Grenade, 40 mm	1 ea	0.50
Hygiene Kit	1 ea	2.00
Intercom system, soldier	1 ea	1.40
Jacket, field	1 ea	3.30
Light Anti-tank Weapon (M72A3)	1 ea	5.50
Light Anti-tank Weapon, AT-4 (M-136)	1 ea	14.80
Light, Aiming, Infrared (AN/PAQ-4C)	1 ea	0.36
Light, Aiming, Target Pointer, Illuminator, Infrared (AN/PEQ-2)	1 ea	0.47
Liner, field jacket	1 ea	0.70
Mess kit	1 st	2.80
Mine, AP (M16A1)	1 ea	8.25
Mine, AT (M-21)	1 ea	18.00
Mine, Claymore (M18)	1 ea	3.50
Mortar base plate—large (M7), 60 mm	1 ea	14.80
Mortar round, 60 mm	1 ea	3.50
NBC suit (JSLIST)	1 st	9.60
Night Vision Goggles (AN/PVS-7D)	1 st	1.80
Night Vision Monocular (AN/PVS-14D)	1 st	0.86

TABLE 9.1 *(continued)*		
Overshoes	1 pr	4.20
Pads, knee/elbow	1 pr	1.70
Pad, sleeping	1 ea	1.30
Polypropylene underwear	1 st	0.50
Poncho	1 ea	1.50
Poncho liner	1 ea	1.60
Precision Lightweight GPS Receiver (PLGR) AN/PSN-11	1 ea	2.75
Protective mask (M-40)	1 ea	4.50
Radio, MBITR (AN/PRC 148) w/battery	1 ea	1.90
Radio, (PRC 6725E) w/battery	1 ea	3.00
Rain suit, Improved	1 st	2.90
Rifle, M16A2	1 ea	7.80
Rucksack, MOLLE (Complete)	1 ea	16.80
Scarf, wool	1 ea	0.40
Shelter half (3 poles, 5 pegs, 1 rope, 1 canvas)	1 st	4.50
Short-Range Assault Weapon (SRAW)	1 ea	21.25
Sleeping bag, Gortex ®	1 ea	3.50
Sleeping shirt	1 ea	0.70
Sling rope, work gloves, snap link	1 ea	1.90
Socks	4 pr	1.20
Spare barrel w/bag for M-240B MMG	1 st	8.00
Telephone, TA-1	1 ea	2.00
Thermal Weapon Sight—Lt (AN/PAS-13): M4	1 ea	3.00
Thermal Weapon Sight—Med (AN/PAS-13): M249 & M240B	1 ea	4.50
Thermal Weapon Sight—Hvy (AN/PAS-13): M2, MK-19	1 ea	5.00
Tripod with T & E mechanism (M122A1) for M240B MMG	1 ea	19.50
Trousers, Gortex ®	1 pr	1.90
Water	1 qt	2.67

(continued)

* Weights derived from various sources to include:

Jane's Infantry Weapons 2001–2002, edited by Terry J. Gander (United Kingdom: Thomson Company, 2002).

U.S. Department of the Army, *Antipersonnel Mine M18A1 and M18 (Claymore)*, Field Manual No. 23-23 (Washington, D.C.: 6 January 1966), 2-1.

U.S. Department of the Army, *Crew-Served Machine Guns 5.56-mm and 7.62-mm*, Field Manual No. 3-22.68 (Washington, D.C.: U.S. Government Printing Office, 21 July 2006), 1-2, 3-2.

U.S. Department of the Army, *Foot Marches*, Field Manual No. 21-18 (Washington, D.C.: June 1990), 5-6, 5-7.

U.S. Department of the Army, *40-MM Grenade Launcher, M203*, Field Manual No. FM 3-22.31 C1 (FM 23-31) (Washington, D.C.: U.S. Government Printing Office, 19 March 2007), 2-3.

U.S. Department of the Army, *Grenades and Pyrotechnic Signals*, Field Manual No. 3-23.30 (Washington, D.C.: 1 September 2000), Chapter 1.

U.S. Department of the Army, *Javelin Medium Antiarmor Weapon System*, Field Manual No. FM 3-22.37 (Washington, D.C.: U.S. Government Printing Office, January 2003), 1-3.

U.S. Department of the Army, *Jungle Operations*, Field Manual No. 90-5 (Washington, D.C.: 16 August 1982), H-1–H-5. Note how the soldier's combat load is stripped down to accommodate more water. In jungle operations, water is the major load concern.

U.S. Department of the Army, *Mortar Gunnery*, Field Manual No. 23-91 (Washington, D.C.: 1 March 2000), Chapter 6.

U.S. Department of the Army, *Rifle Marksmanship M16A1, M16A2/3, M16A4, and M4 Carbine*, Field Manual No. 3-22.9 (FM 23-9) (Washington, D.C.: U.S. Government Printing Office, April 2003), 2-1.

U.S. Department of the Army, *Shoulder-Launched Munitions*, Field Manual No. 3-23.25 (Washington, D.C.: U.S. Government Printing Office, January 2006), 2-2, 5-2.

U.S. Department of the Army, *Soldier's Guide for the PLGR* (Fort Monmouth, NJ: U.S. Army GPS Project Office, January 1995), 4.

U.S. Department of the Army, *Tactical Employment of Mortars*, Field Manual No. 7-90 (Washington, D.C.: U.S. Government Printing Office, 9 October 1992), Chapter 1.

U.S. Department of the Army, *TOW Weapon System*, FM 3-22.34 (FM 23-34) (Washington, D.C.: U.S. Government Printing Office, November 2003), 1-1, 2-4.

U.S. Department of the Army, *United States Army Weapon Systems 2002* (Washington, D.C.: U.S. Government Printing Office, 2002).

Internet Sites

AN-PVS7D: http://www.morovision.com/littonanpvs7d.htm.

AN-PVS14D: www.morovision.com/littonpvs14d.htm.

AT-4 LAW: http://www.fas.org/man/dod-101/sys/land/at4.htm.

Hand Grenades: http://www.fas.org/man/dod-101/sys/land/grenade.htm.

Javelin: http://www.army-technology.com/projects/javelin; www.strategypage.com/fyeo/howtomakewar; http://www.fas.org/man/dod-101/sys/land/javelin.htm.

Land Warrior: www.army.mil/soldiers/jan2000/pdfs/todaysoldier.pdf.

M224 60mm Mortar: http://www.fas.org/man/dod-101/sys/land/m224.htm.

M203 Grenade Launcher: http://www.colt.com/colt/html/a2f27_m203grenade.html.

M240 Machine Gun: http://www.hk94.com/m240b.html; http://www.fas.org/man/dod-101/sys/land/m240g.htm.

M249 SAW LMG: http://www.fas.org/man/dod-101/sys/land/m249.htm.

M4 Carbine: http://www.colt.com/colt/html/a2f20_m4carbine.html; http://www.fas.org/man/dod-101/sys/land/m16.htm; http://www.hk94.com/m4a1.html.

M9 9mm Beretta Pistol: http://www.fas.org/man/dod-101/sys/land/m9.htm; http://www.hk94.com/black2.html.

MBITR Radio, PRC-6725E: http://www2.thalescomminc.com.

Rock Island Arsenal Mortar Team: http://tri.army.mil/LC/Cf/Cft/Cftm/cftm.htm.

U.S. Army Material Command: ttp://aeps.ria.army.mil/aepspublic.cfm.

U.S. Land Warfare Systems: http://www.fas.org/man/dod-101/sys/land/index.html.

If the battalion commander cannot accept a reduction of the soldier's load, the commander should recommend that troop transportation assets be exploited for movement and that dismounted distances be shortened. The commander can also employ nonmission-essential soldiers or local civilians as porters to load mission contingency and sustainment items in rucksacks. In such cases, porters can each carry 120 pounds over 20 kilometers per day.[7] Porters are not to be used in combat, because their fatigue level will render them ineffective.

History is replete with instances of higher leadership burdening soldiers with noncritical mission loads in order to meet every contingency. Numerous casualties and failed attacks are directly attributed to this lack of judgment. Perceiving that higher leadership was not up to this task but also was unwilling to delegate it to lower echelons, soldiers discarded unnecessary equipment early in a campaign only to suffer terribly when the weather conditions changed. The bottom line is that if the command refuses to address the soldier's load, the soldier will make the decision himself—with deleterious results.

NOTES

1. S. L. A. Marshall, *The Soldier's Load and Mobility of a Nation* (Washington, D.C.: The Combat Forces Press, 1950), 20, 26–28, 30, 48–50. Even this upper limit is probably too much. Marshall cites studies that conclude that the tolerable load for road marching is 40–45 pounds.

    The Infantry Soldiers Load (TA 1122-84A). An old mimeographed briefing I obtained at the 7th Infantry Division. This study did not have any further documentation. Its conclusions follow the themes of Marshall, but I have not discovered the source. English, 227. British and German studies concluded that the optimum load is one-third of a man's weight. FM 21-18, 5-3–5-6. "Battlefield stress decreases the ability of soldiers to carry their loads. Fear burns up the glycogen in the muscles required to perform physical tasks. . . ." "The fighting load . . . should not exceed 48 pounds and the approach march load should not exceed 72 pounds." Ellis, 33. World War I loads were as follows: British, 60–77 lbs; Germans, 70 lbs; French, 85 lbs. Ellis repeats the British studies which concluded that the optimum load for a soldier was one-third his weight.

2. Carl von Clausewitz, *On War,* edited and translated by Michael Howard and Peter Paret (Princeton, N.J.: Princeton University Press, 1976), 115. Clausewitz warns that soldier fatigue also weighs heavily on the commander's mind, causing him to slow the effort unduly: "it takes a powerful mind to drive his army to the limit."

3. English, 223. English concludes that fatigue and fear are closely associated, feeding off each other until the soldier is consumed with panic or exhaustion. The soldier's load is a primary contributor to fatigue. Marshall, 22–23, 42, 46–47. Marshall concludes that there is a limit to what a soldier can carry and that no amount of physical conditioning will overcome this

fact. If commanders fail to strike nonessential equipment from the soldier's load, then the soldiers will be so exhausted when contact is made that they are likely to break. Afterward, the survivors will make the decision for the commander regarding what equipment to discard.

4. English, 223; Marshall, 13–14, 18–19. Marshall observed that overloading soldiers with excessive ammunition is a common mistake. During World War II, 80 rounds and 2 hand grenades proved more than adequate. Soldiers will not worry about a lack of ammo but will look around for more if needed. Instances of battles lost for lack of ammunition are virtually nonexistent. The focus should be on replenishing ammunition after the engagement rather than overloading the soldier beforehand.

5. Lt. Col. Bolger conscientiously reduced the load per soldier to 51.2 pounds during his final rotation to JRTC. Bolger, 135. During World War II, the average Soviet soldier's load was 62 pounds in the summer and 78 pounds in the winter; James Lucas, *War on the Eastern Front: The German Soldier in Russia 1941–1945* (Novato, Calif.: Presidio Press, 1991), 57.

6. Luvass, 18. Frederick the Great once made a captain, who was a scholar on ancient warfare, bear the load of a grenadier when the captain commented that the Roman soldier carried more, in order to remind the scholar of the difference between what is written and what is reality.

7. FM 21-18, 5–10.

Chapter 10

Unit Manning Roster

The Unit Manning Roster (UMR) along with the Weapons Capability Charts help commanders "see themselves." The UMR helps company leaders track the strength and pulse of their units. It helps the commander make decisions on platoon taskings, because some taskings require more manpower (assault element) than others (security element). It also helps the commander and first sergeant (1SG) assign replacements to the platoon with the greatest need quickly.

The battle roster number normally comes from the battalion S1 and is used for the rapid, detailed reporting of casualties via radio. Battle roster numbers help the S1 track precise losses and assign replacements for vacancies. A simple battle roster assignment technique is to use the soldier's first and last initials and the last four digits of the soldier's social security number.

> **TTP**
>
> Maintain data in pencil for flexible revisions. For longer wear, acetate it and use alcohol markers.

The remarks column is used to keep track of weapon assignments and personal data. Often, casualties require the cross-leveling of weapons, their redistribution within the unit. So, it is not unusual for team leaders or even squad leaders to arm themselves with an M203 or squad automatic weapon (SAW). The commander and 1SG use it to reflect future leaders or losses depending on the potential of the soldier. Platoon leaders and platoon sergeants use it to track weapon qualification proficiency, additional skill identifiers, award recommendations, and promotion points. Squad leaders and team leaders can use it to track personal history, family issues, and special skills.

Personal data serves to keep track of issues that affect the soldier and his family. Leaders often neglect or forget important matters that are very important to the soldier and his family simply because the chain of command has no management system in either peacetime or war. If a soldier is recommended for an award, the leader is in the best position to collect witness statements and write the award recommendation or to track the process through his subordinate leaders. It also reminds him to write to the next of kin of killed soldiers. It helps the leader keep tabs on the soldier's concerns. Basically, the manning roster is a tool for the leader to track that which he feels is important. Such efforts reap tremendous dividends. If leaders care for the soldier, he will be loyal and dedicated. It builds rapport and mutual admiration. Besides, it's the right thing to do.

The radiation exposure status chart simply helps the commander track the amount of radiation the unit has absorbed in nuclear warfare. It is a mistake to assume that higher headquarters will track this for the unit.

The UMR affords the company leadership great advantages during peacetime. With such a tool, the company leadership can track weapons qualifications, schooling, future losses, awards, special duty, and other items of interest. The commander and 1SG can use the UMR to portray the strength of the company present for duty to battalion HQ in order to argue a case for replacements or to avoid being overtasked. Of course, the S3 and CSM will probably be annoyed with you for it, but it's the best way to take care of the company.

TABLE 10.1 UNIT MANNING ROSTER

Battle #	Name	Rank	DTY PSN	DMOS	Remarks
Company HQ					
NB1805	Napoleon Bonaparte	CPT	CDR	11B00	
HB0216	Hannibal Barca	1LT	XO	11B00	T-3 Eye infection
GK1206	Genghis Khan	1SG	1SG	11B5M	
JC1704	John Churchill	SSG	SUPPLY	76Y30	
GM1944	George Marshall	SSG	COMMO	31G30	
EF1915	Erich Falkenhayn	SGT	NBC	54E20	
BF1914	Al Krupp	SPC	ARMORER	76Y10	
CC1832	Carl Clausewitz	PFC	RTO	11B10	
HJ1862	Henry Jomini	PFC	RTO	11B10	
LM1916	Lord Moran	SPC	MEDIC	36B10	

TABLE 10.1 *(continued)*

FK1792	Francois Kellerman	2LT	FIST OFF	13F00	
GB1918	George Bruchmuller	SSG	FIST NCO	13F30	
Antitank Section					
LD1806	Louis Davout	SSG	SEC LDR	11B30	
JL1809	Jean Lannes	SGT	TM LDR	11B20	WIA—RTD Dec
DE1944	Dwight Eisenhower	SGT	TM LDR	11B20	
PS0202	Publius Scipio	CPL	TM LDR	11B20	
PE1704	Prince Eugene	SPC	AA GNR	11B1C2	
PL1863	Pete Longstreet	SPC	AA GNR	11B1C2	
JP1918	John Pershing	PFC	AA GNR	11B1C2	
JH1914	John Hindenburg	PFC	AA GNR	11B1C2	
EL1918	Erich Ludendorf	PFC	AA GNR	11B1C2	
JS1862	James Stuart	PFC	AA GNR	11B1C2	
TD1864	Tom Devine	PFC	AST GNR	11B1C2	
JB1863	John Buford	PV2	AST GNR	11B1C2	
WS1847	Winfield Scott	PV2	AST GNR	11B1C2	Overweight–CH
Mortar Section					
EM1805	Ed Mortier	SSG	SEC LDR	11C30	
LD1800	Louis Desaix	SGT	SQD LDR	11C20	WIA—RTD12Nov
AW1815	Arthur Wellesley	SPC	MTR GNR	11C10	
PS1864	Phil Sheridan	PFC	MTR GNR	11C10	
AK1943	Alfred Kesselring	PV2	AMMO BR	11C10	
MM0488	Miltiades Marathon	PV2	AMMO BR	11C10	
1ST Platoon					
JC0048	Julius Caesar	1LT	PLT LDR	11B00	
FL1807	Francois Lefebvre	SSG	PLT SGT	11B4G	
HM1872	Helmut Moltke	PFC	MG GNR	11B10	
TJ1863	Thomas Jackson	PFC	MG GNR	11B10	

(continued)

TABLE 10.1 *(continued)*					
EL0371	Epaminondas Leuctra	PV2	ASST MG	11B10	
GZ1942	George Zhukov	PV2	ASST MG	11B10	
BH1954	Basil Liddell Hart	PFC	RTO	11B10	
BM1942	Bernard Montgomery	PFC	MEDIC	36B10	
HF1415	Henry Fifth	SPC	FIST FO	13F10	
FH1757	Frederick Hohenzollern	SSG	1 SL	11B30	
JP1812	Joseph Poniatowski	SGT	A TM LDR	11B20	
EM1943	Erich Manstein	SPC	AR	11B10	
GT1863	George Thomas	PFC	GREN	11B10	Overweight–CH
			RIFLE	11B10	
NS1811	Nicholas Soult	SGT	B TM LDR	11B20	
PS1796	Phil Serurier	PFC	AR	11B10	
HG1940	Heinz Guderian	PFC	GREN	11B10	
			RIFLE	11B10	
ER1942	Erwin Rommel	SSG	2 SL	11B30	
LS1811	Louis Suchet	CPL	A TM LDR	11B20	M203
WC1066	William Conqueror	PFC	AR	11B10	
			GREN	11B10	
			RIFLE	11B10	
CV1800	Claude Victor	SGT	B TM LDR	11B20	
EA1917	Edmond Allenby	SPC	AR	11B10	
JM1808	John Moore	SPC	GREN	11B10	
			RIFLE	11B10	
RL1863	Robert Lee	SSG	3 SL	11B30	OCS packet
PA1807	Pierre Augereau	CPL	A TM LDR	11B20	M203
CB1813	Charles Bernadotte	PFC	AR	11B10	
			GREN	11B10	
			RIFLE	11B10	

TABLE 10.1 *(continued)*

GP1943	George Patton	SGT	B TM LDR	11B20	Next SL
LD1760	Leopold Daun	SPC	AR	11B10	
DE1643	Duc d'Enghien	PV2	GREN	11B10	
GC1503	Gonzalo de Cordoba	PVT	RIFLE	11B10	
2D Platoon					
CB0530	Count Belisarius	PV2	PLT LDR	11B00	
NS1811	Nick Soult	SFC	PLT SGT	11B4G	
MN1805	Mike Ney	SSG	MG GNR	11B10	
HM1944	Hasso Manteuffel	SPC	MG GNR	11B10	
AM1797	Andre Massena	PFC	ASST MG	11B10	
SB1241	Subedei Bahadur	PFC	ASST MG	11B10	
ST500	Sun Tzu	SPC	RTO	11B10	
JB1636	Johan Baner	PFC	MEDIC	36B10	
JF1918	John Fuller	SPC	FIST FO	13F10	
AM0331	Alexander Macedon	SSG	1 SL	11B30	
UG1863	Ulysses Grant	SGT	A TM LDR	11B20	
WS1864	William Sherman	SPC	AR	11B10	
BK1241	Batu Kahn	PFC	GREN	11B10	
			RIFLE	11B10	
HG1066	Harold Godwinson	CPL	B TM LDR	11B20	
JP1920	Joseph Pilsudski	PFC	AR	11B10	
PM0338	Philip Macedon	PV2	GREN	11B10	ADAD
			RIFLE	11B10	
GC1812	Gouvian St. Cyr	SSG	2 SL	11B30	
MR1951	Matt Ridgeway	SGT	A TM LDR	11B20	BS Awd—15 Sep
WS1944	William Slim	SPC	AR	11B10	
BA1777	Benedict Arnold	PVT	GREN	11B10	AWOL 15 Oct
AC1812	Armand Caulaincourt	PV2	RIFLE	11B10	
IY1941	Isoroku Yamamoto	SGT	B TM LDR	11B120	

(continued)

TABLE 10.1 *(continued)*

EG1807	Emanuel Grouchy	PFC	AR	11B10	
JS1944	Joe Stillwell	PFC	GREN	11B10	
			RIFLE	11B10	
JJ1813	Jean Jourdan	SSG	3 SL	11B30	
DM1945	Doug MacArthur	SGT	A TM LDR	11B20	
TY1942	Tomoyuki Yamashita	SPC	AR	11B10	
GH1943	Gottard Heinrici	PFC	GREN	11B10	
			RIFLE	11B10	
GW1781	George Washington	PV2	B TM LDR	11B20	USMAPS Candidate
AM1812	August Marmont	SPC	AR	11B10	
HB1944	Hermann Balck	PV2	GREN	11B10	
			RIFLE	11B10	
3D Platoon					
GA1631	Gustavus Adolphus	2LT	PLT LDR	11B00	
RE1943	Robert Eichelberger	SFC	PLT SGT	11B4G	Wife Due Sep
ET1346	Edward Third	PFC	MG GNR	11B10	
JC1863	Joshua Chamberlain	PV2	MG GNR	11B10	WIA—RTD Nov
QF0217	Quintus Fabius	PV2	ASST MG	11B10	
HB0211	Hasdrubal Barca	PV2	ASST MG	11B10	
AP1868	Ardant du Picq	SPC	RTO	11B10	
CC0801	Charlemagne Carolingien	PFC	MEDIC	36B10	
OC1645	Oliver Cromwell	SPC	FIST FO	13F0	
SM1809	Steve Macdonald	SSG	1 SL	11B30	
JB1808	Jean Bessieres	CPL	A TM LDR	11B20	
CJ1877	Chief Joseph	PFC	AR	11B10	
NM1877	Nelson Miles	PFC	GREN	11B10	
			RIFLE	11B10	
DM1781	Daniel Morgan	SGT	B TM LDR	11B20	

TABLE 10.1 (continued)

WB1799	William Brune	SPC	AR	11B10	
GC1873	George Crook	PFC	GREN	11B10	
			RIFLE	11B10	
DP1795	Dom Perignon	SGT	2 SL	11B30	
NG1780	Nathanael Greene	CPL	A TM LDR	11B20	
AJ1815	Andrew Jackson	PFC	AR	11B10	
PG1709	Peter Great	PV2	GREN	11B10	
			RIFLE	11B10	
JM1799	Jack Murat	CPL	B TM LDR	11B20	
HS1919	Hans Seeckt	PFC	AR	11B10	ETS 15 Dec
MT1919	Mikhail Tuckhachevski	PFC	GREN	11B10	
			RIFLE	11B10	
AM1814	Adrien Moncey	SGT	3 SL	11B30	
JW1899	Joseph Wheeler	SGT	A TM LDR	11B20	
EU1884	Emory Upton	SPC	AR	11B10	Hosp—RTD 30 Nov
GB1879	Gonville Bromhead	PFC	GREN	11B10	
			RIFLE	11B10	
NO1812	Nick Oudinot	SGT	B TM LDR	11B20	
LW1899	Leonard Wood	SPC	AR	11B10	
JC1879	John Chard	PFC	GREN	11B10	
HF1842	Harry Flashman	PVT	RIFLE	11B10	
Attachments					

(continued)

TABLE 10.1 *(continued)*

TABLE 10.2 RADIATION EXPOSURE STATUS CHART

Unit	RES 0: No Exposure	RES 1: $0 < x \leq 70$ cGy	RES 2: $70 < x \leq 150$ cGy	RES 3: $x > 150$ cGy
HQ				
AT SEC				
MORT SEC				
1 PLT				
2 PLT				
3 PLT				
Operational Exposure Guidance				
Negligible Risk: 50 cGy				
Moderate Risk: 70 cGy				
Emergency Risk: 150 cGy				

Source: U.S. Department of the Army, *Fundamentals of Nuclear and Chemical Operations,* PO34 COMPS (Fort Leavenworth, Kans.: U.S. Army Command and General Staff College 1989), 34.

Chapter 11

U.S. Weapons Capabilities

One of the greatest limitations in training is the inability to experience the deadly effectiveness of weapons and munitions. For defensive efforts MILES training (the multiple integrated laser engagement system) gives soldiers a false sense of security, because soldiers in fighting positions or behind foliage cannot be hit by MILES lasers, whereas actual weapons have greater penetration capabilities. Further compounding the problem, observer/controllers (OC) are reluctant or unavailable to assess damage to a position unless it has a MILES harness attached.

Knowledge of a weapon's capabilities is the first step in allowing the commander to "see himself." The first step toward enlightenment is to assemble a table of weapons and their capabilities. This chart helps the leader align specific weapons against specific targets with a reasonable assurance of destruction. Maximum effective range is the range that affords the average firer a 50 percent probability of hitting the target. For planning, maximum effective range is important when assessing the distance of the support position to the target area, but this is not an absolute maximum range. The effective area helps the planner visualize the width and depth of protection afforded by artillery and mortars. Penetration helps the leader assess which munitions will destroy the target without overkill. Basically, this information assists in planning.

The second step is to show soldiers the effects of munitions on specific targets. Scheduled marksmanship training at ranges provides an excellent opportunity for soldiers to observe munitions effects. The building of bunkers and fighting positions at the M203 range, grenade range, and antitank (AT) range is easy to coordinate and execute. Ostensibly, the commander can reward the best marksman the honor of destroying these targets. Soldiers really enjoy these opportunities, because it breaks up the monotony of marksmanship training and it gives them the opportunity to destroy something legally. The range OIC includes this demonstration as part of the training. After the soldiers have completed firing, the OIC leads soldiers down range (once this area has been swept for unexpended rounds) and allows them to see the effects up close. Such an experience remains with the soldier and gives him a sober appraisal of the weapon effectiveness.

The third step is to conduct a firepower demonstration against a strongpoint twice a year. The strongpoint has fighting positions with and without overhead cover (18 inches), entrenchments, triple-strand concertina, tangle foot, apron fence wire obstacles, and minefields (with fake mines). Clothed dummies are placed in all positions. Clothes stuffed with newspapers and balloons filled with dye at kill points on the body give the soldiers a good idea of the level of protection afforded by entrenchments.

The demonstration begins with 105 mm artillery and 81 mm mortar preparatory fire (10 minutes or longer) using a mixture of high explosives (HE) and white phosphorous with variable time (VT), impact, and delay fuses. The commander conducts a walk-through of the area with his soldiers after it is swept for unexpended rounds and shows the effect of artillery and mortars on positions and obstacles. The commander notes the effect on the wire obstacles and minefields. He also notes the effect on the positions and their occupants. He further points out the type of wounds for the occupants of positions with and without overhead cover as well as the occupant's posture in the positions (standing, sitting, prone). The lesson is to show the protection afforded the various positions. You can learn a lot from a dummy.

Next, the fire support element sets up around 150 meters and fires M203 grenades, 60 mm mortars (direct and indirect fire), light antitank weapons (LAWs) and AT-4s, M202 Flash and flamethrowers (if available), and machine guns. The commander again conducts a walk-through of the area and notes the damage to positions and the dummies.

The commander should also highlight the fact that dummies are impervious to the physiological and psychological effects of indirect fire and how long it takes soldiers to recover from an intense bombardment. He also needs to bring out that despite such artillery preparations, history demonstrates that well-entrenched defenders normally withstand short, intense shelling remarkably well. Nevertheless, he should not sidestep the issue entirely regarding the physiological and psychological effects of munitions on the individual. For instance, soldiers forced to remain in place under enemy fire become exhausted, because the adrenaline generated from fear has no opportunity to dissipate through physical activity.

The commander should also impart a good appreciation of the effects terrain and weather will have on weapons and munitions. In heavy forests and jungles, machine gun fire has decreased penetration effects. M203 grenades may not be able to arm because of the dense vegetation. Lack of good observation will prevent forward observers from guiding in indirect fire onto targets.[1] In extreme cold, artillery and mortar fuses freeze, rendering them useless. Deep snow also reduces the effective area of explosions.[2]

During arctic conditions, soldiers must protect equipment and munitions by storing them in heated bunkers or buildings. Machine guns can be kept warm and placed into operation immediately during an alert or can be rotated into the line frequently.[3] Maintaining the extra firing pins in the warmth of a bunker protects them from extreme cold, which can cause them to become brittle and to snap when fired. Storing batteries in the bunker also prevents the warping of their plates. Soldiers can keep additional firing pins for their individual weapons next to their body (waist band) for such contingencies. The squad leader can check weapons more easily for proper maintenance. In extreme cold, soldiers should use only a thin coat of winter lubricants for weapons. If lubricants are not available, the soldiers should remove all oil from the weapon action groups and use a substitute such as finely ground sulfur powder or sunflower oil.[4] Mortar men can keep powder charges dry by stuffing them into their shirts.[5]

A good appreciation of weapon capabilities increases the soldier's confidence and allows him to be more effective in the beginning stages of a conflict. He will be able to destroy targets with minimum firepower needed, saving heavier weapons for heavier targets. His knowledge will allow him to be more creative as he is confronted with tactical problems.

TABLE 11.1 U.S. WEAPONS CAPABILITIES

Weapon	Cal	Type Round	Max Effective RG (Meters)	Min RG (Meters)	Effective Area (Meters)	Remarks
Mortar						
M30 Battery of 4	107mm	M329A1/HE	6,840	770	75 diameter 40x20 150 x 50 160 x 50	1 round Radius FPF Linear
		M328A1/WP	5,650	920	600	
		M335A2/ ILLUM	5,490	400	1500 diameter	Burn Time: 90 sec
81mm Battery of 4	81mm	M374A3	4,789	73	34 diameter 50 100 x 35 120 x 35	1 round Radius FPF Linear
		M375A2/WP	4,737	72	20 diameter	1 round
		M301A3/ ILLUM	3,150	100	1,200 diameter	Burn Time: 60 sec

(continued)

TABLE 11.1 (continued)

M224 Battery of 2	60mm	M720/HE	3,489 1,350	75 75	27.5 diameter 100 x 35	1 round FPF Handheld: Charge 1, trigger
		M302A1/WP	1,629 1,200	75 75	20	— Handheld: Charge 1, trigger
		M83A1/ ILLUM	931 800	75 75	,200 diameter	— Handheld: Charge 1, trigger
Artillery						
M102 Battery of 6	105mm	HE	12,400 15,100 with RAP	—	75 diameter 150 x 50 200 x 50	1 round FPF Linear
		SMK			500	High Cloud
M198 Battery of 6	155mm	HE	22,400 30,000 with RAP	—	100 diameter 200 x 50 300 x 50	1 round FPF Linear
		SMK			600	High Cloud
Antitank						
M220A2 ITOW/ TOW 2B	—		3,750	65/ 288 (ground mounted); 200 (vehicle mounted)		Back blast: 75 m (Water over 1,100 meters wide reduces range)
Javelin	127mm	HEAT	2,000	65 (Direct Atk) 150 (Top Atk)	N/A	Back blast: 25 m

TABLE 11.1 *(continued)*

M72A2 LAW	66 mm	HEAT	Stationary: 200 Moving: 165	10	N/A	Back blast: 40 m
M136 AT4	84 mm	HEAT	300	15	N/A	Back blast: 35 m
SRAW	140mm	HEAT	500	17	N/A	
Rifle						
M4	5.56 mm	Ball	Area: 600 Point: 500	N/A	N/A	Round Length: 2.25 inches
Machine Gun						
M2	.50 in	AP	Area: 1,830 Vehicle: 1,100 Point: 700	N/A	N/A	Grazing Fire w/Tripod: 800m
M240	7.62 mm	Ball	Area: 1,800 Point: 800 Moving: 200 Suppression: 1,800	N/A	N/A	Grazing Fire w/Tripod: 600m Ranges are with tripod; with bipod subtract 200 m
M249	5.56 mm	Ball	Area: 1000 Point: 800 Suppression: 1,000	N/A	N/A	Grazing: 600 m Ranges are with tripod; with bipod subtract 200 m
Grenade Launcher						
M203	40 mm	HEDP	Area: 350 Vehicle or point: 150	14-28	5	Arms: 14-28 m Max Range: 400 m
		Anti-Personnel	35		N/A	Shotgun round
		White Star Parachute	300	N/A	200 m diameter	Burn Time: 40 sec

(continued)

TABLE 11.1 *(continued)*

			Flame Devices			
M202 Flash	66 mm	—	Area: 5-750 Vehicle: 200 Bunker: 50	20	22	Back blast: 40 m
M2A2	N/A	Thickened Fuel	40-50	N/A	N/A	6-9 sec
		Unthickened Fuel	20-25	N/A	N/A	6-9 sec

TABLE 11.2 MINIMUM SAFE DISTANCES FOR EXPLOSIVE ORDNANCE

Weapon	Distance
60 mm	28 meters
81 mm	35 meters
105 mm	50 meters
Hellfire Missile	100 meters
Cluster Bomb Unit	195 meters
Maverick Missle/MK82 Bomb/Hydra-70 Rocket	220 meters
127 mm Naval Gun	410 meters

TABLE 11.3 AT WEAPONS BACK BLAST ALLOWANCE WITHIN ENCLOSURES

Weapon	Room Size (Feet)	Ceiling Height (Feet)	Vent Size (Square Feet)	Muzzle Clearance (Inches)
TOW	15 x 15	7	20 (3 x 7′ door open)	9
Javelin	12 x 15	7	20 (3 x 7′ door open)	N/A
SMAW-D	12 x 15	7	20 (3 x 7′ door open)	6
LAW	12 x 15	8	20 (3 x 7′ door open)	N/A
AT4	17 x 24	8	20 (3 x 7′ door open)	6

Source: FM 3-06.11 (FM 90-10-1), Chapter 7; FM 3-22.37, pp. 1-3; FM 23-25, Chapter 6.

TABLE 11.4 PENETRATION EFFECTS OF MUNITIONS

WPN	Type RD	Range (Meters)	Target Material	Penetr-ation (Inches)	Diameter of Loophole (Inches)	# RDS Required for Loophole (3-5 RD Bursts)
M4	Ball	200	Pine Board	25	N/A	N/A
			Loose Sand	4		
			Concrete	1		
		50-100	Reinforced Concrete	8	7	250 (35—initial penetration)
			Triple Brick	14	7	160 (90—initial penetration)
			Cinder Block with Single Brick Veneer	12	7	250 (60—initial penetration)
			Double Brick	9	7	120 (70—initial penetration)
			Tree Trunk or Log Wall	16	N/A	1-3
			Cinder Block filled with Sand	12	7	35
			Double Sandbag Wall	24	N/A	220 (will not penetrate sandbags at less than 50 m)
			Mild Steel Door	3/8	N/A	1
M240 MG	Ball	25/100/ 200	Pine Board	13/18/41	N/A	N/A
			Loose Sand	5/4.5/7		
			Cinder Block	8/10/8		
			Concrete	2/2/2		
		25	Reinforced Concrete	8	7	100
			Triple Brick	12	7	200
			Concrete with Triple Brick Veneer	12	6	60

(continued)

TABLE 11.4 *(continued)*

M2 MG	AP	200/600/1500	Armor Plate (Homogenous)	1/.7/.3		
			Armor Plate (Face-Hardened)	.9/.5/.2		
			Clay	28/27/21		
			Loose Sand	14/12/16		
			Packed Earth	28		
	Ball	100 200	Reinforced Concrete	2/3/4	N/A	300/450/600 1200/1800/2400
		25 35	Reinforced Concrete	8 10/18	7 12/7	100 50/140
		25 35	Triple Brick Wall	14 12	7 8/26	170 15/50
		25 35	Concrete Block with Single Brick Veneer	12 12	6 /24 10/33	30/200 25/45
		25	Cinder Block (filled)	12	N/A	18
			Double Brick Wall	9	N/A	45
			Double Sandbag Wall	24	N/A	110
			Log Wall	16	N/A	1
			Mild Steel Door	3/8	N/A	1
M203	HEDP	31-400	Armor Plate	2	N/A	N/A
			Brick	6-8		
			Sandbags (Double layer)	20		
			Sand-filled Cinder Block	16		
			Pine Logs	12		

TABLE 11.4 *(continued)*

M224 TOW[a]	HEAT	65-3750	Armor	23.6(+)	
			Homogeneous Steel	11	
			Packed Earth	96	
			Reinforced Concrete	48	
			Steel Plate	16	
			Log Walls	18	
Javelin	HEAT	65-1000	Armor	23.6	N/A
			Homogeneous Steel	8	
			Reinforced Concrete	48	
			Steel Plate	13	
M136 AT4	HEAT	17-300	Armor Armor Plate	14 17.5	
M72A2 LAW	HEAT	10-200	Armor	12	
			Earth	72	
			Reinforced Concrete	24	
			Steel	12	
Improved LAW	HEAT	25-220	Rolled Homogenous Steel Armor	11.8	
TNT	5 lbs	Untamped	Non-reinforced Concrete	12	
	2 lbs	Tamped			
C4	10 lbs	Untamped	Masonry Wall	Complete	Breach hole large enough for a man to enter
M3 A3	Shaped	N/A	Reinforced Concrete	60	N/A
M183; M37	Satchel	N/A	Concrete Wall	36	N/A

Conversions:
1 inch = 25.4 mm
1 meter = 3.28 feet

Sources:

Table 11-1: Technical data for weapon capabilities came from the following sources. Discrepancies between sources were resolved in favor of the manual dedicated to the weapon. Technical Manuals are also excellent sources.

ATIA portal to Reimer Library http://www.adtdl.army.mil/.

AT-4 LAW http://www.fas.org/man/dod-101/sys/land/at4.htm.

M203 Grenade Launcher http://www.colt.com/colt/html/a2f27_m203grenade.html.

M224 60mm Mortar http://www.fas.org/man/dod-101/sys/land/m224.htm.

M240 Machine Gun http://www.hk94.com/m240b.html; http://www.fas.org/man/dod-101/sys/land/m240g.htm.

M249 SAW LMG http://www.fas.org/man/dod-101/sys/land/m249.htm.

M4 Carbine http://www.colt.com/colt/html/a2f20_m4carbine.html; http://www.fas.org/man/dod-101/hsys/land/m16.htm; http://www.hk94.com/m4a1.html.

M9 9mm Beretta Pistol http://www.fas.org/man/dod-101/sys/land/m9.htm; http://www.hk94.com/black2.html.

Hand Grenades http://www.fas.org/man/dod-101/sys/land/grenade.htm.

Javelin ttp://www.army-technology.com/projects/javelin; www.strategypage.com/fyeo/; owtomakewar ttp://www.fas.org/man/dod-101/sys/land/javelin.htm.

Rock Island Arsenal Mortar Team http://tri.army.mil/LC/Cf/Cft/Cftm/cftm.htm.

U.S. Army Material Command http://aeps.ria.army.mil/aepspublic.cfm.

U.S. Department of the Army, *Browning Machine Gun, Caliber .50 HB, M2*, Field Manual No. 23-65 (Washington, D.C.: U.S. Government Printing Office, 19 June 1991), Chapter 1.

U.S. Department of the Army, *Combined Arms Operations In Urban Terrain*, Field Manual No. FM 3-06.11 (90-10-1) (Washington, D.C.: U.S. Government Printing Office, 28 February 2002), Chapter 7.

U.S. Department of the Army, *Crew-Served Machine Guns 5.56-mm and 7.62-mm*, Field Manual No. 3-22.68 (Washington, D.C.: U.S. Government Printing Office, 21 July 2006), 1-2, 3-2.

U.S. Department of the Army, *Fire Support For The Combined Arms Commander*, Field Manual No. 6-71 (Washington, D.C.: U.S. Government Printing Office, 29 September 1994), Appendix A, Appendix B.

U.S. Department of the Army, *40-MM Grenade Launcher, M203*, Field Manual No. FM 3-22.31 C1 (FM 23-31) (Washington, D.C.: U.S. Government Printing Office, 19 March 2007), 2-3.

U.S. Department of the Army, *Javelin Medium Antiarmor Weapon System*, Field Manual No. FM 3-22.37 (Washington, D.C.: U.S. Government Printing Office, January 2003), 1-3.

U.S. Department of the Army, *Rifle Marksmanship M16A1, M16A2/3, M16A4, and M4 Carbine*, Field Manual No. 3-22.9 (FM 23-9) (Washington, D.C.: U.S. Government Printing Office, April 2003), 2-1.

U.S. Department of the Army, *Shoulder-Launched Munitions*, Field Manual No. 3-23.25 (Washington, D.C.: U.S. Government Printing Office, January 2006), 2-2, 5-2.

U.S. Department of the Army, *Tactical Employment of Mortars*, Field Manual No. 7-90 (Washington, D.C.: U.S. Government Printing Office, 9 October 1992), Chapter 1.

U.S. Department of the Army, *TOW Weapon System*, FM 3-22.34(FM 23-34) (Washington, D.C.: U.S. Government Printing Office, November 2003), 1-1, 2-4.

Table 11-2:
Daniel P. Bolger, *The Battle For Hunger Hill* (Novato, CA: Presidio Press, 1997), 205.

Table 11-3:
FM 3-06.11 (FM 90-10-1), Chapter 7; FM 3-22.37, 1-3; FM 23-25, Chapter 6.

Table 11-4:
FM 3-06.11 (FM 90-10-1), Chapter 7; FM 23-65, Chapter 1; FM 3-22.31 C1, 2-9; FM 3-23.25, 2-2, 5-4.

FM 3-06.11 (FM 90-10-1), Chapter 7. AT weapons should not be used to breach walls for loopholes or mouse holes since this requires too many rounds. To maximize spalling on enemy positions in buildings or bunkers, aim six inches below or to the side of the weapon aperture.

U.S. Department of the Army, *Engineer Field Data*, Field Manual No. 5-34 (Washington, D.C.: U.S. Government Printing Office, 14 September 1987), 6-8–6-9. Good reference for breaching calculations. The 2005 manual is unfortunately classified.

NOTES

1. PAM 20-269, 206.
2. Lucas, 81.
3. Timothy A. Wray, *Standing Fast: German Defensive Doctrine on the Russian Front during World War II,* Combat Studies Institute Research Survey No. 5 (Fort Leavenworth, Kans.: U.S. Army Command and General Staff College, September 1986), 82.
4. PAM 20-269, 246. Lucas, 81, 93. Intense cold on the eastern front could make "metal as brittle as glass." Wray, 75. The Germans made use of Russian villages to keep weapons and equipment warm because "the extreme cold made gunmetal brittle and weapons kept outside tended to jam or malfunction due to broken bolts and firing pins."
5. Doubler, 222.
6. FM 90-10-1, chapter 8. AT weapons should not be used to breach walls for loopholes or mouse holes because this requires too many rounds. To maximize spaulding effects on enemy positions or bunkers, aim six inches below or to the side of the weapon aperture.

Chapter 12

Passage of Lines

Passage of lines requires extensive preparation, because the company is extremely vulnerable during this critical event. The company cannot assume that the unit through which it is passing (the stationary unit) is prepared for a passage of lines. The platoon leader (OIC) coordinating the passage of lines might need to prompt the stationary unit leadership for essential information and may need to lead this unit's leaders through the company procedures. Following the checklist, the platoon leader can exchange information rapidly with no ambiguity. The platoon leader also alerts the stationary unit of company reconnaissance conducting a passage of lines in front of the company. If the stationary unit has little knowledge of the area to its front, the route recon team will need to ensure that the immediate area of the passage point is clear and dispatch an element back to the company to report on the security situation.

The passage of lines briefing should be as visual as possible, using a sketch map. Unit radio-telephone operators (RTOs) will copy the overlay for their platoon or section for added specific information. If the company elects to have multiple passage lanes (as would be used for an infiltration mission), the platoon leader meets with platoon or squad guides to provide in-depth information on the passage of lines. If time permits, these guides should accompany the platoon leader on the initial coordination. Following the operations order (OPORD), the recon teams conduct the passage of lines with no further guidance.

> **TTP**
>
> For efficiency, the platoon leader's platoon sergeant (PSG) is responsible for the foot march annex, thereby ensuring continuity of effort for two closely related tasks.

At the coordinated time, the stationary unit guides link up with the company at the attack position and lead the unit through the passage lane. At the attack position, the primary land navigation platoon assumes the point, but the passage of lines OIC continues to control the passage of the company up to the security-listening halt.

For reentry passage of lines, the company should dispatch a patrol for coordination, especially if the reentry point is through a different stationary unit. This precaution is less hazardous for the company, because the soldiers of the stationary unit are less likely to fire to their front if alerted to the expected presence of a friendly unit in the near future. It is also easier for a small recon element to find the reentry point as opposed to a company groping around in front of friendly lines.

Once the company passes through the lines and closes in an assembly area, the commander contacts battalion headquarters to receive a fragmentary order (FRAGO). The main concern of the commander is to establish an assembly area that is out of the stationary unit's sector to avoid congestion and to avoid indiscriminate enemy indirect fire.

TABLE 12.1 PASSAGE OF LINES	
Task	**Responsibility**
1. Upon receipt of the tasking, report to the stationary unit for the initial coordination:	OIC/RTO
a. Provide stationary unit representative (CDR/XO) with the following:	
1) Unit identification.	
2) The time of the passage of lines.	
3) Approximate number of personnel conducting the passage of lines.	
4) The number of passage points required.	
5) The unit's general area of operation.	
6) Company frequencies and call signs.	
7) Time company recon conducts passage.	
8) Company contingency plans:	
a) Enemy contact before passage of lines (from the attack position to the passage point): Return to attack position and contact battalion HQ for guidance.	

TABLE 12.1 *(continued)*	
b) Enemy contact during passage of lines (from the passage point to the release point): Return to attack position and contact battalion HQ for guidance.	OIC/RTO
c) Enemy contact after passage of lines (from the release point to the phase line denoting the stationary unit's maximum fire support line): Continue mission. Request stationary unit fire support.	
b. Obtain the following from the stationary unit representative:	
1) Unit frequencies and call signs.	
2) Challenge and password.	
3) Running password.	
4) Locations of company primary and alternate attack positions.	
5) Enemy situation.	
a) Known or suspected positions.	
b) Likely ambush sites.	
c) Latest enemy activity.	
d) Known or suspected obstacles.	
e) Source of the above information (higher intelligence or own recon patrols). How old is the information?	
6) Stationary unit situation:	
a) Unit **Sector Sketch.**	
b) Depth of no-man's-land.	
7) Description of surrounding terrain.	
8) Primary and alternate passage points: make overlay with start point, lane, release point, and maximum fire support phase line.	
9) Possible fire support, aid and litter teams, reactionary forces, guides, water, and food.	

(continued)

TABLE 12.1 *(continued)*	
10) Casualty evacuation plan, casualty collection OIC/RTO points, and the company aid post.	OIC/RTO
11) Time of final coordination between commanders.	
12) Link up with stationary unit guide(s). Conduct joint recon of the primary attack position and the passage lane(s). Arrange for guide(s) to link up with the company at the final coordination meeting at the stationary unit CP.	
2. Prepare passage of lines plan for OPORD:	OIC
a. Back-brief commander on initial coordination.	OIC
b. Disseminate stationary unit sector sketch and passage point overlay.	RTO
c. Brief passage of lines at OPORD.	OIC
d. Link up with platoon/section guides, move to attack position, and prepare it for occupation.	OIC/CO GUIDES
3. Occupy attack position:	CDR
a. Attend final coordination meeting with stationary unit commander:	CDR/OIC/PL/SEC LDR
1) Receive tactical situation updates.	CDR/OIC/PL/SEC LDR
2) Provide changes to the initial coordination.	OIC
3) Provide the exact number of personnel conducting the passage of lines.	OIC
4) Provide exact time of departure for each subelement.	OIC
5) Coordinate post mission reentry plan (if applicable):	
a) Date/time of return.	
b) Primary and alternate reentry lanes.	CDR
c) Projected frequencies and call signs.	
d) Reentry rally point.	
e) Linkup point with stationary unit guide.	

TABLE 12.1 *(continued)*	
f) Final rally point of company after reentry of lines.	CDR
g) Actions on enemy contact contingency plan CDR for reentry.	
1] Before linkup with guide: request fire support, break contact, and move to alternate reentry lane rally point.	
2] During reentry of lines: move to final rally point, form perimeter, and inform battalion.	
6) Link up with guides and return to attack position.	CDR/OIC/PL/ SEC LDR
4. Execute passage of lines:	CDR
a. Inform stationary unit CP and request that it alert its subunits of the departure.	CDR
b. Accompany guide at the head of the company/ platoon/ section to the release point, verify the number of personnel with the guide as the unit passes through, and tell the guide to wait 15 minutes in case of enemy contact.	1SG/PSG/SEC TM LDR
c. Establish security-listening halt once the unit is outside of the stationary unit's maximum fire support phase line.	CDR/PL/SEC LDR
5. Execute reentry of friendly lines through a stationary unit:	CDR
a. Occupy reentry rally point 500 meters from stationary unit's FEBA.	CDR
b. Establish radio contact with stationary unit, confirm linkup point/time for guide.	CDR
c. Exchange 5 Point Contingency Plan with commander and conduct leader's recon for linkup point.	OIC
d. Link up with stationary unit guide and coordinate the following:	
1) Number of personnel in the unit and number OIC of EPWs.	OIC
2) Need for medical evacuation support for casualties.	

(continued)

TABLE 12.1 *(continued)*	
3) Location of final rally point.	OIC
4) Verification of reentry contingency plan.	
5) Food and water support.	
e. Radio main body to come forward or send back guides.	OIC
f. Advance to the front of the unit and verify the identity of each soldier as he passes through the reentry point. Inform guide when the last man has passed through.	1SG/PSG
g. Conduct reentry of lines and occupy final rally point.	CDR
h. Take accountability of personnel.	PSG/1SG
i. Brief stationary unit commander of the immediate tactical situation to his unit's front.	CDR
j. Contact battalion HQ for further instructions.	CDR

Source: ATEP 7-10-MTP, 5-118–5-124. "Perform Passage of Lines" (7-2-1040) forms the foundation document for this portion of the SOP. FM 7-71, 3-28.

Chapter 13

Air Assault

Similar to an infiltration, an air assault is a type of maneuver. However, at the company level it is not an end in itself but a means to an end. In this regard, it is planned as a movement unless the company is assaulting the objective directly from the air. An air assault directly on the objective is an extraordinary exception to normal operations because the benefits rarely exceed the risks.

An air assault requires extensive coordination and preparation. It is a fragile operation in the sense that the mission can be aborted or mission success severely endangered at any point of the process. The mission is influenced by weather, terrain, helicopter readiness, and pilot proficiency. All these factors require the company to begin planning as soon as it learns that an air assault is in the making.

Alert Procedures. Once the commander learns of a planned air assault (warning order or a "heads up" from the staff), he designates a platoon to coordinate and prepare for the air mission. For planning, the designated platoon is in the last lift serial because of its pickup zone (PZ) duties (if conducted at company level; otherwise, PZ preparation is a battalion task). The designated platoon leader becomes the pickup zone officer in charge (PZ OIC), and the platoon sergeant becomes the PZ NCOIC. The PZ OIC gathers information to assist the S3 Air with the air mission briefing (AMB) and submits it while the battalion staff is planning. The total number of passengers (PAX) allows the supporting helicopter battalion to allocate the proper number of aircraft for the mission. The total number of anticipated chalks (groups of soldiers for transport by helicopter) is not critical, but it does reinforce the number of aircraft and lifts required.

The AMB may be held immediately after the battalion operations order (OPORD), so the PZ OIC may need to accompany the commander to the OPORD. Once the primary PZ is identified, the PZ NCOIC recons, prepares, and secures the primary PZ. If time permits, the PZ OIC assists the NCOIC by conducting the PZ recon and sketching the PZ. Unless it is a battalion PZ, the battalion S3 Air selects the PZ through a map recon but should be amenable if the company identifies and recons a suitable PZ before the OPORD. Before any work is done on the PZ, the company gains approval from the S3.

At the battalion OPORD, the commander agrees to the selection of the landing zone (LZ) and voices reservations to it if he feels that a better alternative exists. In this regard, the PZ OIC and the commander review the area around the objective area for advisable LZs at least one terrain feature away from the objective. Without a compelling reason (accessibility), planning an LZ on or within sight of the objective is imprudent. The company must have a chance to organize itself for the attack instead of assaulting helter skelter.

PZ Preparation. The essential criteria for a company-sized PZ is its size, proximity, and horizontal surface. All other matters (obstacles, ground conditions, and security) can be resolved through preparation. The PZ NCOIC ensures that the PZ meets the essential criteria and has the work detail prepare the PZ for the helicopter lift. He also identifies and marks the locations for the PZ command post, the landing point of the lead aircraft, the assembly area, the chalk-holding areas, and the bump plan rally point for soldiers unable to load with other chalks. The designated guides accompany him from the assembly area to the anticipated, respective chalk-holding areas and PZ posture chalk points. The PZ NCOIC bases the number of chalk points on the number of helicopters that can land on the PZ at one time. This can be adjusted once the AMB reveals the number of aircraft per lift. At a minimum, a lift of one platoon (+) (three to four Blackhawks for training and two for combat conditions) is needed. The PZ NCOIC then makes a PZ sketch. Upon departure, the NCOIC leaves a three-man surveillance team with a radio to watch the PZ.

The Battalion Air Mission Brief. The PZ OIC takes as much information as available with him to the AMB. The AMB should only take 15 to 30 minutes, depending on how often the helicopter battalion supports the battalion. The PZ OIC uses the checklist to ensure that both the helicopter battalion and the company have coordinated properly for the mission.

The Air Assault Annex Briefing. Upon return from the AMB, the PZ OIC back-briefs the commander and receives additional guidance before preparing the air assault annex. The PZ NCOIC also meets with the OIC to learn of changes to the allowable cargo load (ACL) and the number of aircraft available per lift. He then relays this information to the chalk leaders, who, in turn, submit their manifests on a 3 X 5 card to the OIC. The 3 X 5 cards allow the OIC to organize the company elements for the lift quickly and with flexibility. Cards are easier to handle and manipulate than varied scraps of paper. He can also use these cards on a sand table (if time allows) when briefing the PZ occupation and lift plan. Each company section submits the preferred subdivision of its section for cross loading (leadership and key weapons). The OIC adds the subdivided section to each chalk. Depending on the time allotted, the PZ OIC can go into lesser or greater detail for the air assault phase of the mission. Such details as code words, LZ rules of engagement (ROE), scatter plan, the lift schedule, and the suppression of enemy air defenses (SEAD) plan are of less import to the mission itself and may be briefed later to the commander and chalk leaders in order to free up more preparation time for the platoon and sections.

The air route plan loses value if the distance to the LZ is great and the flight is executed by night. From a practical point of view, a long distance will require a small-scale map or a series of bulky large-scale maps per leader. The smaller the scale, the more difficult it is to navigate and pinpoint progress along the route. At night, navigation is nigh impossible without a Global Positioning System (GPS). We need to recognize this fact of air assault life and plan accordingly. It is ludicrous to expect everyone to know his location along the route. Most likely, the commander and platoon or section leaders will have small-scale maps of the route. Presented with this problem, each chalk leader needs to don a flight helmet and track the helicopter's progress by communicating with the pilot, who in turn informs the chalk leader of checkpoints, significant terrain features, and the reading on the helicopter's GPS position of these points along the flight. In this manner, the chalk leader will have an idea of his location should the aircraft crash or be forced to land early.

Occupation of the PZ. Upon occupation of the PZ assembly area, the guides lead the chalk groups to their respective chalk-holding areas. The chalks of lift 1 continue to move to their PZ posture chalk points. At this time, the chalk leader inspects each soldier's equipment and conducts a safety briefing. He reminds those personnel designated for the bump plan (should be the last two soldiers in the chalk, who also load the rucksacks on the helicopter) and shows everyone the bump plan rally point on the ground. He also goes over LZ procedures and the LZ assembly area map location and marking. The chalk rehearses (even without a helicopter present) loading and off-loading procedures to include rucksacks. At this point, the seating arrangement is verified. Once the chalk leader feels confident that the soldiers are ready, he goes over the mission subtasks with them again. The chalk leaders in the last chalk serial are responsible for establishing the perimeter of the chalk-holding area. Once the other chalks in the holding area have conducted their inspections, the perimeter chalk leader conducts his inspections and rehearsals.

The PZ NCOIC is responsible for PZ security. He designates which chalk conducts security patrols of the PZ and establishes observation posts (OPs) on avenues of approach into the PZ. The degree of security is largely dependent on the general threat picture. If the PZ is located deep in friendly territory, the establishment of perimeters at each chalk-holding area will suffice. The chalk guides are responsible for monitoring the arrival of each lift and alerting the next chalk to move to the PZ posture chalk point.

The PZ OIC and NCOIC ensure that the PZ is properly marked upon arrival. Once the PZ OIC has established contact with the lead aircraft, he relays the landing heading or the aircraft alignment markings on the PZ, marking of the lead touchdown point, chalk-loading side, A/C landing formation, and PZ landing conditions and local security status.

Knowledge of the coordinating checkpoint (CCP) helps the PZ OIC anticipate the approach azimuth of the aircraft. (A CCP is an easily recognizable terrain feature the aircraft use to navigate to the PZ.) He guides them to the PZ once he establishes visual contact (that is, "I have you visual. PZ is at your 9 o'clock") or, if at night, can perform a long or short count on the radio. The OIC informs the Battalion S3 Air of the departure of each lift and when the company is PZ clean. With the next-to-the last lift, the NCOIC informs the OIC of the number of chalks in the last lift. The OIC passes this information to the lead lift pilot as a reminder that one more lift is required. The PZ NCOIC ensures that all bumped personnel, security, and control elements are manifested and moved to a PZ posture chalk point(s).

For each lift that lands, the PZ posture chalks load while the chalks for the next lift move to the PZ posture chalk points. Once loaded, each chalk leader gives the pilot or flight chief the manifest and a 3 X 5 card with the PZ and LZ grid coordinates and a place for the pilot to write the GPS reading upon landing.

The Landing Zone. If a company-sized assembly area is designated, the lead aircraft chalk is responsible for establishing the LZ assembly area. Otherwise, platoon sized assembly areas are established by their respective lead chalks. The chalk leader(s) organize the perimeter IAW.

Occupy Simple Perimeter. The chalk leader also marks the assembly area (night/day) to guide on-the-ground and follow-on chalks. Assembly areas are located on the edge of the LZ and should be prepared to meet enemy attacks immediately. Once the company has assembled and conducted its personnel and equipment check, the commander gives the battalion S3 a SITREP and continues with the mission.

TABLE 13.1 AIR ASSAULT	
Task	**Responsibility**
1. Warning order: Upon notification gather the following information from the responsible personnel for the initial planning of the air assault:	PZ OIC
a. Total number of passengers (PAX) to be lifted.	1SG

TABLE 13.1 *(continued)*

b. Anticipate type of aircraft and allowable cargo load (ACL); submit tentative number of chalks and their chalk leaders. Use the following for planning: *Training* *Combat* UH-60 11 20 UH-1H 8 12 CH-47 33 48	PSG
c. Location of primary PZ.	S3 AIR
d. Location of alternate PZ (map recon).	S3 AIR
e. Time and location of air mission brief.	S3 AIR
f. Whether soldiers will carry rucksacks.	CDR
g. Abort criteria. Minimum number of aircraft needed to complete mission.	CDR
h. Internal or external load requirements for equipment/supplies.	1SG
i. Primary and alternate LZ.	CDR/S3 AIR
2. Verification and preparation of the PZ:	PZ NCOIC
a. Organize PZ recon force consisting of security force, guides, and PZ preparation work detail.	PZ NCOIC
b. Secure PZ for recon.	Security NCO
c. Conduct recon of primary PZ and determine suitability:	
1) Ground slope:	
a) 0-6 percent: Excellent; aircraft land upslope, never downslope.	
b) 7-15 percent: Fair; aircraft land sideslope. Soldiers must load on downslope side of aircraft.	
c) < 15 percent: Unusable.	PZ NCOIC
2) Usable PZ size: 10:1 ratio obstacle clearance along approach and departure axis of the PZ. Example: if the tree lines intersecting the axis of approach and departure are about 10 meters high, the landing area for the helicopters must be 100 meters from the tree line. Plan 35 meters between aircraft in tandem and 50 meters to each side.	

(continued)

TABLE 13.1 *(continued)*	
3) Obstacles: whether obstacles make PZ unusable.	PZ NCOIC
4) Wind direction: Aircraft should land and depart facing into the wind.	
5) Ground: Vegetation knee level or lower; firm with little loose sand, dust or snow; good drainage.	
6) Make sketch of PZ using the above criteria and dimensions. Return and back-brief PZ OIC immediately.	
d. Prepare PZ:	PZ NCOIC
1) Remove or mark all obstacles with VS-17 panel (day) or red light (night) secured firmly to obstacle.	WORK DETAIL NCO
2) Ensure each landing area is cropped of excess vegetation and that loose sand or snow is packed down.	WORK DETAIL NCO
3) Designate and mark PZ assembly area or chalk release point.	PZ NCOIC/GUIDES
4) Designate and mark chalk-holding areas and PZ posture chalk points; assign guides and brief them on their responsibilities.	PZ NCOIC/GUIDES
5) Designate and mark bump plan rally point.	PZ NCOIC
6) Designate and mark PZ control post.	PZ NCOIC
7) Designate and mark location of landing point for lead aircraft.	PZ NCOIC
8) Upon completion of PZ prep, establish surveillance point for a three-soldier surveillance team (from security team) and return to company assembly area.	PZ NCOIC
3. Attend air mission brief (AMB):	
a. Provide S3 Air and supporting helicopter unit S3 with the following:	
1) Number of PAX to be lifted.	PZ OIC
2) Whether soldiers will carry rucksacks.	
3) Location of primary and alternate PZ.	
4) Sketch of PZ with all pertinent data.	

TABLE 13.1 *(continued)*	
5) Marking of the PZ for day and night: -PZ control -A/C approach and departure -Lead A/C point -PZ posture chalk points	PZ OIC
6) PZ control frequency and call sign.	
7) External and internal load requirements for equipment/supplies.	
8) Primary and alternate landing zones (LZ).	
b. Obtain the following from S3 Air:	
1) Verify the type of aircraft and ACL.	PZ OIC
2) Serial configuration per lift (number of aircraft on the PZ per lift).	
3) Number of lifts. Ensure the planned ACL, and that the number of aircraft and lifts are sufficient to lift the company.	PZ OIC
4) Aircraft CCP and approach azimuth to the PZ.	
5) Aircraft approach heading into the PZ.	
6) Aircraft frequency and call sign.	
7) Turnaround time of each lift.	
8) Primary/alternate air route overlay with easily recognizable checkpoints/rails from the air identified.	
9) Suppression of enemy air defense (SEAD) plan.	
10) Medevac plan: Location of casualty collection point at LZ.	
11) Primary and alternate extraction PZ (if applicable).	
12) Downed aircraft procedures.	
13) Enemy air scatter plan to include rally point.	
14) Aircraft ROE into LZ.	
15) Map requirements.	
16) Aircraft laager site.	

(continued)

TABLE 13.1 *(continued)*	

17) Code Words for:
  a) Hot PZ
  b) Hot LZ
  c) First liftoff
  d) PZ clear
  e) Abort air mission
  f) Conduct air extraction
  g) Conduct ground extraction
  h) Conduct enemy air scatter plan
	PZ OIC

18) Time of final AMB (usually two hours before first lift).	
19) Request 1–3 aircraft to be at PZ 3 hours before first lift (H-3) for the purpose of chalk loading and unloading rehearsals.	
4. Prepare **air assault annex** (sketch) to OPORD:	PZ OIC
a. Brief CDR on changes as a result of the AMB.	PZ OIC
b. Disseminate changes to ACL as a result of the AMB to plt/section leaders.	PZ NCOIC
c. Submit manifests on 3 X 5 card to PZ OIC in the following format: *Number sequence Name Rank SSN Type Wpn Spc Equip*	CHALK LEADERS
d. Devise unit bump plan:	
1) Verify that key weapons and personnel are not designated for bump on manifest cards. Bump personnel are located at the bottom of the manifest card with a "B" suffix on the number sequence (i.e., B12).	PZ OIC
2) Identify which chalk in each lift is the bump chalk. Normally the last chalk in each lift in case the anticipated aircraft does not show. Verify that no key weapons or leaders are in the bump chalk.	
e. Prepare PZ security and fire plan.	PZ OIC/FIST OFF

f. Prepare sketch or sand table of PZ with the following:
  1) Dimensions of PZ
  2) Wind direction
  3) PZ assembly area
  4) PZ chalk-holding areas
	PZ NCOIC

TABLE 13.1 *(continued)*	
5) PZ posture chalk points 6) Location and marking of obstacles 7) Method of marking PZ 8) Bump plan rally point 9) Aircraft landing direction 10) Notes on peculiarities of PZ (slope, surface conditions)	PZ OIC
5. Brief **air assault annex:**	PZ OIC
a. Attend air assault annex brief along with normal contingent of leadership.	CHALK LEADER
b. Link up chalk guides with chalk leaders at end of briefing	PZ NCOIC
6. Occupy PZ assembly area (H-4):	CDR
a. Establish security at main avenues of approach.	PZ NCOIC
b. Recon route, chalk-holding area, PZ posture chalk point, bump plan rally point, and aircraft landing area. Chalk leaders in the last lift serial are responsible for establishing the security perimeter of their chalkholding area.	CHALK LEADER/GUIDE
c. Occupy PZ control post and aircraft landing marker if rehearsal aircraft are due. PZ control party consists of PZ OIC, RTO with two radios, and signalman (day).	PZ OIC
d. Inspect PZ for final marking of obstacles, landing point, bump plan rally point.	PZ NCOIC
7. Occupy chalk-holding areas (H-3):	CHALK LEADER/GUIDE
a. Inspect soldiers for the following: 1) Loose gear tied down. 2) Short-whip antennas on radios folded down and secured. 3) Precombat checks of equipment to include radio checks. 4) Weapon on safe, magazine not in weapon. 5) Patrol cap stowed or helmet chinstrap snapped.	TM LDR
b. Establish seating plan and loading plan: assign seats to each soldier. If floor loading, assign general position of each man.	CHALK LEADER

(continued)

TABLE 13.1 *(continued)*	
c. Issue rucksack loading and unloading plan: Soldiers drop rucksacks at aircraft entry point while loading. Last two soldiers pass rucksacks into aircraft, including their own. Rucksacks lay on soldiers' laps.	CHALK LEADER
d. Issue safety brief:	CHALK LEADER
1) Approach and depart helicopter from front or side, never from or to the rear.	
2) Never walk to the rear of the helicopter; if required to load on the opposite side of the helicopter, circle around to the front.	
3) Approach and depart helicopter hunched over. Do not chase any gear that flies away.	
4) Once seated, ensure weapons are on safe with muzzles pointed down toward the floor of the helicopter.	
5) When offloading, take three steps and take the prone position. Depart LZ once all helicopters have lifted off.	
6) If aircraft lands on a slope, everyone offloads on the downward slope side of the helicopter.	
7) If required to offload from a hover, await signal from chalk leader, throw out rucksack, ensure your landing area is clear, land on both feet, and roll to left or right. Move forward with rucksack three steps and assume prone position.	
e. Conduct loading and offloading rehearsals (with/ without rehearsal aircraft).	CHALK LEADER
f. Physically show the entire chalk where the bump plan rally point is located.	CHALK LEADER
8. Assume PZ posture (H-2):	CDR
a. First lift occupies PZ posture chalk points.	CHALK LEADER
b. Supervise flow of chalks to chalk points.	GUIDES
c. Occupy bump plan rally point and establish commo with PZ OIC and security element. Organize new chalks from bump personnel and guide them to chalk points. Chalk leader of bump chalk provides PZ NCOIC with manifest.	PZ NCOIC/RTO
d. Conduct final inspection of chalk personnel (equipment/weapons).	TM LDR

TABLE 13.1 (continued)

9. Execute PZ:	PZ OIC
a. Make contact with approaching helicopters, advise of tactical situation and PZ conditions.	PZ OIC
b. Once visual contact is made, guide helicopters to PZ, until they have visual contact with guide or landing marker.	PZ OIC
c. Once helicopter lands, guide chalk to helicopter, and supervise loading. Enter last and provide crew chief with manifest and a 3 X 5 card with PZ grid location and LZ grid location for the GPS. Don crew flight helmet and establish contact with pilot. Instruct crew chief to return card upon landing with the grid coordinate of the LZ as given by the GPS to verify the actual LZ location.	CHALK LEADER
d. Hook up radio to helicopter antenna outlet. Establish radio contact with higher command.	PLT LDR/CDR
e. Radio first liftoff and PZ clear to battalion S3 section.	PZ OIC
f. Bring in security team and guides to the bump plan rally point once the next-to-the-last lift departs. Verify chalks and lead them to the chalk points.	PZ NCOIC
10. Execute LZ:	CDR
a. Upon landing, the first lift moves to the LZ assembly area and secures it for follow-on lifts. A guide for each subsequent lift and a security team moves to a linkup point on the LZ and signals (chemlight/red lens flashlight, VS-17 panel) follow-on lifts. At night challenge and password is exchanged. Each guide leads his lift to the assembly area.	PLT LDR
b. If LZ is hot, advise battalion and execute contingency plan.	CDR
c. Once the company has consolidated in the assembly area, leaders conduct and report personnel and sensitive items checks. Company informs battalion of situation and continues with mission.	CDR

Source: ATEP 7-10-MTP, 5-97–5-102. "Perform Air Assault" (7-2-1036) forms the foundation document for this portion of the SOP.
U.S. Department of the Army, *Air Assault Operations,* Field Manual No. 90-4 (Washington, D.C.: March 1987), A-1–A-6, B-6–B-14, C-20–C-27, appendix D, appendix E. Provides an overview of planning checklists and preparation of the PZ.

FIGURE 13-1

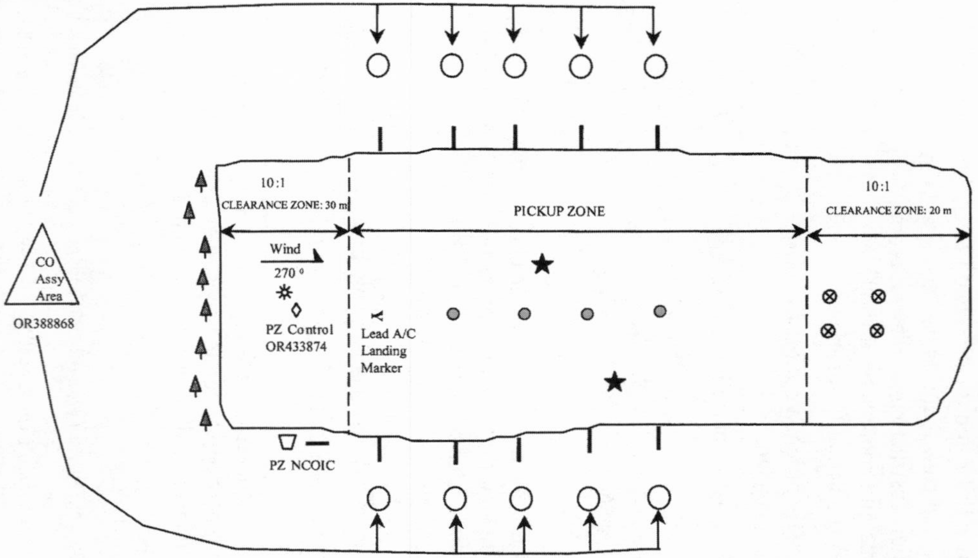

Codewords

Hot PZ: Volcano
Cold PZ: Eagle
Hot LZ: Cherry
Cold LZ: Ice
Lift # 1 Clear: Hammelburg
Lift # 2 Clear: Melbourne
Lift # 3 Clear: Schweinfurt
PZ Clear: Topeka
Abort Air Msn: Popcorn

Legend

△	PZ Assy Area	✳ Direction of Departure Marker
○	Chalk-Holding Areas	⊗ Direction of Approach Marker
─	PZ Posture Chalk Points Bump Plan Rally Point	★ Obstacle
		● Subsequent A/C Landing Points

Data

PZ Grid: OR456897
PZ Control Grid: OR433874
A/C Callsign: Black Widow
A/C Frequency: 37.70
A/C per Lift: 5
PZ % Slope: 0%
ACL: 20
H Hour: 2300

CLEARANCE ZONE: 20 m

10:1

PICKUP ZONE

CLEARANCE ZONE: 30 m

10:1

Wind
270°

PZ Control
OR433874

PZ NCOIC

Y Lead A/C Landing Marker

CO Assy Area
OR388868

Lift Plan

TIME	LIFT#	SER 1	SER 2	SER 3	SER 4	SER 5
H-30:30	1	1/1	2/1	CDR/M1	3/1	AT1
H-20:15	2	1/2	2/2	1SG/AT2	3/2	M2
H-10:20	3	1/3	2/3	XO/AT3	3/3	AT4

Pickup Zone Planning Sketch

Chapter 14

Linkup Operations

A linkup operation may involve the coalescence of company elements conducting an infiltration, or it may involve the meeting of two separate units during the course of a larger operation (battle handover, passage of a mechanized unit forward, and so forth). The main goal is to join together for the larger mission while preventing fratricide. Gaining confidence with linkup operations increases the options for higher command to execute riskier operations that would otherwise be a gamble with less-trained units.

The objective of the linkup site is to conduct a swift linkup without enemy interference. Selection of an easily recognizable terrain feature is crucial, particularly for night linkups. A prominent knoll or draw is distinct even at night. A site with cover and concealment, and away from natural lines of drift and likely enemy activity, decreases the likelihood of enemy interference. A defensible site with multiple access routes promotes security and survivability. With this in mind, a linkup site in a canyon or depression is not a practical choice.

The contingency plans give the linkup officer in charge (OIC) guidance when confronted with problems with the linkup. For an infiltration maneuver, the alternate site will be either closer to or at the objective rally point (ORP). For a larger operation, the alternate site may be closer to the moving unit or may need to shift laterally.

Time is the greatest adversary to linkup operations. Infiltration by squads could take two hours to get the company reformed. Because of the time factor in an infiltration and attack mission, the commander may be compelled to attack with 70 percent of the force and calculate that the remainder will join it on the objective. For a larger operation, the burden of timely linkup lies with the moving unit. The stationary unit continues to make contact until successful, even if substantial time elapses. In this case, only the commander can cancel the linkup operation for his subordinate element.

Coordination with the moving unit for larger operations should be made personally, but this generally is difficult to do below battalion level. The company commander may only have the opportunity to coordinate through the battalion S3. In this case, the commander gives the S3 a copy of the linkup operations SOP along with the crucial coordination information. The SOP gives the moving unit the required information to set the conditions for a successful linkup.

The company scouts are normally detailed to prepare the linkup, especially during an infiltration. Otherwise, a squad performs the task. The linkup element is split into a linkup site security team and the signal site overwatch team. The linkup site is different from the far recognition signal site in case the signal site is compromised. Because the signal site has a marker pointing to the overwatch position, enemy discovery means only that the overwatch position is compromised. The signal is pointed in the direction of the expected friendly approach. Infrared (IR) chemlights or a flashlight with an IR lens are very effective and secure devices for signaling the moving unit that has night observation devices. Once the signal site is prepared, the stationary unit attempts to make contact at regular intervals (every ten minutes). To improve line of sight, the squad leader establishes a radio site on high ground. If blessed with a Global Positioning System (GPS), the linkup OIC verifies the position. The linkup site team prepares it for occupation referring to Occupy Simple Perimeter SOP.

The moving unit halts short of the signal site and detaches a team equipped with a GPS to make initial contact. Once the contact team spots the far recognition signal, it responds with its far recognition signal in toward the overwatch position. This prompts the overwatch team to approach for the exchange of the challenge and password. Once positive contact is established, the overwatch team informs the linkup site security team, which in turn informs the company command post (CP). The overwatch team then accompanies the contact team to the moving unit and guides it to the linkup site. This procedure is repeated for linkup operations involving multiple elements, as in an infiltration. Once all infiltrating elements have completed the linkup, the linkup site element guides them to the ORP. For larger operations, the linkup site element guides the moving unit through the passage point or lane in cooperation of the effort.

TABLE 14.1 LINKUP OPERATIONS	
Task	**Responsibility**
1. Identification of linkup site using map recon for OPORD using the following guidelines:	CDR
a. Located on terrain feature readily identifiable at night.	
b. Provides cover and concealment.	

TABLE 14.1 *(continued)*	
c. Away from natural lines of drift.	CDR
d. Defensible.	
e. Has multiple access routes.	
f. Not located near likely enemy active areas (key terrain, intersections, etc.).	
2. Review of contingency plans:	SL
a. Enemy contact before or during linkup: break contact and proceed to alternate linkup point. Gain contact with moving unit and provide SITREP.	
b. Linkup not effected: gain radio contact for SITREP. If linkup not possible after 30 minutes of planned linkup time, proceed to alternate linkup site.	
3. Coordination with moving unit (for larger operations):	CDR
a. Exchange frequencies and call signs.	
b. Pertinent code words.	
c. Location and marking of the linkup site.	
d. Far and near recognition signals.	
e. Fire coordination measures (restrictive fire line).	
f. Review contingency plans.	
4. Stationary unit preparation of linkup site:	SL
a. Secure linkup site and establish overwatch position near signal site.	
b. Mark signal site with far recognition signal (chemlight, VS-17 panel). Mark site with directional marker (pile of rocks with pointing stick) pointing toward overwatch site.	
c. Establish radio site on high ground and contact moving unit.	
5. Moving unit procedures:	OIC
a. Establish security halt and establish radio contact with stationary unit.	
b. Dispatch a contact team to conduct the linkup with the stationary unit.	

(continued)

TABLE 14.1 *(continued)*

6. Linkup procedures:	SL
a. Contact team spots signal site and gives its far recognition signal (flashlight with colored lens or VRS panel) in the direction of the overwatch position.	SL
b. Stationary unit overwatch team approaches and exchanges challenge and password.	SL
c. The stationary unit overwatch team informs higher CP of initial contact, returns with contact team to moving unit, and leads moving unit to linkup point.	
7. Once linkup operation is complete, continue with mission.	CDR

Source: AT&EP 7-10-MTP, 5-108–5-116. "Perform Linkup" (7-2-1060) forms the foundation document for this portion of the SOP. FM 7-71, 3-35–3-38.

Chapter 15

Occupy Simple Perimeter

Occupying a simple perimeter is normally a nocturnal activity because of security and concealment requirements. It is also exponentially more difficult to execute at night than during the day. It behooves the company to practice this procedure at every opportunity, because an unorganized occupation degenerates into chaos quickly and is difficult to set right without great noise and consternation, alerting enemy soldiers in the general vicinity.

For the company and its elements, occupation of a simple perimeter is the geometric task associated with establishing an objective rally point (ORP), a patrol base, attack position, assembly area, or even a hasty perimeter while awaiting the execution of another phase of an operation.

Before occupying the perimeter, the company executes a security halt. The security halt is a prudent measure, because it allows time for a quartering party (Q-party) to clear and prepare the prospective area for a quick and smooth occupation. In enemy territory, a quartering party is less likely to compromise the mission if an enemy element happens to be in the area. Nevertheless, the main body is located within close proximity (within 300 meters) in case the quartering party runs into trouble.

The quartering party leader is a squad leader for company perimeters. For ORPs and patrol bases, he executes a 90 degree turn (sometimes called a dogleg) about 35 meters from the tentative perimeter. This precaution is aimed at catching enemy patrols that are tracking the company. At the dogleg, the Q-party leader establishes a two-man team to serve initially as a navigational aid for the main body, then as a listening or observation post (LP/OP) once the company occupies the perimeter. Placing a machine

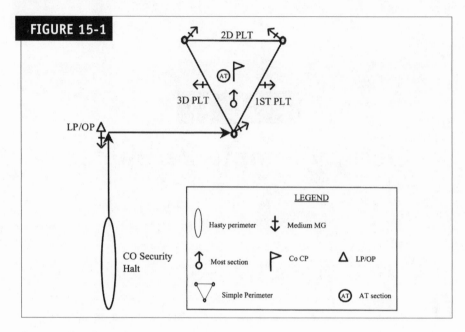

FIGURE 15-1

Occupy Simple Perimeter

gun with the LP/OP is a lethal means of eliminating enemy trackers who run into the LP/OP. There should be no misunderstandings for the action of the LP/OP and company sector—immediate fire response. It is better to compromise a position by aggressive fire and then to move to another location than it is to suffer a hasty enemy attack at close quarters simply because everyone is afraid that firing will give away the position. Part of establishing moral superiority is not being afraid to engage the enemy. If such mishaps with the enemy degrade the element of surprise, it is a small price compared to losing an aggressive edge. The company wants the enemy forces to be afraid every time they leave the safety of their lines.

The Q-party leader halts the Q-party at the tentative company CP and dispatches clearing teams with return time instructions (15–20 minutes). At the same time, he designates and marks the platoon sectors of the triangle-shaped perimeter. At night, a triangle perimeter is the easiest to prepare and occupy, because it is easier for all to conceptualize without ambiguity. Any weakness that a triangle perimeter has is mitigated by the short duration of occupation and hence should be regarded as ideal for military operations. The Q-party leader marks the apexes of the triangle only, using markers that can be recognized only from short distances (for example, engineer tape, an open Lensatic compass, or a directional chemlight on a tree or stake). He also ensures that the legs of the triangle are long enough for each platoon to occupy (30–60 meters, or 1–2 meter intervals between soldiers). From a practical point of view, the clearing team informs the Q-party leader if natural or man-made obstacles hinder occupation. The team does not begin clearing away obstacles, because it is far easier to move the perimeter to another location. This applies to chemical warfare contamination as well.

Once the perimeter is cleared and marked, everyone meets back at the tentative company CP within the perimeter to back-brief the Q-party leader. The Q-party leader places platoon and section guides at the six o'clock position (the entry point of the perimeter) to guide in their respective platoon or section. He also dispatches a 2- to 3-man team to return to the company and guide it into the perimeter.

As the company elements enter the perimeter, the guides lead their platoons into sector by tracing the leg until the lead soldier approaches the proper marked apex. The platoon then expands to fill the leg. It is important for the lead soldier of each platoon or section to alert his guide of his element's arrival at the entry point.

Once in place, the company begins preparation for the mission (ORP) or priority of work (patrol base, defense sector assembly area). Sector sketches and fire plans are only needed for a patrol base, because the length of stay warrants such activity.

TABLE 15.1 OCCUPY SIMPLE PERIMETER

Task	Responsibility
1. Security halt:	
a. Halt company 200–300 meters from tentative perimeter (small assembly area, patrol base, attack position, ORP) and form a cigar-shaped perimeter.	CDR
b. Mortars set up in hand held mode, trigger 1, for 360 degree coverage of hasty perimeter.	MORT SEC LDR
c. Organize quartering party: Consists of leader with RTO, one guide of each platoon and company section, and at least a squad for clearing and security.	Q-PARTY LDR
d. Exchange **Five Point Contingency Plan** with CDR: 1) The task. 2) Return time (about 30 minutes). 3) Company action if Q-party is late. 4) Q-party actions on enemy contact. 5) Company actions on enemy contact.	Q-PARTY LDR
e. Upon departure of quartering party, adjust perimeter and reestablish 360 degree tactical posture for main body.	1SG

(continued)

TABLE 15.1 *(continued)*	
2. Quartering party activities:	Q-PARTY LDR
a. For ORPs and patrol bases, make a 90 degree turn (dog-leg) approximately 35 meters (by day, 100 meters) from tentative perimeter and establish an LP/OP here. NOTE: If the commander decides to occupy a perimeter in force, the unit still makes a dogleg and establishes an LP/OP.	Q-PARTY LDR
b. Establish initial security at tentative Co CP.	Q-PARTY LDR
c. Clear designated sectors using clearing technique (box, zigzag, cloverleaf, etc.) and return to tentative CP in 15–20 minutes.	CLEARING TMs
d. Designate flanks of sectors and supervise method of marking of triangle formation. Sector markers (open Lensatic compass, directional chemlight, engineer tape) are established at the triangle apexes.	Q-PARTY LDR
e. The AT section weapons are placed along the perimeter, covering possible vehicular avenues of approach. If no avenues of approach exist, the AT section selects a location within the perimeter and becomes the company reserve.	AT SECTION LDR
f. Check area for NBC contamination if battalion HQ indicates a threat. If contamination exists, inform quartering party leader to evacuate the area.	NBC NCO
g. Check for obstacles. If mines are located, inform quartering party leader and evacuate the area.	CLEARING TMs
h. Upon clearing and marking, all guides report to tentative CP for back-brief and then move to perimeter entry point (normally six o'clock). Two guides return and guide company to perimeter release point.	Q-PARTY LDR
3. Occupy perimeter.	CDR/PLT/ SEC LDRS
a. Guides pick up respective units and guide into location.	QP LDR/GUIDES
b. Once perimeter is established, leaders adjust perimeter as needed and emplace LP/OPs.	PLT LDR
c. Begin mission preparations or priorities of work.	ALL

Chapter 16

Actions from Objective Rally Point to Objective Positions

This event is most critical to the mission, because this phase is the culmination of all prior planning and preparation. Too often, simply making it to the objective is an end in itself for companies. In combat, a sloppy attack will likely be repulsed with a concomitant heavy toll in casualties. The savvy commander uses every ploy and device to give the company an edge over the enemy in the initial period of the attack. Much of the activity occurs simultaneously but is still controlled by the senior leader in the area.

Priority of Work. Before final preparations begin, the NCOs ensure that the perimeter is secure with interlocking fields of fire. It is also reassuring to the leader's recon that each soldier knows the challenge and password as well as the running password in case the recon runs into trouble.

The final check of weapons, ammunition, and equipment assures that everything functions as expected. A fast cleaning of ammunition and the functioning parts of weapons practically assures that the weapons will not foul. Machine gun ammunition is particularly susceptible to filth. The belt must be thoroughly brushed and carefully folded in its container (the two-quart canteen pouch or butt pack are good durable containers). Wrapping the machine gun belt around the body or allowing it to hang from the machine gun without support increases the likelihood that the machine gun will malfunction. Noteworthy is the assistant machine gunner, who not only helps the gunner select targets but also monitors the feeding and expenditure of ammunition into the machine gun. He signals the gunner to stop firing when the belt nears its end, hooks up a fresh belt, and signals the gunner to resume firing. Good assistant gunners can do this without a cessation of fire.

For those mounting night observation devices (NODs) and donning nightvision goggles (NVGs), a final battery and functions check must be performed in the objective rally point (ORP), because a nonfunctioning NOD will not be corrected during the attack. It is better to know beforehand that it is not functioning than to have a false sense of security. Because the NVG head harness is difficult to don in the dark, the user will need time and sometimes assistance to get it in place properly (helmet harnesses are far superior for NODs). NODs are best used with the fire support and security personnel. They are less useful for assaulting personnel because three-to-five-second rushes and proximate flash of munitions and explosions tend to disorient the user. Leaders also verify that their laser pointers are functioning by using NODs.

The final cleaning of radio terminals is necessary, because dampness is a common cause of malfunction. It does no harm to change out the battery even if the old battery has a few hours of life in it. Sometimes that little extra level of power means the difference between radio contact and static.

Special demolitions and breaching equipment are also accounted for and prepared in the ORP, because time and stealth may not permit this later on. This also permits the immediate supervisors to check the preparation without haste.

Each soldier also takes the time to dry off, powder, and replace his T-shirt and socks. After any dismounted distance, soldiers carrying a standard load perspire a great deal. They often lie and wait for hours in position before H hour. Conserving body heat lessens fatigue and maintains marksmanship accuracy. The final application of camouflage on exposed flesh completes the individual preparation for the attack.

The mortar squad leader informs each platoon sergeant of the mortar round collection point, supervising the assembly of all the rounds. The transfer of mortar rounds to the fire support element is best done by having the carriers deposit the rounds at the designated fire support element position in the ORP. The mortar squad leader makes a count and informs the mortar section leader of the total rounds available. The mortar section leader informs the company commander of the tally and the amount of fire (rounds per minute and for how long) his section can provide during the attack.

Finalize Plan. The objective recon leader back-briefs the primary leadership on the objective area, referring to his sketch. He identifies observations that affect the plan and recommends alternatives. Of particular importance are the locations of minefields and observed locations of registered fire. Among its other tasks, the objective recon always clears lanes through minefields at night from the assault position if possible—carefully replacing the camouflage. The commander considers whether these revelations are significant enough to deviate from the original plan. Ultimately, it is the commander's call. The details of the surrounding terrain or the enemy situation may turn a good plan into a mediocre one. But the old adage that a poor plan executed well is better than a good plan executed poorly is still valid. The commander needs to weigh whether a deviation of the plan increases or jeopardizes the odds of success. In this, the time available and the adaptability of the company are paramount.

The commander then conducts a leaders' recon with the principal element leaders. The executive officer (XO) adheres to the contingency plan and makes plans to continue the attack, conferring with the alternate element leaders for continuity. The leaders' recon links up with the surveillance team initially for updates. To save time, the leaders' recon may split up at a release point to recon respective positions. A guide from the objective recon leads each element to its respective position. The release point must be a clearly identifiable piece of terrain, and the commander dictates when everyone meets back for the return to the ORP. The leaders' recon should return to the ORP no later than one hour before H hour. The commander recons the assault position first with the assault element leaders and the objective recon leader and then goes to the fire support position and confers with the fire support element leader on how to best support the advance of assault element. If time allows, the commander and the recon leader may recon more of the objective area.

TTP

If a leaders' recon is not possible, the commander relies on the expertise and judgment of the objective recon leader, finalizes the plan on the spot, and issues the fragmentary order (FRAGO). The company forms into its task-organized positions automatically upon closure of the company. The first sergeant (1SG) supervises the reorganization.

Each element leader should take the opportunity to mark the initial positions of his element if the danger of compromise is low. This is particularly important for the fire support element, because the leader needs to emplace the element in the most favorable position. Additionally, a marked position serves as a limit of advance, thereby increasing the soldiers' confidence that they will not blunder into the enemy. It is important to note that enemy contact does not automatically mean that the mission is compromised and that the commander must abort the mission or worse, commence a pell-mell

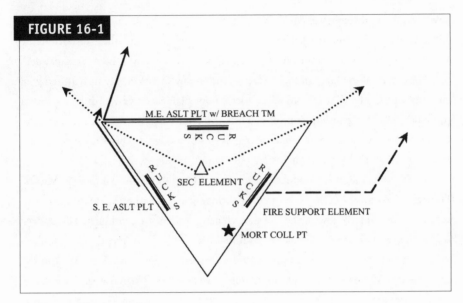

FIGURE 16-1

Task Organizing in ORP

attack. The enemy will never be sure whether the contact is just with a reconnaissance unit or even is a false alarm. Even if the timing and point of attack is compromised, the commander can confer with the battalion commander and recommend a change in either or both.

If the objective recon element also comprises the security element, there is no need for the security team leader to conduct a leaders' recon of each security position. This is done as part of the objective recon. If the commander deems necessary, the security element may deploy to its positions during the leaders' recon, because emplacement of the security element consumes the most time.

Prepare to Deploy. Once the leaders' recon returns to the ORP, the 1SG directs and supervises the company into its task-organized elements. (See figure 16-1.) The locations of the task-organized elements are standardized to avoid confusion: the sector between two and six o'clock is designated for the fire support element. The mortars move to a point next to the fire support leader's command post (CP) and establish the mortar rounds collection point. The mortar section leader supervises the distribution of the rounds to the fire support personnel for transportation to the mortar firing position. The fire support element personnel place these rounds in their butt pack or rucksack for transportation to the mortar fire support position. The assault elements occupy sectors

10 to 2 and 6 to 10, respectively (maintaining platoon integrity as much as possible). The security element locates within the perimeter near the company CP. Rucksacks are placed neatly in line at the respective element CPs. Soldiers continue the priority of work and supporting mission back-briefs until the unit begins deployment to assigned positions.

The commander completes the plan and issues the FRAGO. Once alerted, the element guides link up with their respective elements to lead them into place.

Deploy to Assigned Positions. Depending on the time, distance, and available cover and concealment to the assigned positions, the fire support team (FIST) officer and commander may decide to begin the preparatory fire or harassing fire. Such fire masks the movement of the unit and suppresses the enemy forces while the elements deploy into position. However, the preparatory fire phase may expire before the unit is ready for the attack, allowing the enemy forces to recover and steel themselves for the attack. If the enemy forces are well entrenched, the effects of suppression will be degraded. Generally, starting the preparatory fire once the unit is poised for the assault allows the FIST officer to adjust fire and allows the breaching team to reduce obstacles.

> **TTP**
>
> The fire support element leader may deploy the 60 mm mortars as part of the base fire support element in order to use them in the direct fire mode.

If the security element does not deploy during the leaders' recon, the commander should dispatch it about 30 minutes before the company occupies the assault and fire support positions. If convenient, security element personnel transport mortar rounds to the mortar firing position in transit to their security positions.

The fire support element task organizes into the fire support base element (machine guns), hunter-killer teams (M203, antitank [AT] weapon, squad automatic weapon [SAW]), and the mortar section. (See figure 16-2.) The mortar section should be near the front of the fire support element in the order of movement. The mortar section turns off to its selected firing position, with the mortar squad leader remaining at the turnoff point to establish the ammo point. As the fire support element continues toward the fire support position, the personnel deposit the mortar rounds with the mortar squad leader with no disruption in the movement.

The fire support element stops short of the fire support position and deploys into the team positions slowly. The fire support leader and platoon sergeant ensure that all

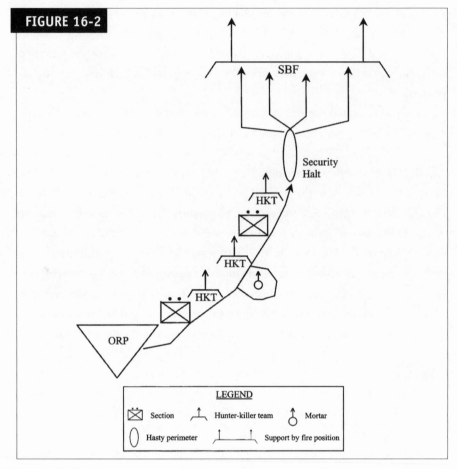

FIGURE 16-2

SBF

Security
Halt

HKT

HKT

HKT

ORP

LEGEND

⊠ Section   ⊥ Hunter-killer team   Mortar

◯ Hasty perimeter   Support by fire position

Movement of Fire Support Element

weapons are in place with general sectors of fire as well as orienting the crews to the location of the assault position and general scheme of maneuver on the ground. Squad and team leaders remind everyone of the signals for lifting or shifting of fires. If tanks are to link up with the fire support element, the fire support leader along with the guide recon the tank primary and alternate firing positions, ensuring that obstacles do not block the access route. The fire support leader's primary choice is defilade positions. The guide goes to the linkup point and guides the tanks into position. The guide also acts as the observer for the tank, using the field phone to communicate targets to the crew. Depending on the risk, the guide can either position himself behind the turret or lie beneath the tank in the prone position.

TABLE 16.1 ACTIONS FROM ORP TO OBJECTIVE POSITIONS

Task	Responsibility
1. Begin priority of work in ORP:	1SG
a. Check security of assigned perimeter sector (LP/OP, sectors of fire, running password, and challenge/ password).	PSG
b. Final check of weapons (lightly oiled, locked and loaded, on safe).	IAW [CH 5] **MISSION PREPARATION**
c. Final inspection of ammunition (clean, properly carried).	IAW **MISSION PREPARATION**
d. Change radio battery and clean terminals. Make final commo check.	IAW **MISSION PREPARATION**
e. Prepare demolitions for obstacle breaching.	BREACH TM LDR
f. Final inspection of NODs (mounted and tested).	IAW **MISSION PREPARATION**
g. Ensure subordinates have dried off perspiration and changed T-shirts and socks.	TM LDR
h. Inspect camouflage of soldiers.	TM LDR
i. Account for mortar ammo carried by subordinates and supervise transportation to ORP collection point.	TM LDR
2. Finalize plan:	CDR
a. Back-brief CDR and principal leaders of objective area, using sector sketch.	OBJ RECON LDR
b. Organize leaders' recon: CDR, RTO, OBJ Recon Ldr, Fire Support Element Ldr, Assault Element Ldr, Security Element Ldr, and guides (OBJ Recon Tm).	CDR
c. Issue contingency plan to XO:	CDR
1) If leaders' recon does not return by H-30 mins, deploy elements to assigned positions and commence attack.	
2) If leaders' recon is compromised, it withdraws to the ORP, using running password.	

(*continued*)

TABLE 16.1 *(continued)*

3) If ORP is compromised, XO defends initially and withdraws to alt ORP.	CDR
d. Upon return from leaders recon, complete plan and issue FRAGO.	CDR
3. Prepare for deployment to assigned positions:	1SG
a. Form into task organized elements.	PSG
b. Rucksacks placed on line at each element CP.	PSG
b. Review assigned tasks with subordinates.	SL/TM LDR
c. Fire support element collects mortar rounds at ORP collection point.	FS PSG
d. Guides link up with assigned elements.	OBJ RECON LDR
4. Deploy to assigned positions:	CDR
a. Begin harassing or preparatory party fire.	FIST OFF/CDR
b. Deploy security element first (CDR waits 10 minutes).	SEC SL
c. Deploy fire support element into assigned positions. Mortar section is second in the order of march. Mortar round carriers deposit rounds at mortar firing position.	FS PLT LDR
d. Deploy assault element into position. Breach team is second in the order of march.	CDR/ASLT PLT LDR
e. Establish company casualty collection point.	CO MEDIC

Source: ATEP 7-10-MTP, 5-130–5-133. "Occupy Objective Rally Point" (7-2-1041) forms the foundation document for this portion of the SOP.

The assault element need not follow the same route as the fire support element. Generally, it is better for it to take its own path because it has a greater distance to go. The breaching team follows the lead squad under the supervision of the main effort platoon leader.

The company medic establishes the company aid post in a covered position near the assault element and a road leading to the rear. A road is essential for quick ambulance evacuation.

Leader Positions. The location of the company commander is a reflection of the level of proficiency and experience of the company. It does commanders little good to read in manuals that their position depends on where they can best influence the battle. That knowledge is self-evident. It is better to discuss the pros and cons of where he decides to establish his CP than to ignore the issue.

S. L. A. Marshall concluded that the commander was most effective a little farther back from the front line in order to ensure that every asset available to the company was brought into the fight.[1] This would mean that he would need to position himself where he could request and link up with an advancing weapon system (tank) or special combat unit (engineers) or simply grab them as an opportunity. He could also take advantage of high ground to oversee the battle and request fire or air support. Additionally, his value to the sustainment of the company is too great to take a chance on becoming a casualty. The soldiers understand and accept this. In fact, a commander who exposes himself needlessly could, in fact, cause such consternation among the troops that they could worry more about his well-being than about completing the mission. The problem with a rearward position is that the leader can quickly lose touch with the battle, particularly in close terrain or at night. It is impossible to maintain a clear picture of the situation via radio. The commander who attempts to command from the rear rarely is able to find a good spot that allows him both to track the battle and to maintain good communications with higher headquarters for the purpose of requesting support. In this sense, he is robbing Peter to pay Paul.

The commander can place himself in the middle of the action—the decisive point—in order to monitor the pulse of the battle. In this sense, he is following the traditions of ancient great captains. Alexander the Great always led from the front and epitomizes this style of combat leadership. Wherever he was positioned, that was the decisive point. From such a position, the commander can immediately exploit an opportunity or correct an error without relying on subordinates to relay detailed information for him to act on. He can also request air or fire support more immediately if his radio link is good. In this sense, he creates the dynamic of decisive action—pushing weapons and manpower to critical points to break the coherence of the enemy defense, and once the enemy defense begins to crumble, maintaining pressure on him until he is destroyed or routed. Young Lieutenant Erwin Rommel was very good at this sort of thing during World War I.

The shortfalls of this style of leadership are numerous, though. The probability of the commander becoming a casualty is significantly higher, and his influence on the battle and the well-being of the company abruptly ends. In fact, his death or injury could cause the whole momentum of the attack to cease as the soldiers become paralyzed by his loss. Even if they are able to carry on, a certain amount of time will pass as the new commander establishes control. Leaders who lead from the front also tend to lose the sight of the larger picture—tunnel vision syndrome. Subordinates have a harder time maintaining communications with him to keep him aware of the overall situation. He likely misses opportunities elsewhere on the battlefield and will rarely be in a position to request support from higher headquarters. He in a sense becomes a platoon leader or even a squad leader, depending on the terrain, instead of orchestrating the entire effort.

The commander is best placed where he can see the progress of the assault and also the effects of the fire support element. This is easier to do in more open terrain. For closer terrain, he needs to be closer to the assault element and keep the fire support element informed of the progress, because platoon and squad leaders tend to forget to report during the fray. In really close terrain, such as deep woods or jungles, the commander probably will not have a separate fire support element, electing to distribute these weapons among the platoons. So, in this sense, he can be closer to the assault element.

During the attack, the commander's location constantly changes as the assault progresses. He may initially be near the breach point and then follow behind the assault element. From this perspective, he can see how the attack is progressing and how effective the fire support element is supporting the assault. This style of command requires great energy, ensuring that the assault and fire support elements are complying with his intent. Such urgency requires him to move to multiple points in the area, checking the progress of the breach, then the positioning of the fire support element, then the assault itself—almost like a bee pollinating flowers. Napoleon Bonaparte in his prime was extremely active on the battlefield because he knew that execution was marred by friction. The differences between his earlier and later campaigns is not in the planning (which remained brilliant to the end of his career) nor in the number of mistakes he and his subordinates made; the key difference lay in his energy in correcting mistakes, resolving misunderstandings, and generally ensuring that his subordinates remained focused on his intent. In point of fact, every one of his defeats can be traced to his increased lethargy as he became older, less fit, and, most important, in ill health.[2] The guiding principle for the commander is to destroy the enemy in detail by concentrating overwhelming firepower at distinct enemy centers of resistance and finish each with maneuver. If he can translate this concept into action, he usually will be successful.

The XO should be located near the fire support position, where he can see most of the objective. He monitors the progress and takes over if the commander becomes a casualty. He can relay requests for higher headquarters support and keep the battalion commander apprised of the situation. He can also keep tabs on the progress of the advancing assault and advise the fire support leader of needed shifts in fire or positions, if applicable.

The 1SG should be located near the assault position, because this is where casualties and enemy prisoners of war will flow. During consolidation and reorganization, he goes to the linkup point for the company trains and guides them into place.

The security element squad leader is located at the security position along the most likely enemy avenue of approach. Since an enemy attack along this avenue would be deemed the most dangerous to the mission, his leadership is most effective there.

Conclusion. Rarely will the movement, deployment, and attack go as planned. A thousand unforeseen things will happen that could degrade the power of the attack. Degradation does not mean failure. Each leader should expect problems and maintain mission focus. Subordinates need to accept the idea of friction and maintain their faith in their leaders. Through discipline, individual initiative, and perseverance, the company will bring off success even if the plan falls apart.

NOTES

1. S. L. A. Marshall, *Men against Fire* (Novato, Calif.: Presidio Press, 1947), 187–99.
2. Chandler, 733, 763, 777, 788–89, 799, 804, 823, 860–61, 940, 1046–47.

Chapter 17

Consolidate and Reorganize

The company is most vulnerable during consolidation and reorganization (C&R). C&R planning is critical in both the defense and attack. But, since it requires greater planning and execution in the attack, this discussion will focus on C&R in the attack.

Clausewitz's ever-present friction requires the company to consolidate and reorganize, or risk becoming combat ineffective. The task organization for the attack may require squads and fire teams to operate separately from their parent units. During the course of the attack, assaulting squads and fire teams also become separated from their parent units. Casualties and breakdowns are likely to claim some key leaders and key weapons. The soldiers who constitute the assault element rarely end up in their assigned zone for the consolidation and reorganization phase and probably haven't a clear idea where they are much less where the assigned zone is. The greater the objective area and resistance of the enemy, the greater the amount of chaos.

To add to the confusion, the soldier's adrenaline surge wanes once the immediate source of danger disappears. He is exhausted and parched, and since there is no identifiable event that signals the transition from the attack to C&R, a soldier will generally stay in place to recover his strength or assume a state of lethargy.[1] The idea of moving from his current refuge makes less sense now than it did during the operations order (OPORD). Even when he does move, it is slow and uncertain. This is the moment the enemy wishes to exploit for a counterattack.

Consolidation. The commander's greatest challenge is getting the unit to follow through. The risky place to stay is directly on the objective area, because the most likely enemy reaction is predesignated artillery on the objective itself. If time and terrain allows, the company should establish a hasty defense off of the objective area.[2] At this point, platoon integrity has less significance than the establishment of a hasty defense. The commander and platoon leaders direct squads to the nearest predesignated sectors without considering their parent designation for the moment. Platoon leaders maintain their original assigned sectors and establish the initial defense with whatever squads are available. Naturally, this will result in some sectors being weaker than others, but this is better than wasting time attempting to sort out who goes where.

The members of the fire support element retain their cohesion and can rapidly displace to their assigned sectors without too much confusion. It is best for the platoon sergeant to guide these teams to the platoon command post (CP) because he knows the location and can use his authority to move them quickly.

The security element squads and fire teams revert to the listening and observation posts (LP/OPs) and recon patrols in order to extend the eyes and ears of the company as well as establishing contact with units on the flanks. The security element contributes more in this role than adding to the manpower of the immediate defense. The question remains what to do with the platoon leader and platoon sergeant who have contributed their squads to the security effort. The commander has several options: he can generate a powerful reserve from the remnants of the platoon; he can incorporate the platoon's key weapons into the defense and have the platoon leader assist the executive officer (XO) or first sergeant (1SG) during C&R; or he can reconstitute the platoon by having the company scouts take over the patrol mission, leaving the platoon responsible for local security.

Reorganization. While the commander and platoon leaders concentrate on establishing and improving the defense, the 1SG, XO, and platoon sergeants concentrate on casualty evacuation and resupply. Many of these tasks occur simultaneously.

Team leaders collect the status of their teams for the squad leader using the acronym ACE (ammo, casualties, and equipment). Ammo status for rifles is best stated in terms of magazines even if most of the magazines are not complete. Machine gun ammo status is best accounted by boxes or belts. Casualties are reported by name for the platoon sergeant (PSG) to verify. Key weapons and equipment status is best reported by what is missing or destroyed rather than by what is present. This also alerts the platoon sergeant and squad leader to recover the missing equipment quickly. Detailed reports can come later. The PSG submits his ACE report to the XO in terms of number of personnel per squad, losses in key weapons, and average number of rounds per weapon. The XO quickly records this information on an acetated ACE report chart and passes this on to the battalion tactical operations center (TOC). It is quite unrealistic of the XO to attempt to track the ACE status of the company during the heat of battle. Certainly, the junior leaders should not attempt to take an accounting of their losses while fighting. The only time to make this assessment and report is during consolidation and reorganization.

Rapid and efficient casualty evacuation is essential not only to those who are casualties but also to the mental well-being of their comrades. The commander wants to avoid the immediate attrition of strength resulting from soldiers carrying their wounded buddies to the company aid post when their immediate presence is needed for the fight. To avoid this, the company must have an established system that gives the soldiers confidence that everything is being done to save their friends.

The casualty evaluation (CASEVAC) system seeks to aid and evacuate casualties quickly without sacrificing the accuracy in accountability. Upon discovery of a casualty, the soldier applies first aid, using the casualty's first aid packet (not his own) and alerts a medic. At this point, he continues the attack. The alerted medic performs additional aid if needed and guides or brings the casualty to the company aid post. If the case is urgent and requires a litter, the medic is authorized to enlist the assistance of two soldiers to move the casualty to the aid post. If the case is not urgent but requires a litter, the medic continues with other casualties and alerts an aid and litter team to evacuate him. During C&R, aid and litter teams sweep the area for casualties to evacuate. The platoon sergeant visits the aid post in search of his soldiers. For each casualty, the platoon sergeant removes the DA Form 1156 (Casualty Feeder Report) from the soldier's first aid pouch and completes it. Once he has accounted for all his soldiers either with the 1156 or 1155 (Witness Statement), he submits them to the 1SG, who in turn updates the **Unit Manning Roster.** The 1SG then submits the packet of forms to the aid evacuation NCO for submission to the battalion aid station. The platoon sergeant ensures that crew-served weapons are returned to the front and that only the soldier's personal weapon is evacuated with him. The status of personal weapons is always a point of great debate concerning evacuation between the company and the forward support battalion medical company. The soldier's personal weapon remains with the soldier. This is so important that the commander should authorize the use of force to ensure that this happens. The medical community does not call the shots on this issue and woe be the medic who attempts to refuse a soldier's weapon. As for the soldier's rucksack, the platoon sergeant transfers it to the supply sergeant for safekeeping in the company trains.

The company medic assists the aid evacuation NCO with coordination of evacuation. Cases are evacuated by priority (Urgent, Priority, Routine) via ambulance or aerial medevac. For the latter, the company medic prepares a pickup zone (PZ) nearby if there is no battalion coordinating checkpoint (CCP). During daylight, a soldier can guide in the helicopters; at night, the company medic can mark the PZ with chemlights, using the inverted "Y" marker. For speed, the company medic should prepare a marker with premeasured 550 cord (7' X 7' X 14' legs) for quick setup.

The process and evacuation of enemy prisoners of war rarely receive attention in training. The company is normally left to evacuate enemy prisoners of war (EPWs) to the rear with no thought of the strain EPWs have on the company's manpower and logistics. Because of the emotional state of the soldiers after a battle and the strain on the company's limited resources, soldiers may possibly commit atrocities on these prisoners.

The company commander designates an EPW collection point in the OPORD. As the enemy surrenders, slightly wounded soldiers can be designated as escorts and guards. If none are available, the fire team to which the enemy has surrendered escorts them to the collection point, and a section of the fire support element is detailed as guards. During C&R, the designated EPW team processes the EPWs rapidly and arranges for their transportation to the rear. If transportation is not available, the EPW team marches them to the designated EPW point at the combat trains, submits the prisoner manifest to the S1, and transfers responsibility.

TTP

Soldiers never leave their position to approach surrendering enemy—the enemy must approach the friendly lines with no weapons and their hands in the air and pass through to the EPW point. Soldiers are to continue firing at other enemy soldiers in the vicinity that are not part of the surrender group.[3]

Leaders must be attuned to the psychological makeup of the EPW team leader. The commander does not want someone in charge who is likely to commit or tolerate atrocities. The odd thing about war is that even ethical people can be moved to rationalize murdering prisoners; so the commander must be attuned to changes in his subordinates' attitudes to make sure that those soldiers don't have the opportunity to make such a rationalization.

The destruction of enemy equipment is normally a function of a raid. Unless there is a high likelihood that the enemy will recapture the area after an attack, there is no need to destroy it.

Resupply is the responsibility of the supply sergeant even though few supply sergeants feel useful in the field. This must be so because they are so reluctant to deploy to the field during exercises. (And in truth, their duties keep them quite busy in the garrison.) Now, to make the supply sergeant feel useful, the commander needs to give him plenty of responsibility and challenges. The main challenge is to have him coordinate, organize, and deliver supplies at the earliest opportunity. The supply sergeant oversees the resupply packages IAW **Resupply Planning.** The supply sergeant coordinates the method of delivery with the S4, depending on the mission. Air assault and infiltration attacks favor the use of helicopter resupply if there is no air defense threat. Otherwise, the supply sergeant can use the company vehicle to deliver supplies. If the commander assesses that the threat is low, the supply sergeant delivers the supplies to the platoon and section CPs. Otherwise, he establishes the company trains at a point to the rear of the company and notifies the chain of command of supplies for pick up. The 1SG may be able to form a resupply detail comprising profiles and headquarters personnel under the control of the supply sergeant to deliver supplies forward. To keep the soldier's load down and based on mission, enemy, terrain and weather, troops and time available

FIGURE 17-1

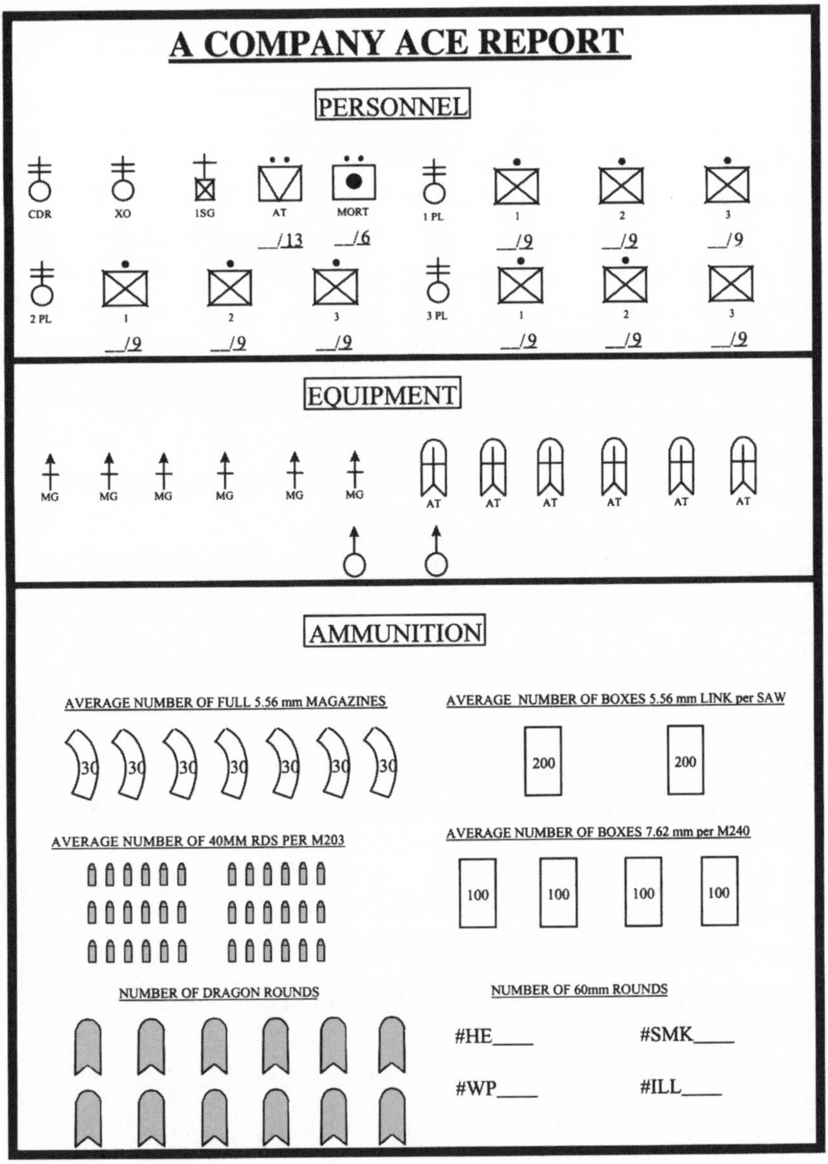

ACE Report Tracking Sheet

(METT-T), the commander may even dedicate a squad to carry supplies and extra equipment forward. After delivery of supplies, the XO submits logistical status and other reports to the supply sergeant for delivery to the battalion combat trains.

Once there is a lull in activity, the commander submits the unit status to the battalion commander. The purpose is to give the battalion commander a clear idea of the combat effectiveness of the company. Lastly, he informs the battalion commander when the company will be ready for further operations.

The company does not wait for the commander to issue formal orders for preparations for the defense; it begins the priority of work automatically. In the interim, the commander prepares for immediate movement by planning the movement to contact, because this is the most versatile of plans.

Consolidation and reorganization is just as applicable for the defense as for the attack. The majority of activity is devoted to repairing positions and obstacles as well as evacuating casualties and prisoners. Fortunately, C&R normally takes less time and effort for the defense, but that does not mean it can be ignored.

TABLE 17.1 CONSOLIDATE AND REORGANIZE

Task	Responsibility
1. Establish hasty defense:	CO CDR
a. Establish company or separate platoon perimeters 100–200 meters past the objective (toward enemy) or on nearby key terrain that controls the objective area.	CO CDR/PLT LDR
b. Move to fire support element position and guide subordinate elements to parent platoon/section CPs.	PSG
c. Hasty positions complete.	TM LDR
d. Fire protective line established on likely counter-attack approach.	CO CDR/FIST OFF/MORT NCO
e. Displace company aid post to friendly side of objective area IAW OPORD.	CO MEDIC
2. Establish security:	
a. Adjust security element positions to provide early warning as OP/LPs IAW OPORD.	SECURITY SL
b. Dispatch recon patrols (1 km max). Establish contact with units on flanks.	SECURITY SL
c. AT weapons positioned on most likely mounted avenues of approach.	XO/AT NCO
3. Reorganize:	1SG

TABLE 17.1 *(continued)*	
a. Submit **ACE** report (ammo, casualty, equipment status) to PSG: on-hand **A**mmo reported in terms of average number of full magazines per man, boxes of ammo per machine gun, average number of M203 rounds per grenadier; **C**asualties reported by name; **E**quipment losses reported by type.	SL
b. Submit **ACE** report to XO: average of on-hand ammo per type of weapon; personnel remaining per squad/section; mission essential equipment losses.	PSG
c. Submit **ACE** report to battalion TOC.	XO
d. Crew-served weapons accounted for and emplaced with sectors of fire.	PLT LDR
e. Chain of command redesignated.	ALL LDRS
f. Ammunition cross-leveled by magazine and belt, not loose rounds.	TM LDR
4. Medical evacuation and reporting:	1SG/CO MEDIC
a. Perform first aid on casualty and inform nearest medic.	ALL
b. Evacuate casualties to company aid post.	MEDIC/AID & LITTER TM
c. Collect and/or complete DA form 1156 (Casualty Feeder Report) and/or 1155 (Witness Statement) [located in first aid pouch] on each casualty. Submit to 1SG.	PSG
d. Referring to the DA forms 1156 and 1155, update **Unit Manning Roster.** [Ch 10] Submit forms to aid evacuation NCO (ambulance sqd).	1SG
e. Submit DA forms 1156 and 1155 to battalion adjutant in combat trains.	AID EVAC NCO
f. Inform the nearest medical treatment team (MTT) of the number and category of casualties and estimated time of evacuation to the designated casualty collection point (CCP).	AID EVAC NCO
g. Transport casualties to CCP, transfer to MTT, and return to Co aid post.	AID EVAC NCO

(continued)

TABLE 17.1 *(continued)*	
h. Killed in actions (KIAs) consolidated away from Co aid post. Inform S4 (combat trains) of the number and location using roster numbers. S4 back-hauls to hasty GRREG point.	1SG
i. Ensure that crew-served weapons are not evacuated with casualties (only individual assigned weapon).	PSG
5. Prepare pickup zone for medevac:	
a. Prepare PZ at or near CCP if feasible. Mark PZ with Apanels or chemlights: inverted "Y" is preferred marking.	ID EVAC NCO
b. Use aerial medevac for urgent cases. Evacuate priority and routine cases via ambulance.	
6. Process enemy prisoners of war (EPW):	XO
a. During assault: designate immediate escort to EPW point IAW OPORD.	SL
b. During C&R: EPW and search teams process EPW and noncombatants, using the 5 Ss:	EPW TM LDR
1) Search: Buddy team method. With flashlight, search from boots to head; remove bootlaces and belts and place in EPW pockets. Mark dead by pulling shirts over head. Collect all weapons/equipment/ documents and tag. Contact the battalion S2.	EPW TM LDR
2) Silence: No talking. Gag persistent EPWs.	EPW TM LDR

3) Segregate: By rank, sex, and nationality. Tag with following information:
    a) Date and location of capture
    b) Name
    c) Rank
    d) Serial number
    e) Unit
    f) Capturing unit
    g) Circumstances of capture
    h) Equipment/weapons/documents found
	EPW TM LDR

4) Safeguard: Protect from populace and unreliable guards.	XO
5) Speed to the rear: Contact S4 for back-haul to bn EPW point.	XO

TABLE 17.1 *(continued)*	
c. Submit EPW manifest (number by rank, sex, nationality) to S1.	EPW TM LDR
7. Destroy enemy equipment (if applicable):	CDR
a. Demolition team returns to objective and destroys designated enemy equipment.	DEMO TM LDR
b. Once demolitions are in place, announces "FIRE IN THE HOLE" three times to alert all personnel to clear the area. On the final iteration, the demo team ignites the fuses.	DEMO TM LDR
8. Conduct resupply:	SUPPLY SGT/PSG
a. Establish trains IAW OPORD. Organize resupply items into platoon and section packages for rapid and organized delivery: ammo, water, medical supplies, and food. (chapters 6 and 18)	SUPPLY SGT
b. Dispatch detail to pick up supply package.	PSG/SEC LDR
c. Dispatch detail to retrieve rucksacks from ORP.	PSG/SEC LDR
d. Submit LOGSTAT to supply sgt for delivery to bn combat trains.	XO
9. Submit unit status to bn cdr: Situation, unit location, and available strength, ammo, and weapon status.	CDR
10. Begin **Priority of Work in the Defense** or conduct movement to contact. (chapter 19)	CDR

Source: ATEP 7-10-MTP, 5-228–5-230, 5-214–5-216. "Consolidate and Reorganize" (7-2-1047) and "Process Enemy Prisoners of War/Captured Material" (7-2-1049) form the foundation document for this portion of the SOP.

NOTES

1. Doubler, 22–23. Commanders learned that once the stress of action was over, soldiers became "docile and careless," requiring leaders to force them into any activity just to break the inertia.
2. Ibid., 25. Since the Germans routinely had registered fire on their positions, commanders always pushed beyond the objective to avoid artillery fire. Consequently, the Germans normally abandoned plans for an immediate counterattack.
3. Patton, 328.

Chapter 18

Defense Operations Order

The offense operations order (OPORD) covers most of my thoughts on the use of the OPORD format in general. In this regard, this chapter addresses those elements that pertain to the defense. One item that requires emphasis is the use of visual aids to disseminate the plan. Using the task organization, planning, and fire plan matrices with the sketch or sand table significantly enhances the transmission and clarity of the plan.

The task organization matrix helps the commander array his forces without ambiguity or confusion. The matrix reveals generic position assignments and tasks.

Situation. The S2 provides the majority of the information on the enemy forces. Intelligence gaps will exist, however. Filling in the gaps requires the use of assumptions. As more intelligence comes in, the commander validates or revises his assumptions. This mental process is important, because it helps the commander anticipate enemy actions. For instance, if the enemy is a mechanized force, located 10 kilometers away, the probability of the enemy conducting a movement to contact and hasty attack is high. Enemy methods are revealed through past contact with the enemy. If the commander can determine the enemy's tactical modus operandi, he can plan accordingly. The duke of Wellington's observation at Waterloo of the French attacking in the same old way is a case in point.

Information on friendly forces gives everyone the bigger picture. The battalion mission and the brigade and battalion commander's intent are significant. If the battalion mission, higher intent, and company mission contain contradictions or ambiguities, the company commander seeks clarification from the battalion commander, executive officer (XO), and S3. Remember, the battalion commander and his staff are working quickly and under pressure. It is not unusual for ambiguities to arise under these conditions. Company commanders are part of the quality control process and address issues at the end of the battalion OPORD, back-brief, or rehearsal.

The commander and XO brief on the effects of terrain and weather on the friendly and enemy operation, respectively, referring to terrain and weather analysis for the offense (enemy) and defense (friendly) in Mission Preparation. (See chapter 5.)

The final section is the most probable enemy course of action (COA). Often the COA is tied to the primary avenue of approach into sector. Although the enemy may not strike from this direction, securing the avenue of approach for subsequent echelons is likely to be the ultimate objective.

Mission. The restated mission statement stands alone as the focus of the company's activities. It is the product of the commander's analysis of the battalion mission, the battalion commander's intent, and the specific and implied tasks of the company. Most of the time, the subunit instructions of the battalion OPORD provide the commander with the core of his mission statement. Because the mission is so important, the commander reads it twice and ensures that everyone has copied it before moving on. It may be useful to write the mission on a placard near the sketch or sand table.

Execution. While referring to the sketch or sand table, the commander lays out the plan. FM 100-5 defines the commander's intent as "a concise expression of the purpose of an operation, a description of the desired end state, and the way in which the posture of that goal facilitates transition to future operations."[1] The purpose explains how the company mission is linked to higher headquarters plans. The desired end state translates to denying the enemy key terrain and avenues, destroying a critical unit, or delaying his time schedule. Repulsing the attack or eroding his strength is a means to this end. The subsequent posture describes how company success expedites the plans and opportunities of higher headquarters. The commander's intent is extremely important, because it is the basis of the operational art of war. By instilling the concept of linkage with higher objectives, our officers and NCOs gain an understanding of two principles of war—objective and unity of effort. The intent, however, must be concise and clear, and must not digress into a dissertation on how the commander foresees the enemy forces' attack.

The basic plan describes the basic dispositions to accomplish the mission—forward edge, reverse edge, or open terrain. The commander should resist the temptation to go into the details yet. He wants to disseminate the plan in stages for better subordinate consumption.

The planning matrix provides direction and a sense of urgency to subordinates. It complements the Priority of Work in the Defense (see chapter 19) and also helps the commander determine if the defensive preparation is on schedule or not.

The detailed plan and subunit instructions provide subordinates with the specific defensive positions, deception plan, and fire plan. Here, the commander discloses how he plans to distribute fires to defeat the enemy attack. The sketch or sand table with graphic control measures and the fire plan matrix clearly portray how the details of the plan come together. Although he devotes the majority of his briefing to the most likely enemy course of action, he also touches on secondary sectors of fire and realignment of the defense (limited withdrawal to alternate and supplementary positions and deployment of the reserve) in order to address enemy penetrations or flanking maneuvers.

The deception plan seeks to enhance security and surprise. The commander wants to obscure the true dispositions, size, and force array and delude the enemy commander. The deception plan is covered in greater detail in Priority of Work in the Defense. (See chapter 19.)

The counterreconnaissance plan focuses on how the company will deny the enemy information on the details of the defense. The element leader provides a detailed counterrecon plan to the commander once the company occupies the defensive sector. The counterreconnaissance effort is unlikely to deny completely enemy reconnaissance access into the sector. The counterreconnaissance effort strives to guide the enemy into identifying the deception positions only and harrying the enemy's more obtrusive efforts.

The fire plan matrix helps the commander develop a coherent defense with synchronized fires. The components of the fire plan explain the details for designating and distributing fires in the matrix.

As already mentioned, supplementary and alternate positions are designed to realign the defense once the attack has commenced. They provide the commander with the flexibility and depth to meet an enemy attack from any direction with a coherent defense.

The overrun rally point is an emergency control measure in case the enemy succeeds in breaking the defense. This rally point is designated one to two kilometers behind the company sector for the company to consolidate and reorganize and prepare for the deliberate counterattack if ordered. The company is the best unit to guide the counterattack unit into sector since it is familiar with the terrain and avenues of approach.

The immediate counterattack plan is merely the counteraction to an enemy penetration. The commander orders a company element (reserve or out-of-contact squads) to deploy to designated positions covering the penetration. A reserve can react the fastest to a breach. So it is prudent for the commander to designate one for this contingency. Once the reserve is deployed, the commander designates a new reserve from another part of the sector and realigns the defense (refuse a flank, implode positions, or employ economy of force). Supplementary or alternate positions with hasty positions already prepared make excellent firing platforms for the counterattack element to use. To hold the shoulders of the breach, personnel in untenable fighting positions move laterally to another position on either side of the breach.

The rest of the OPORD concerns recurring information that requires no further amplification here. The bottom line of the OPORD is to impart the plan quickly and clearly. If it appears that the complete OPORD cannot be briefed in the original time allotted (because of an accelerated time schedule), the commander briefs the mission and detailed plan only and saves the rest for later.

TABLE 18.1 DEFENSE OPORD WORKSHEET

Company element	Task Main effort	Organization Supporting effort	Other
1st Platoon			
2d Platoon			
3d Platoon			
AT Section			
MORT Section			
Attachment:			
Attachment:			

Main effort: strongpoint, battle position

Subordinating effort: battle position, sector, hide position, strongpoint

Other: counterreconnaissance, outpost, reserve

I. Situation (from battalion OPORD):

  a. Enemy forces:

    1) Likely type of contact: MTC and hasty attack; deliberate attack; mounted, dismounted attack, or air assault (or combination).

    2) Expected composition and size: mechanized/dismounted infantry, armor, air assault; Co, Bn, Rgt.

    3) Strength and capabilities:

      a) Projected strength percentage and readiness. A factor of the number of continuous days enemy unit has participated in continuous combat.

      b) Location, projected arrival, and size/composition of second echelon.

      c) Organic and probable artillery and mortar support for attack.

      d) Air and NBC capabilities and likely use.

TABLE 18.1 *(continued)*

4) Peculiarities/habits: favored time of attack, use of reconnaissance, use of supporting fires.
5) Probability that our sector lies in zone enemy main effort.
b. Friendly forces:
1) Battalion mission and Bn and Bde Cdr's intent.
2) Mission of unit to the left.
3) Mission of unit to the right.
4) Mission of unit to the front.
5) Mission of unit to the rear (reserve).
6) Available DS and GS Fire Support (FA, CAS, AT, Tanks, ADA).
c. Detachments: gaining unit POC, reporting time and location for linkup.
d. Terrain and weather (refer to map, sketch, sand table when briefing):
1) Terrain effects on operations for friendly and enemy forces (refer to **Mission Preparation,** chapter 5: Make a tentative plan (METT-T)):
a) Distance for **O**bservation and fields of fire.
b) **C**over and concealment available.
c) Natural **O**bstacles.
d) Identification of **K**ey terrain.
e) Identification of **A**venues of approach.
2) Current data available (from BN OPORD or S2):

Wind direction/speed	Sunset	Temp high/low	Moonrise	EENT
Percentage of precipitation	Sunrise	Percentage of illumination	Moon set	BMNT

3) Weather effects on operations for friendly and enemy forces (refer to **Mission Preparation,** chapter 5: Make tentative plan (METT-T)):
a) Relative mobility in the defensive sector:
b) Visibility in meters:
c) Effects on use of NBC and Smoke:
e. Probable enemy course of action: entry into sector—MTC with a hasty attack, deliberate assault; disposition (main and supporting

(continued)

TABLE 18.1 *(continued)*

points of attack)—rear, flank, front; array (assault formation)—linear, zone, penetration:

II. Restated mission (from mission analysis) [Read twice]:

III. Execution:

a. Concept of the operation (refer to map overlay, sketch, or sand table):

1) Commander's intent and basic plan:

a) Purpose of operation—linkage with higher HQ plans:

b) End state signaling success:
- Retain/control/neutralize specific key or decisive terrain
- Repulse key enemy unit
- Deny an avenue of approach to the enemy

c) How success facilitates objectives of higher HQ:

2) Movement to defensive positions:

a) Order of march:

b) Technique (traveling or traveling overwatch):

c) Formation (wedge, column, vee, echelon left or right) for specific events:

d) Planning matrix for movement and key priority of work goals:

Event or Planning Graphic	NLT Time
1. Depart assembly area:	
a. Start point (SP)	
b. Release point (RP)	
2. Occupy defense assembly area:	
3. Conduct leaders recon of sector.	
4. Depart defense assembly area to positions:	

TABLE 18.1 *(continued)*	
a. SP	
b. RP	
5. Begin priority of work.	
6. Overhead cover packages staged.	
7. Obstacle wire packages staged.	
8. Obstacle mine packages staged.	
9. Complete defensive positions.	
10. All defensive preparations complete.	
11. Target turnover	
12. Battle handover	
13. Expected time of attack.	

b. Detailed plan and subunit instructions:
1. Emplace units:
a) Subunit missions: Task/purpose of main effort and supporting efforts (battle positions, strong points, redoubts, defensive sectors, ambush sites, and hide positions):
b) Designate reserve by composition, size, position, and orientation.
c) Critical LP/OPs.
d) AT positions—primary, secondary.
e) Mortar position(s).
2. Deception plan (dummy positions and obstacles):
3. Counterreconnaissance plan: commander's intent, actions after enemy contact.
4. Fire plan (refer to fire plan matrix):
a) Designate and assign kill zones for the purpose of massing fires:
1) Sectors of fire: Primary and secondary, mutually supporting range fans with defined left and right limits (TRPs and/or terrain features).
2) Engagement area(s) (EA):

(continued)

TABLE 18.1 *(continued)*
• Select decisive point (trigger point where greatest concentration of fires occurs). • Dispose platoons to cover EA—overlay multiple sectors of fire. • Integrate use of natural and man-made obstacles to canalize advancing enemy thrusts toward the decisive point. • Designate TRPs for corners of EA and the decisive point.
b) Designate distribution of fire:
1) Cross fire: Engage enemy on flanks simultaneously.
2) Fire in depth: Divide sector or EA in half and assign responsibility for the far sector to specific weapons/elements and the near sector to others (also called near-half/far-half).
3) End points to center: closest key weapons engage select targets on end points of enemy formation and then shift fires toward the center of the enemy formation.
c) Designate fire control measures: TRP, trigger point, trigger line.
d) Designate priority of fires and when to lift/shift fires.
e) Incorporate artillery/mortar fire support plan.
f) Assign priority of targets by weapons available:
1) Engineer vehicles: TOW, M47, Arty, AT4, and field expedient AT device.
2) Tanks: TOW, M47, Arty, AT4, and field expedient AT device.
3) APC/AFV: LAW/AT4, M203, M202, M-60 MG with armor-piercing rounds, or field expedient AT device.
4) Wheeled vehicles and dismounted infantry: artillery, mortars, rifle, squad automatic weapon (SAW), and M203.
5. Designate alternate and supplementary positions designed to add depth to the defense.
6. Designate overrun rally point.
7. Immediate counterattack plan:
a) Identify event that triggers counterattack (penetration into main position, loss of key position).
b) Identify counterattack element (reserve) and general route to fire position.

TABLE 18.1 *(continued)*

c) Identify counterattack firing position (normally supplementary positions covering enemy penetration).
c. Consolidation and Reorganization (chapter 17): Remind subordinates to review. Assign responsibility for contact points once R&C is accomplished.
d. Coordinating instructions:
1) Target turnover responsibility.
2) Obstacle plan responsibility.
3) NBC MOPP level: • MOPP 0: mask carried • MOPP 1: CPOG worn • MOPP 2: CPOG, boots worn • MOPP 3: CPOG, boots, mask worn • MOPP 4: CPOG, boots, mask, gloves worn
4) CCIR: • Priority intelligence requirements (from S-2): • Friendly forces information requirements (from Cdr): • Enemy friendly forces information requirements (from Cdr):
5) Constraints and restrictions.
6) Brief-back times and rehearsal times (if not designated in warning order).
7) Withdraw route (primary and alternate) to overrun rally point and beyond.
8) Stand-to for BMNT and EENT reminder.
9) Annexes:
a) Air assault into sector.
b) Foot march.
c) Truck movement.
d) Land navigation plan.
e) Relief in place
10) ADA alert status: Red: air attack is imminent or in progress.

(continued)

TABLE 18.1 *(continued)*

Yellow: air attack is probable. White: air attack is not probable. *Weapons Hold": do not fire except in self-defense. *Weapons Tight": engage only positively identified hostile aircraft. *Weapons Free": fire at any aircraft not positively identified as friendly.

11) Rules of engagement (ROE)—RAMP (OOTW only)[2]
  • **R**eturn fire with well-aimed fire
  • **A**nticipate attack (Hand SALUTE)
    Hand—what is in his hands?
    Size—how many?
    Activity—what are they doing?
    Location—are they within range?
    Uniform—are they in uniform?
    Time—how soon before they are upon you?
    Equipment—if armed, with what?
  • **M**easure the amount of force used (VEWPRIK)
    Verbal warning
    Exhibit weapon
    Warning shot
    Pepper spray
    Rifle butt stroke
    Injure with bayonet
    Kill with fire
  • **P**rotect with deadly force only human life and property designated by commander

IV. Service support:
a. Location of combat trains.
b. Location of field trains.
c. Location of company trains.
d. **Resupply Plan** to include battalion LRP.
e. Location of battalion aid station (if not collocated with combat trains).
f. Location of company aid post.
g. Casualty evacuation plan.
h. Location of battalion CCPs.
i. Location of company and battalion EPW point during C&R; guard and transportation plan.
j. Meal cycle and time of water resupply.

TABLE 18.1 *(continued)*

k. Trace of battalion main supply route (MSR).
l. Tentative LZ/PZ locations for resupply and casualty evacuation.
V. Command and signal:
a. Command:
1) CP location.
2) Chain of command.
3) Second in command's location.
4) Battalion TOC and TAC location.
b. Signal:
1) Priority of communication: radio, wire, messenger, visual, and sound.
2) SOI duration and DTG of change.
3) Radio encryption secure fills: Time and location of fill update; duration of each fill and the corresponding positions for each duration.
4) Code words or signals used during mission (as applicable):
a) Initiate fire (signal/code word).
b) Move to supplementary position (signal/code word).
c) Move to alternate position (signal/code word).
d) Lift fire (signal/code word).
e) Shift fire (signal/code word).
f) Switch to alternate frequency (signal/code word).
g) Conduct immediate counterattack (signal/code word).
h) Counterattack successful (code word).
i) Initiate, consolidate, and reorganize (signal/code word).
j) Consolidation and reorganization completed (code word).
k) Move to overrun rally point (signal/code word).
l) Close lane in obstacle or execute target turnover (signal/code word).
m) Execute emergency TRP on defensive position (plan VT) (signal/code word).
n) Destroy all COMSEC/zero out secure devices.

(continued)

TABLE 18.1 *(continued)*

5. Challenge and passwords used throughout mission.
6. Running password: for counterrecon teams and LP/OPs.
7. Number combination password: as backup to challenge and password.
8. Actions to take if SOI compromised (drop down 20 kHz [second set of numbers] on radio).
9. Method of marking target for CAS.
10. Method of alerting unit of enemy contact (normally for perimeter defenses):
a) In-contact element's code name.
b) Clock direction of enemy and distance of enemy from position.
11. Method of marking casualty collection point at night.

NOTES

1. FM 100-5, glossary-2.
2. Bolger, 98–100. LTC Bolger gives an in-depth explanation of the use of RAMP and its successful use at JRTC and Haiti.

TABLE 18.2 FIRE PLAN MATRIX

Combat Event	Fire Control Measure	Indirect Fire	Position/ Unit	Action	Purpose	Engagement Priority	Type Fires	Fire Pattern

(continued)

TABLE 18.2 (continued)

Key
Combat event: Enemy prebattle actions (combat patrols, advance guard, phased fires), enemy main battle actions (phased fires, fire line/SBF activated, breaching effort, penetration)
Fire control measure: TRP, trigger line, trigger point, engagement area
Indirect fire: Company mortar section, battalion mortar platoon, supporting artillery battery
Position or unit: Battle position, strongpoint, redoubt, sector, hide position, reserve, or specific unit
Action: Acquire, track, engage, and deploy to new position, counterattack into firing position
Purpose: Destroy, suppress, neutralize, fix, realign defense
Engagement priority: Engineer, tank, IFV, and dismounted infantry
Type of fires: FPF, interdiction, harassing, suppression, volley, individual, scheduled
Fire pattern: Cross-, depth-, and end points to center, point, and area

Chapter 19

The Defense

THE STRONGER FORM OF COMBAT

Using the Clausewitzian approach to the study of warfare, why is the defense the stronger form of combat, even at the small unit level? The chief advantage lies in the influence of friction, physical exertion, and time.

Friction. Friction lessens the effectiveness of both the attack and the defense. But the influence of friction is greater on the attack because there are more moving parts. The attacker must plan, prepare, and rehearse under a cloak of incomplete knowledge of enemy dispositions. Resupply in the midst of a campaign is often a critical factor. Extended lines of communication result in delays and the need to prioritize classes of supply. A dearth of ammunition, water, and fuel can stop an attack just as surely as a resolute defender can. Transportation may arrive late or not at all, desynchronizing the attack. Attacking forces can become lost during movement, suffer mechanical breakdowns and injuries, and run into unexpected obstacles, all resulting in a piecemeal attack.

The defender knows where he shall defend and can focus on improving the defense. His lines of communication are shorter, and he is more likely to receive critical supplies—a dearth of ammunition, water, and fuel rarely stops the defender from defending. Since he moves little to get into posture, the defender does not incur this type of friction.

Physical Exertion. Attacking forces have less opportunity for rest, particularly the night before an attack. His dismounted forces must advance over various terrain and carrying burdensome loads. Weather conditions, such as cold, precipitation, or heat further drain the soldier's strength. If he lacks food and water, the effects are even more pronounced. With all these cumulative factors, even physically fit soldiers will attack with less energy.

Other than digging positions, the defender has more opportunity to rest. He is less likely to suffer from the effects of the elements. Since he defends from a position, he does not need as much physical strength to fight. He just needs to remain in place to be effective.

Time. Time is constant, but that does not mean it is neutral. The attacker is often beset by delays as a result of friction and physical exertion. Arriving on the objective is often so difficult that commanders equate success with attacking on time rather than defeating the defender. Success for the attacker depends on the quick and decisive defeat of the defender. Possessing the initial advantage, the attacker wants to make contact as quickly as possible, defeat the defender, and maintain the advance before the defender can recover. The defender wants to delay this clash until he is strong enough to repulse the attack, or at least to delay and erode his strength with successive defenses until the attack culminates. Even though he does not feel he ever has enough time to prepare a proper defense, every minute in which he is not attacked increases his chances of success. Time therefore favors the side that can take better advantage of it.

The priority of work in the defense is designed to give the commander the opportunity to strengthen the defense in a methodical manner. It helps him set track of progress. But he must not think of the priority of work as an end in itself. He must understand the dynamics between the attack and the defense. The priorities of work ensure that subordinates are actively employed in preparing a coherent defense. The commander focuses on denying the attacker any advantage that will set the conditions of a successful attack. This means that the commander will need to give higher priority to some activities because mission, enemy, terrain and weather, troops, and time available (METT-T) factors demand it. Rarely can all activities become completed, so the commander sets the priority of work.

OCCUPATION OF THE DEFENSIVE SECTOR

At the conclusion of the company warning order, the commander dispatches a squad from the lead platoon (in the order of march) to recon the defensive sector and to ward off enemy recon elements.[1] The recon squad's main objective is to become familiar with the area to assist the leaders' recon. It identifies an initial place (viewing area) for the leaders' recon to view the defensive sector.

The linkup point between the recon squad and the lead squad of the main body is at the defense assembly area. If linkup is not made, the lead squad of the main body informs the commander (who may or may not establish a hasty perimeter) and secures the assembly area. The company then occupies the defense assembly area in force.

The assembly area is located on the decisive terrain that will form the backbone of the defense. This ensures that the most important portion is defended immediately against enemy advance elements before the complete occupation of the sector. The members of the leaders' recon immediately report to the company command post (CP). The commander then departs with the leaders' recon to the viewing area to provide initial guidance to the leaders.

During the leaders' recon of the sector, the first sergeant (1SG) adjusts the perimeter to cover the main avenue of approach with key weapons while maintaining the integrity of the perimeter. The intent is to provide immediate defense of the key terrain and not the entire sector.

The purpose of the leaders' recon is to make adjustments to the plan, not to rewrite it. Time does not allow this, so leaders must avoid the temptation. At the viewing area, the leaders discuss the proposed decisive point, the geographic trace of the sectors of fire or engagement areas, the integration of obstacles, and the location of trigger points and target reference points (TRPs) for command and control. After the initial discussion, the platoon and section leaders disperse to their respective sectors for personal recons.

The purpose of the personal recons is to verify key weapon locations, adjust the general trace of squad sectors, verify left and right flanks, and establish command posts. The commander accompanies the antitank (AT) section leader, ensuring that AT positions support the plan and are protected by the platoon positions.

At the directed time (no longer than an hour), the leaders regroup and conduct a brief-back. From this the commander sketches a rough portrayal of the defense two levels down, displaying key weapons. The platoon and section leaders also provide the commander with the grid location of their CPs.

Upon return from the leaders' recon, each platoon or section leader immediately guides his unit to his CP, where it forms a security perimeter. He then conducts a leaders' recon with his squad leaders. During their absence, the platoon sergeant checks the security of the perimeter.

The purpose of the platoon leaders' recon is to verify flanks, the location of key weapon positions, and the general trace of squad sectors. The weakest part of any defense is along the seams. Enemy recon seeks to find those seams. The platoon leader must personally conduct the liaison to ensure that the unit on his flank knows the location of his positions, his fields of fire, and obstacles. For this purpose, a copy of his sector sketch is ideal. He should obtain the same information from the flanking unit too. The platoon leader also assigns responsibility for each listening and observation post (LP/OP) but leaves the exact placement to the discretion of the squad leader. The platoon sergeant (PSG) verifies these positions once the priority of work begins.

PRIORITY OF WORK IN THE DEFENSE

A systematic, detailed priority of work is absolutely essential for a deliberate defense. The immense number of tasks and supporting tasks requires close management. The priorities of work reflect the commander's priorities and the time line allotted for completion of tasks. The commander must view the preparation of the defense as an ensemble of security, position preparation, and kill zone preparation. The commander must strike a balance among the three areas, because emphasis of one over the others can set the conditions of defeat. Using METT-T as the guideline, the commander sets the focus of effort.

Establishing Security. During preparation, the company cannot operate in a vacuum vis-à-vis the enemy, an attitude so often adopted during training. The company must constantly conduct reconnaissance and security operations (LP/OPs, counterreconnaissance, and clearing patrols), maybe with as much as a third of the company, to deny the enemy the details of the defense. Counterreconnaissance is an essential part of the priority of work, because success for the enemy depends on pinpointing the defender's center of gravity (the keystone of the defense). If the enemy cannot pinpoint the defense, his chances of success are greatly diminished. He cannot formulate an appropriate plan of attack with any confidence; he cannot determine where to focus his weapons at a key point of the defense; and he cannot synchronize his or his higher command's efforts. In short, he is forced to conduct a movement to contact, culminating in a hasty attack.

An intelligent adversary will not attack under such conditions if he can avoid it. He must recon the route, pinpoint the main line of defense and key weapon positions, and ascertain the size and composition of the defending force. He will devote enough forces to provide him with this information and will be prepared to reconstitute the reconnaissance effort with additional forces, if needed. If his reconnaissance effort fails, he is likely to divert his main attack to another sector. Even if the enemy is forced to forgo detailed reconnaissance in favor of a movement to contact, the defender, set in well-disposed fighting positions with weapons sighted and with synchronized fire plans, will likely inflict intolerable casualties.

With so much at stake, the commander is justified to divert considerable resources for the counterreconnaissance at the temporary expense of position preparation.

The main purpose of security is to protect the force. The company adopts passive measures (cover, concealment, and deception) and active measures (early warning and patrols) in order to confuse, delude, and disrupt the efforts of the attacker.

Listening and observation posts (LP/OP). Leaders often give lip service to LP/OPs and probably would not employ them if they were not an integral part of training evaluations. Unimaginative use and improper placement of LP/OPs result in poor early warning for the defense. To make them effective, the commander must determine the best location of each OP and LP, and he must be able to articulate how much early warning an LP/OP will provide the company in terms of time.

First, not every platoon sector needs an OP or LP. If the frontline positions have uninterrupted observation of their sector out to a kilometer, the OP is superfluous. Second, OPs are established far enough out during the day to provide early warning to the company. That is, soldiers must have enough time to occupy their positions before battle is joined. LP/OPs are located at the most likely covered and concealed avenues of approach and maybe at points where the enemy is exposed during his advance (choke points, major streams and roads perpendicular to his route of advance, and open areas). At night (or during reduced visibility), the OP is withdrawn to the LP position, about 35 to 50 meters from the frontline trace. The LP is positioned at the likely assault position or line of deployment and fire support positions.

Although wire communications are preferred, the distance may preclude their use. In such cases, the LP/OP must have a radio set to the company frequency. This increases the odds that platoons will receive the alarm, because they must also monitor the company net.

Even if a platoon sector requires no LP/OP, the platoon must maintain local security. The platoon sergeant can either have his soldiers switch off between digging and guarding or designate some guard positions 50 to100 meters in front of the positions. Local security is not an LP/OP. It is not recorded on the squad or platoon **Sector Sketch** (see chapter 20), because this would give a false picture to the commander.

The platoon sergeant (PSG) manages both local security and LP/OPs. He ensures that the local security rotates every hour and the LP/OP rotates every four hours. A good method of managing an LP/OP is to assign a six-man shift to each two-man LP/OP. In addition to their LP/OP, the six soldiers prepare two three-man fighting positions located nearest to the assigned LP/OP. The LP/OP team rotates between manning the LP/OP and preparing the fighting positions. When attacked, the LP/OP occupants withdraw to their assigned fighting position as the third man.

The OP consists of two hasty fighting positions reinforced by sand bags and camouflaged. The LP is also similarly constructed, but it is also reinforced by a single strand of concertina wire between 35 and 50 meters directly to its front. This obstacle allows the LP to withdraw without the attacker following in close pursuit.

The LP/OP team consists of a guard and a challenger. Whenever the LP/OP hears or observes movement to the front, the challenger averts his head and loudly instructs the intruder to stop moving with "Halt!" Averting the head prevents the intruder(s) from pinpointing his position and also alerts nearby soldiers, who in turn can sound the alarm if the intruders turn out to be enemy. The guard maintains vigilance with his weapon to protect the challenger. With grenades (or claymore mine) at the ready, the challenger demands the identity of the intruder with "Who goes there!" If the intruder does not respond immediately, he again demands identification. If this elicits no response, the challenger throws his grenade (or detonates the claymore), awaits the detonation, and both the challenger and the guard suppress the general area with automatic fire as they withdraw to their assigned fighting positions. Upon the sounding of the alarm or detonation of the grenades or claymore, soldiers occupy their positions, assuming 100 percent security. The company remains at 100 percent security until attacked or while a clearing patrol is dispatched. Once the area is cleared and the LP/OPs are reoccupied, the company stands down in accordance with the commander's alert status.

If the intruder claims to be "friendly" (U.S. or allied) and responds immediately and loudly with his name and rank, the challenger directs him to come forward with "Advance to be recognized," then "Halt!" for the exchange of the challenge and password, followed with "Pass through." Once the situation is resolved, the LP/OP provides the PSG with a SITREP via radio or field phone.

There are two disadvantages to this system. First, nearby enemy recon forces will likely probe for LP/OPs hours or days before the attack; or they may monitor LP/OP challenges to friendly troops and have a good idea of the LP/OPs' general locations. The danger of compromise is mitigated by shifting each LP/OP every 18 to 24 hours. It is impractical to move them immediately after the possible compromise. Second, the LP may shoot at friendly troops who wander into the company area and do not know the procedures. Well, war is dangerous—the alternative is worse. Destruction of a company as a result of an overcautious LP is unforgivable. Too often, LPs, afraid to give their position away, say nothing or whisper challenges too low for the intruder to hear. The intruder is upon the LP before he realizes that he is being challenged. If he is enemy, he is in a good position to bypass or overrun the LP before the company is alerted. Too many defensive positions have been wiped out this way!

Counterreconnaissance. Counterreconnaissance is an integral part of the defense. The size and composition of the counterrecon element depends on the composition of the enemy. If the enemy has only light armored vehicles, the company AT section reinforced with a squad is the best counterrecon element (and reserve for that matter). If the enemy is fully or partially mechanized/armored, the AT section must focus on preparing its AT positions. The commander should assign the effort to a reinforced squad or even a platoon. If manpower is short, the commander must have several two-man patrols continuously in the combat outpost sector. Training soldiers to operate in two-man teams allows the unit to control larger areas with minimum manpower. They hone their land navigation and field craft skills, increasing their self-confidence and making them a lethal commodity on the battlefield.

The counterreconnaissance element comprises squad automatic weapon (SAW) gunners, M203 grenadiers, a radio-telephone operator (RTO), and riflemen. Additionally, the element carries twice the basic load of grenades per soldier, an M18 claymore per fire team, and two AT4 light antitank weapons (LAWs) per squad. With this size and firepower, the counterrecon element can engage large enemy forces with deadly effect and still be able to disengage quickly. Upon enemy contact, the counterrecon conducts cloverleaf tactics: striking, withdrawing, and striking again at another location. Its sole purpose is to aggressively seek out and engage the enemy. Aggressive counterreconnaissance is key both tactically and psychologically in the domination of the combat outpost sector.

Tactically, the enemy recon element's main task is to gather information. It will avoid contact so as not to compromise its mission. The mere presence of patrols harries the enemy recon, forcing it to rush its recon or keeping it at such a distance that it cannot accomplish its task. The effort complements the deception plan, because the counterrecon allows the enemy to observe the dummy positions only. The enemy recon element cannot afford a decisive engagement, because it is far from its support. Its problems multiply when it must evacuate casualties. The enemy recon element can ill afford more than a few casualties before being forced to retire. If one of its soldiers is captured, the enemy higher command is faced with a possible compromised plan. If the enemy forces withdraw after contact, the counterrecon should not vigorously pursue directly. The odds of ambush are great and the counterrecon achieves little more than chasing the enemy forces away temporarily. The counterrecon will derive more benefit from searching captured and dead soldiers for intelligence and alerting higher command immediately.

If the counterrecon element happens to make contact with the enemy advance guard or main body, it has the opportunity to throw off the whole rhythm of the attack. In a deliberate attack, the enemy wants to have all his key elements (security, fire support, and assault) in place before launching the attack. Contact with the counterrecon could force him to conduct a hasty attack before surprise is completely lost. Against a prepared defense, such an attack will suffer heavy casualties.

Psychologically, control of the covering force area or no-man's-land establishes moral superiority. Soldiers feel more confident when allowed to patrol. It gives them a feeling of control over their fate. Soldiers lose that confidence, and the feeling of isolation is multiplied, when they are forced to shroud themselves within the confines of the perimeter. Once the soldiers are gripped by the fear of the unknown, they are ruled by panic and rumor. A soldier on patrol becomes familiar with the surroundings and can rest easier once the fear of imminent attack is not gnawing at him. Additionally, this moral superiority gives the defender an edge and enhances his opportunity to seize the initiative during the course of the engagement.

Counterreconnaissance is rarely conducted in training, because the operation force's (OPFOR's) attack is scripted, and the OPFOR rarely sends out reconnaissance more than 24 hours before the attack. Counterreconnaissance becomes a distraction to the main priorities of work and therefore is ignored.

If counterreconnaissance is not conducted during training, it will be ignored in combat. Higher command will be apprehensive to allow it for fear of fratricide. This fear is well founded, because units do not practice sending out patrols. The mechanisms for such activities must be worked out during training, because the odds in favor of fratricide are high if soldiers are not accustomed to them.

The counterrecon plan is a company-level responsibility. The company commander approves the plan and supervises its execution. The counterrecon force leader draws up the plan and briefs the commander at N + 1. The plan covers probable enemy recon surveillance points, objective rally points (ORPs), linkup points, and choke points on dismounted avenues of approach. Counterreconnaissance extends no more than one kilometer, because anything beyond this is the responsibility of higher headquarters reconnaissance forces. Success is measured in denying the enemy information on the defensive dispositions and in uncovering the attacker's plan. As mentioned above, the attendant objective of the counterrecon is to establish psychological dominance over no-man's-land. The counterreconnaissance effort seeks to force the enemy to worry more about the counterrecon force than his mission.

When considering possible enemy avenues of approach, consider his perspective. The main concern of the enemy commander is to maintain accurate navigation without being compromised. He will likely use terrain rails as guides (ridgelines, streams, roads and trails, railroads, power lines, and so forth). He is also likely to choose a prominent terrain feature (hill, ridge) for an ORP and linkup point with his recon element.

The counterrecon element bears in mind that the enemy recon element is reconnoitering the main body's route, ORP, assault, and fire support positions. The counterrecon element focuses its efforts on uncovering these. First, the counterrecon element searches for signs of enemy presence. The most common are signs of enemy movement along trails and streams or the crossing of a trail or stream. In such case, footprints (count them to determine the number of soldiers) or erasure marks from tree branches are the most apparent. If such a spot is located, the counterrecon shoots an azimuth along the line of movement as a reference to follow. If the enemy has obliterated the footprints with a tree branch, locating the branch will generally divulge the direction of enemy movement.

Before the counterrecon element picks up the trail, the leader checks his map to see where the line of azimuth falls. Prominent terrain features that fall on or near this line are therefore targeted for recon. Additionally, the leader immediately informs the company commander of any discovery of enemy activity. Of note, there is bound to be plenty of man-made activity, friendly or otherwise, around the area of responsibility. Hence, the counterrecon leader relies on his intuition and experience when judging signs of enemy activity. Additionally, enemy route markers (chemlights, markings on trees, chalk marks, a pile of rocks with a direction stick, and so forth) verify the enemy's route. The counterrecon element does not erase or disturb these markings without the commander's approval. It is better to uncover the enemy's plan and set a trap rather than force him to alter his plan. If the counterrecon does discover the enemy ORP (if it is marked), the counterrecon leader places it under surveillance and reports the grid location to the company CP for indirect fire targeting.

The commander and platoon leaders accompany counterrecons on a rotational basis to give them a feel for the surrounding terrain and to prepare them psychologically for the imminent engagement. All soldiers feel more assured and confident whenever they engage in aggressive activity.

No one should assume that no-man's-land automatically belongs to the defense. The attackers may quite possibly conduct a reconnaissance in force or a movement to contact. Even though alarming to the defending company, such an operation alerts higher headquarters of the enemy's intentions and allows it to prepare countermeasures.

Under such circumstances, the attacker normally has had little time for preparation and has forfeited surprise, rest, and mass for the benefit of an immediate attack against an unprepared defense. Provided the defense does not panic, this gives it the opportunity to bloody the attackers' nose.

Clearing patrols. Conducted during stand-to, the clearing patrol recons likely assault and fire support positions. Clearing patrols should be well planned and executed without deviation because of the danger of fratricide. The patrol reports to the company CP for final instructions before stand-to (15 to 30 minutes). The chain of command alerts everyone of the patrol, and the company CP receives acknowledgment before releasing the patrol. At stand-to, the patrol departs the designated departure point and seeks to uncover the enemy, strike, and rapidly withdraw into the perimeter at either the designated departure or reentry points. The intent, besides alerting the company of an imminent assault, is to throw the attacker off balance and encourage him to make a hasty decision. The reentry of the clearing patrol is hazardous. The patrol uses unique far recognition signals (flashlight on, chemlight, or engineer tape on front of load-bearing equipment—LBE) as well as near recognition signals (running password). If reentry is not possible, the patrol withdraws into the attacker's immediate rear and conducts harassment, spotting, or raid activities IAW the commander's special orders.

Early warning devices. The company cannot depend on technology too heavily as a means of early warning, because they are no substitute for sentries. They are used to supplement the security plan. Mechanical devices such as the Platoon Early Warning System (PEWS) are still unreliable. At best, PEWS indicate that something—not necessarily what is displayed on the screen—is in the area. Another failing is the plethora of small components that are vital to the operation of the PEWS. It will not be long before one or more parts are lost, making the system totally ineffective. Trip flares are useful if they are not prematurely tripped off by friendly patrols or animals. For some reason, this is a common occurrence. Field expedient measures, such as tin cans filled with a few pebbles and hung on obstacle wires, are still effective devices. At the very least, they hamper the enemy's ability to approach covertly.

Hasty Positions. Upon assignment of a position, soldiers dig hasty fighting positions. The proper method for digging with an entrenching tool is to fold it and use it as a pick. In normal ground, a soldier can dig an 18-inch-deep position in 15 minutes. Current hasty fighting position designs (a rectangle 18 inches deep) fail to account for

FIGURE 19-1

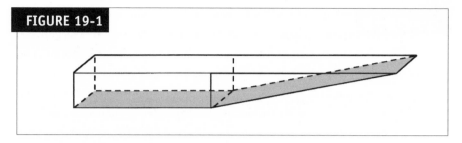

Ergonomic Hasty Position

the physical characteristics of the body. The spine makes it very difficult for the soldier to place his chest and elbows to the front of the position while resting his torso and legs at the bottom of the fighting position—12 to 18 inches below ground level, and still maintain accurate fire. The position itself should slope upward toward the front to accommodate the natural movement of the body in the prone position, thereby making it ergonomical. (See figure 19-1.) This allows the soldier to increase and decrease the exposure of his body as he fires and allows him to adjust the height of his field of fire. The protective berm should be greater toward the front of the position to protect the soldier's head and shoulders from direct fire.

Hasty overhead cover consists of a couple of overhead supports (pickets) and a poncho cover secured firmly over the hasty position, covered with about one-inch layer of loose dirt or sand. During World War II, Soviet soldiers used this technique with good effect. Such cover will stop secondary fragments or 105 mm fragments and will slow down other fragments.

A section of an old camouflage net (6' X 4') provides excellent hasty camouflage. Soldiers should not waste too much time camouflaging the hasty position if a deliberate fighting position is to be constructed too.

As the squad leader assigns the sector of fire for each soldier, the soldier assumes the prone position and aligns himself so as to be able to scan his weapon between his left and right limits with ease. The squad leader stands over his body and records the azimuth of the left and right limit. The soldier marks the outline of the hasty position by fanning his legs and marking the rear of the position with his toes while his elbows mark the front and the arms mark the general outline of the body width. Before digging, the soldier emplaces aiming stakes.

Soldiers clear fields of fire by thinning out the vegetation within their sectors of fire. They clear only enough vegetation to allow them to acquire enemy soldiers. Clearing too much vegetation away may alert enemy soldiers to the layout of the fields of

FIGURE 19-2

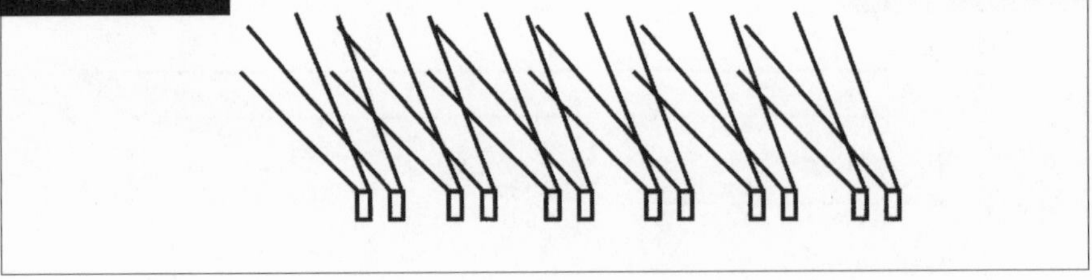

Parallel Hasty Positions

fire. Worse, they may also allow the enemy to pinpoint the fighting positions. In heavy forests or jungles, soldiers should only clear about 18 inches in height, making it difficult for enemy soldiers to spot them. Resembling tunnels, these fields of fires will allow soldiers to wound enemy soldiers' legs initially, but will result in higher casualties as the enemy attempts to maneuver in the prone position or recover wounded comrades.[2]

To assist everyone in determining known distances on the battlefield, the platoon leader identifies landmarks or has objects (wrecked vehicles, rocks, or debris) moved to known distances (200, 400, 600, 800 meters) out to 1,000 meters (3,750 meters, if tube-launched, optic-sighted, wire-guided missile systems [TOWs] are attached). This way soldiers will not engage the enemy until he is within range, especially at night. Additionally, the platoon leader should designate a trigger line (IAW the commander's intent and fire plan). He can either use a linear feature (road, telephone lines) or the known distance markers as the trigger line. Trigger lines help the defense mass fires on the attacker at a single moment. In this sense, the psychological impact on the enemy can be devastating, particularly if he is allowed to close within 100 meters of the main line of resistance.

A bird's-eye view of the hasty positions (which form the foundation of the two-man fighting position) reveals that they should not be parallel as so often portrayed in manuals. With two right-handed (or left-handed) firers, they cannot be parallel without degrading interlocking fires. (See figure 19-2.) The end result of parallel positions is a left-oblique, sectors-at-fire pattern, which leaves the right flank exposed.

This glaring defect is not obvious to leaders for various reasons. The squad leader may assign sectors of fire that *do* intersect, but the leader does not check whether the soldier can physically (or comfortably) fire into this sector from his assigned position. The soldier is forced to attempt to fire into his assigned sector firing without proper stability, or he ignores his selected sector and shifts his fire position until he can fire naturally. Soldiers are want to voice the impossibility of firing into their assigned sector

FIGURE 19-3

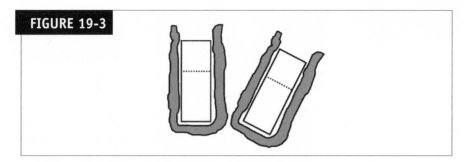

Vee Pattern Hasty Positions

for fear of incurring the wrath of their leaders (who, by the way, are eager to demonstrate their knowledge of a fighting position and will not tolerate deviations from the norm if the problem offers no immediate solution, particularly if the position is already prepared). The soldier briefs his sector, pleases his superiors, and does what comes naturally once in action.

Another factor of self-delusion lies in imprecise sector sketches. If done improperly, the degree of error between the sketch and reality becomes great. The discussion in chapter 20 entitled "The Sector Sketch" virtually eliminates this discrepancy.

To correct the oblique-left fire pattern, the leader must angle the individual hasty positions. When properly emplaced, the pairing of two hasty fighting positions will form a Vee pattern. (See figure 19-3.) The selection of sectors of fire ultimately should govern the angle of the paired fighting positions. From a bird's-eye view, the proper distribution of fire becomes more apparent. (See figure 19-4.)

If the soldier requires sandbags to provide cover for the hasty position and natural camouflage is not available, the 1SG provides field-expedient material. For example, in snowy conditions, white and black paint on sandbags is fast and effective; in the desert, sand paint. Applying natural materials found in the area onto wet paint, oil, or grease

FIGURE 19-4

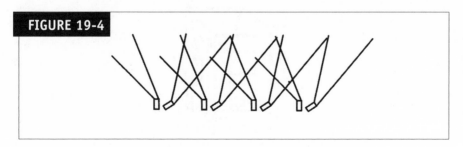

Distributed Fire Pattern

(any sticky substance) enhances the intent. Naturally, the best camouflage for the deliberate fighting position is sod.

Care must be taken not to strip the areas around each position of natural camouflage, because this assists the enemy in locating the main line of resistance or individual positions. It is not enough to direct that soldiers take such care. Soldiers are often in a hurry to complete their positions and will not travel far to acquire camouflage. Squad and platoon leaders should use the areas designated for dummy positions from which soldiers can gather camouflage.

If sandbags are to be used, the 1SG can assist the defensive efforts by organizing a company sandbag filling detail. At first glance, this proposal seems silly, because individual soldiers can fill their own sandbags on location without the need to drain soldiers away from more pressing needs. This view is correct if each position requires only a few sandbags. But if large quantities are needed because defensive positions are located on rocky, loose sand, frozen ground terrain, and so forth, a work detail maximizes time.

Filling sandbags is time consuming, particularly if performed by one soldier. Normally, it is a two-soldier task: One digs and fills, the other holds, ties it off, and emplaces it. Filling sandbags can be streamlined with the aid of a sandbag filler device called a sandbagger. (See figure 19-5.) The sandbag is held upright by the use of a wooden sleeve inserted into another sleeve, allowing one soldier to fill sandbags without assistance.

FIGURE 19-5

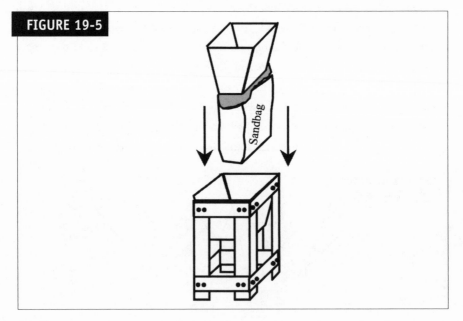

Sandbagger

FIGURE 19-6

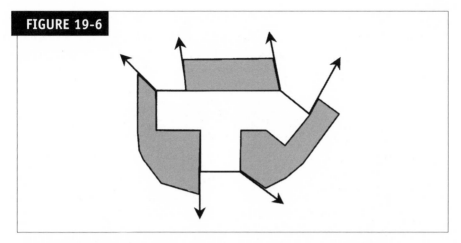

Three-Man Fighting Position

The 1SG can have about twenty of them made and stored in the company trains. With such a device, the company can fill a large number of sandbags quickly.

Fighting Positions. Soldiers excavate fighting positions from the hasty fighting positions—by simply joining them together. Leaders must ensure that soldiers establish protective berms along all sides of the fighting position without obstructing the fields of fire and that the berms have a gradual slope to allow them to blend in with the surrounding terrain as much as possible.

Determining the size of positions. The decision to construct two-man or three-man positions requires a weighing of the advantages and disadvantages of each. One-man positions are not practical, because they tend to create a sense of isolation that can lead to panic or inaction. The soldier is less likely to do his duty if he has no one nearby to voice his concerns to (sometimes a great relief in itself). Because the empty battlefield is a natural occurrence, the soldier is likely to conclude he has been abandoned and flees to the rear. Only very seasoned veterans with full confidence in the command leadership can withstand such urges.

Three-man fighting positions have two distinct advantages. (See figure 19-6.) First, each position has a 360-degree field of coverage, making it difficult for the enemy to attack it from the flank or rear. Moreover, the third soldier can engage penetrating enemy soldiers from the rear; second, the unit can assume 30 percent security without weakening overall security—every position always has a soldier on guard and the rest plan is more manageable; third, with fewer positions, the squad leader has more effective control; fourth, each squad can establish an LP/OP without significantly detracting from the sleep or security plans; and fifth, the LP/OP team can withdraw to a position as the third soldier.

But there are major disadvantages too. First, if a position is destroyed, the squad loses one-third of its strength in one fell swoop and will probably have a major gap in its sector; second, the decreased number of positions means that there will be larger gaps in sector, thus increasing the chance of infiltration, depending on the nature of the terrain; third, such positions tend to become larger or more prominent and thus are easier for the enemy to pinpoint; and fourth, the unit cannot maximize its firepower to the front because of the third soldier facing to the rear.

Two-man fighting positions have three distinct advantages. First, they can cover a wider sector with all available fire and fewer gaps; second, they are less prominent and easier to conceal; and third, interlocking fires from mutually supporting positions mitigate the loss of any one position.

The leader must consider the following disadvantages. First, security measures under 50 percent will mean that some positions will be vacant or off alert status. Attempts to occupy all positions degrade the sleep plan. Second, if the enemy can penetrate to the rear or flank of a position, a bunker-killer team can easily destroy it. Penetrations make the main line of defense untenable. Third, numerous positions are more difficult to control. And fourth, LP/OP requirements are more burdensome and make withdrawal to assigned positions more difficult.

Generally, the unit with fewer available soldiers is better off using three-man fighting positions. To maximize the advantages of each, the leader should consider a mixture of both types.

The squad bunker. The establishment of a squad bunker provides many advantages. When incorporated in the platoon defensive plan, it serves as the squad supplementary position. When forced to abandon their primary positions, soldiers can rally around the squad bunker into prepared positions to continue the defense. (See figure 19-7.)

Squad bunkers also help keep the primary positions less conspicuous. Enemy reconnaissance pinpoints primary positions by spotting activity or signs of activity in suspected areas. To thwart enemy reconnaissance, the defender must be disciplined. Throughout the priorities of work and during the fight, primary fighting positions are used only for security watch and fighting—nothing else. All sleeping, eating, and similar activities (smoking, cooking, brewing coffee, writing letters, reading, purifying

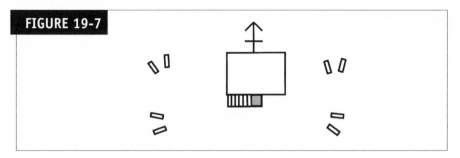

Squad Bunker Defense

water, drying off, conducting personal hygiene, and so forth) are done at the squad bunker. The squad bunker offers the soldier shelter and warmth from harsh weather (rain, snow), because it will have a warming stove, hot coffee, a sleeping area for quality rest, and a place to stow rucksacks. Likewise, equipment not used for position preparation (rucksacks), construction material not immediately installed (sand bags, overhead cover), and unused tools (picks, shovels) are stored at the squad bunker. By having soldiers live in or around the squad bunker in hasty positions, the area around the fighting position remains unspoiled, and thus not as easy to spot. Using squad bunkers to hold the majority of the squad back from the primary positions until needed also reduces the effect of successful enemy penetrations. If the enemy is particularly skillful at stealthily approaching right up to or through the main line of defense without alerting the security forces, the squad is in a better position to repulse the attack from the bunker rather than having individual positions overwhelmed by the initial enemy rush.

The squad leader can also control guard duty and stand-to more effectively. From the bunker he ensures that all soldiers going on guard or stand-to are *awake,* properly *dressed,* and in position *on time.* This system also makes it easier for higher-level leaders to check these activities from a distance.

Each bunker has hasty fighting positions 360 degrees around bunker as supplemental positions. The squad leader determines the disposition of the bunker defenses within the intent of his platoon leader. As general guidance, hasty positions are at least 35 meters from the bunker for dispersion. Supplementary positions may need to be placed farther back in accordance with the company defensive plan. In such cases, the squad bunker is positioned closer to the primary positions for sustainment and not tactical considerations.

The major disadvantage of a squad bunker is the risk of enemy indirect or direct fires. A chance artillery shell can destroy a squad if the bunker becomes a gathering place. Likewise, suppressive fires from direct and indirect weapons could fix the squad before it can occupy its fighting positions.

The disadvantages can be mitigated by improvements to the bunker to increase the survivability of the bunker. To keep the size down, the bunker should be big enough to accommodate only six men at a time.[3] If suppressive fires fix the squad, it simply fights from its supplementary positions. Normally, proper security measures provide ample early warning for the occupation of primary positions. Over time, the squad bunker is connected to the primary positions by communications trenches, turning the squad sector into a resistance nest. Lastly, the supplementary positions make excellent positions for deployment of the reserve to the threatened sector.

Terrain and weather effects on preparation of positions. Preparation of positions is the most time-consuming and strenuous of the priorities of work. Leaders need to experience this drudgery at least once during every exercise to gauge the difficulty of the assigned sector. The Home of the Infantry, Fort Benning, cannot provide a firm basis of experience, because the digging of positions there is a relatively simple task. The standards there are pretty easy to attain. Elsewhere, the realities of rock, clay, and high water tables prevent the excavation of positions within a reasonable time. Leaders should bear this in mind when inspecting positions for standards.

To prepare a full two-man fighting position rapidly, the soldier needs a pickax and a spade (preferably not a shovel). If all he has is an entrenching tool, digging is much slower. Moreover, digging is a two-person job. One soldier breaks the ground with the pickax down one foot. The other shovels out the loose debris and squares the sides of the position. As the work progresses, the pickax breaks up earth, rock, and clay and cuts roots, while the spade scoops them out. To ease fatigue, soldiers switch duties. Leaders must recognize these factors as they discuss the progress of a position with the soldiers. If progress is stopped at a certain level, the leader needs to use judgment regarding the continuance of efforts. If he distrusts the soldiers' explanations, he should try it himself and compare his progress with their previous effort. Ultimately, he must determine when to stop digging and begin overhead cover and camouflage procedures.

In winter conditions, where ground frost makes pioneer tools (picks and shovels) useless, cratering charges or field expedient tamped explosives (TNT, C-4, grenades) are essential. Bonfires can be built to thaw frozen ground enough to allow excavation.[4] If this is not possible or if the layer of snow too thick, soldiers fortify positions by digging out the snow and creating a berm of packed snow and ice. This process requires alternate layers of packed snow and ice (pouring water on the packed snow, allowing it to freeze, and repeating the process to the desired thickness (about 18 inches).[5] White sheets are also effective in camouflaging the firing apertures.[6]

In loose sand or lava ash, sandbags and revetments (poles driven into the floor of the position with wood [timber or stripped branches] placed between the earthen wall and the poles) are necessary. (See figure 19-8.) Even in terrain with soil, positions will eventually need revetments because of erosion.[7] In rocky terrain, sandbags provide the best overall protection. Relying on rocks and boulders is dangerous, since ricochets and spalls from incoming rounds are as lethal as the rounds themselves. Sandbags, filled with sand, provide the best absorption of incoming rounds.

In heavy woods, the company should make unit signs and place them on the roads leading into the company's sector from the rear in order to assist personnel in finding the company. Signs also alert personnel that the front line is near, precluding their wandering across enemy lines.

Engineer support. Leaders do not wait for the arrival of engineers before beginning to dig. Allocated engineer support often is late or becomes unavailable. Infantry companies are usually on the bottom of the priority list, outranked by the battalion and higher command posts, obstacles, and other jobs. Although the common soldier may hold such prudent measures in contempt, experience shows that CPs are often a high-priority target by the enemy and must be hardened. Leaders must explain this battlefield fact to the soldier, because such contempt detracts from the effort.

If the engineers do arrive, the leader should understand their limitations. Digging machines require near-level ground. Positions on forward or rear slopes are generally

FIGURE 19-8

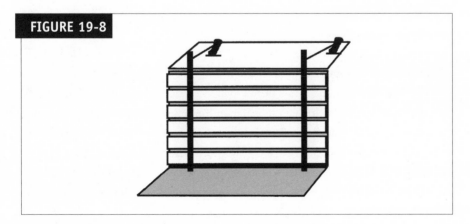

Revetments

beyond the capabilities of such equipment because of the degree of slope. Additionally, these machines create gashes in the ground and destroy the natural surrounding camouflage (engineering excavations are not surgical). The benefit of quick but obvious cover (from the enemy's viewpoint) rarely surpasses the cost in time and exertion as the soldiers repair and strengthen fighting positions and camouflage the scarred earth. Bearing this in mind, the commander allocates the engineer priorities to preparing strong points and squad resistance nests, the company aid post, company and platoon CPs, and the company logistics point. The commander explains to his men that rarely can they expect engineer support for the digging of their positions. The commander assigns a responsible leader (for example, the executive officer—XO) to supervise the engineer effort and does not rely totally on the managerial and tactical expertise of the assigned engineer to prepare the site. The responsible leader prepares the excavation site before the engineers arrive. If the task is a strongpoint, the leader traces out the trench and bunker positions using engineer tape and stakes and determines the order of excavation. This orderly approach saves immense time and frustration and keeps existing trenches from interfering with new ones (yes, unsupervised engineers will box themselves out if left on their own). A sketch of the proposed task is extremely helpful, because it shows the engineer leader the end result and allows him to input his expertise into the plan.

Overhead cover. Overhead cover is as low to ground level as practical. A common mistake is to make the front berm and overhead cover too high. Even with perfect camouflage the enemy can spot positions because of their prominence. An overhead cover of 18 inches includes both the support material and layer of earth or sand bags. Two layers of sandbags suffice when using 4 X 4 inch support beams and plywood. Bear in mind that overhead cover is meant to protect the occupants from small arms fire and shrapnel. Whether overhead cover can sustain someone's weight is irrelevant as long as it supports its own weight. A common mistake is to place too many sandbags on weak support material, resulting in eventual cave-ins.

Overhead cover can be partial or full. Full overhead cover (bunker) affords better protection and allows the defender to maintain continuous fire on the attacker despite the use of 105 mm or smaller caliber air bursts. In fact, the use of air burst munitions (VT fuse) is an excellent defensive tactic against attacking dismounted infantry. Full overhead cover has the disadvantage of limiting the occupants' field of vision and tends to make the occupants feel isolated and vulnerable from the blind spots. It also requires more sandbags and support material and makes the position more prominent. Leaders must keep a watchful eye on positions with unwieldy overhead cover. Soldiers will use whatever materials are available even if not suitable. Since construction material is normally limited, the commander and platoon leaders establish the priority of distribution: key weapon positions, squad bunkers, aid post, and CPs. Leaders closely supervise the construction of overhead cover. A common problem is ceiling clearance from the inside. Often the ceiling is so low that the soldier cannot properly fire his weapon with his helmet on, or he compensates by lowering the weapon platform so much that the field of fire is obstructed. The leader personally checks the ceiling height by physically occupying the position with the soldier's weapon. If the position is unsatisfactory, the soldier removes the overhead cover and starts over—no questions. This greatly upsets the soldier, but he will be more meticulous the next time.

Partial overhead cover (normally positioned over the center of the position) does not provide full protection and forces the defenders to remain undercover until the enemy lifts his suppressive or preparatory fires. Psychologically, they feel less secure (exposed) and tend to shrink to the bottom of their positions, completely ignoring their fields of fire. The danger is that the occupants may become so unnerved that they remain in a fetal position after the attacker has lifted or shifted his supporting fire, allowing the enemy assault elements to close on their position without resistance. Partial overhead cover makes the position more vulnerable to hand grenades once the enemy closes. Nonetheless, partial overhead cover allows the occupants a better field of fire and vision and makes them feel less isolated once they occupy their positions. Grenadiers can more easily wield their M203s while in the open portion of the position. Leaders must encourage open communication during the enemy attack to prevent soldiers from withdrawing into themselves and also to alert them when to take up their positions. Another technique is to position the partial overhead cover to the front of the position, under the berm. This position offers better protection from preliminary artillery and mortar fires, because shrapnel tends to project forward. The open fighting position gives the soldier better observation and allows him to use hand grenades more effectively.

Construction guidelines. Leaders need to be careful not to limit the ingenuity of their soldiers. Soldiers should be encouraged to experiment as long as the finished product satisfies the basic requirements:

- 18 inches of berm and overhead cover.
- Good fields of fire.
- Low profile.
- Camouflage that precludes detection outside of 50 meters.

If sufficient construction material is available, the company standardizes the amount of material per position by establishing a fighting position construction package: 50 sandbags, 6 4 X 4 inch supports, and 1 sheet of plywood. This method ensures that defensive positions are uniformly strong. That is, it precludes one position from being strong at the expense of the other positions. Soldiers, if left to themselves, will attempt to transform their positions into Fortress Europe. These positions become a vortex, sucking down hundreds of sandbags and other construction material. If they could, they would pour in concrete and install plumbing. This phenomenon has two origins. First, U.S. defensive doctrine states that a defensive position is never finished. Actually, doctrine should focus on the defensive sector rather than the individual positions. This makes more sense, because the longer the defender occupies a position, the greater the probability of enemy discovery, and the greater the enemy's employment of heavier caliber weapons to reduce the defense. The defense therefore needs to strengthen the sector with obstacles, field works, and weapons as a means of compensation. But the fighting positions themselves must have an ending point to enable the improvement of the other facets of the defense. Second, soldiers become quite attached to their positions and immediately begin making their new abodes more comfortable. This is inappropriate, because the fighting position should only be used for fighting, not living. Such amenities belong in the squad bunker. Once living conditions become comfortable, soldiers are less inclined to abandon their "homes" even if the tactical situation warrants a withdrawal. So commanders should discourage permanent fixtures the soldiers cannot take with them.

Machine gun positions (excluding SAW). Machine guns are the backbone of the defense. They receive the greatest support and attention. The entire platoon helps to dig these positions in order to complete them as quickly as possible. The platoon sergeant manages the rotation schedule, using two soldiers at a time in 20-minute intervals. The position should be completed in one or two rotations, within a minimum of time and diversion. Soldiers continue work on their own fighting positions when not working on the machine gun position.

In the meantime, the machine gun crew digs the gun platform (12 to 18 inches deep), lays in the gun with tripod, determines dead space for grazing fire, and completes the range card.

As soldiers are apt to do, there will be some complaining about machine gun crews not sharing equally in the preparation. To preclude such talk, make it company policy to assign these malcontents as machine gun crews. The problem will disappear when they realize that the life of a machine gunner is difficult.

The machine gun position is built as a three-man position, with full overhead cover. The machine gun (MG) team emplaces the overhead cover and completes the camouflage. At least one machine gun per platoon is manned at all times. The MG crew trains up the third member of the team so that he can operate it while on guard duty or if one of the crew becomes a casualty.

Machine guns fire only during an enemy deliberate or hasty assault. A clever enemy uses probes to pinpoint crew-served weapon positions and to eliminate them during the actual assault. The machine gun crew uses its experience and judgment to distinguish a probe from an assault. Of course, the gunner may have no choice if the attacker decides to trade lives for this information. Once the MG position is compromised, the crew trades with another fighting position within the platoon leader's plan or withdraws to a supplementary position in or near a selected squad bunker. After stand-to, the platoon prepares another MG position and the other position becomes a dummy position.

Antitank positions. Antitank positions are the fastest to construct, because they are only one meter deep (waist level). The AT crew constructs a position with overhead cover to the rear of the fighting position. The AT weapon and additional rounds are stored here to protect them from indirect fire effects. The AT section constructs numerous supplemental firing positions within the company fire plan, each with two missiles. Each gunner fires one or two missiles from each position and then moves before the attackers can pinpoint him. To maximize their effectiveness, AT weapons should be positioned in pairs, with 35 to 50 meters separation. It is often more effective to place AT positions in the depths of the defense in order to increase survivability. Rather than choose a position with a panoramic view, the weapons should be positioned where the terrain provides cover from the front. A good technique is to angle them across the frontage with sufficient cover to the front. In this manner, they can engage the enemy armor in detail without the enemy pinpointing their position and destroying it. Such a technique is like a window; only the armor in the field of view can be engaged—and vice versa. Window tactics reduce the enemy's ability to destroy the AT systems with his overwatch element and gives sustainable strength to the defense. Dummy AT positions are placed on the battlefield in such a manner as to encourage enemy tanks to expose their flanks to AT fire if they focus on the dummy positions.[8]

Antitank defensive techniques. Units can supplement their main AT weapons with other measures to defeat armor attacks. The first task is to separate the infantry from the tanks, using artillery, mortar, antipersonnel mines, and then small arms.[9] Small arms positions allow tanks to pass through while focusing on infantry and light armored vehicles. AT weapons (and attached tanks) positioned in depth engage the tanks as they enter and pass through the main defensive positions.[10] The commander may form his AT weapons into tank destroyer teams, which, operating under the control of one leader, fire volleys at individual tanks to assure destruction.[11]

Commanders must inoculate their soldiers from tank panic. A proven training technique is to have tanks drive over them while they are in a prepared position and to practice firing light antitank weapons and planting demolitions on them while they are moving.[12] To prevent tanks from surging against the AT weapons, infantrymen stock a few antitank mines at their positions and place them into the path of the tanks. A technique is to tie a wire to an AT mine and place it a distance away from the fighting position. The soldier then pulls the mine into the path of penetrating tanks from the relative safety of his position. The company can form tank hunter-killer teams, which are trained to remain in positions or craters and destroy tanks by placing a demolition charge or AT mine between the turret and the body of the tank.[13]

Immobilizing tanks may prove easier if the defense lacks AT munitions. Firing light antitank weapons (LAW) into the rear or road wheels of tanks; destroying the tracks with demolitions, mines, LAWs, or shape charges; or throwing Molotov cocktails on the engine compartment will turn a tank into an immobile bunker.[14] For this technique to work, enemy infantry must be suppressed and enemy armor must be buttoned up. Nonetheless, close assault of tanks is dangerous. A skilled attacker will echelon his tanks so as to protect the tanks in front with machine gun fire from infantry assaults. The development of tactical intuition concerning when and how to assault a tank comes from discipline and training.

Camouflage. Camouflage is an integral part of cover, concealment, security, and deception. It is an ensemble of variables. Lower one or more and the effectiveness of the whole is reduced. Soldiers conceal not only their individual positions but also the surrounding area they have scarred preparing the position. Sodding of the position is the best because it lasts the longest. Soldiers use the sod where the ground is broken for the position. They can get additional sod from dummy positions. As already mentioned, the position profile is as low as possible to limit the protrusion of the position. To conceal the firing aperture, the soldier constructs a rear berm of the position to preclude light from shining through from behind and creating a halo effect.[15] Soldiers cover the aperture with a section of camouflage net, cloth or canvas strips, or a cloth with slits for foliage weaved in. The aperture cover remains in place until the defender takes his firing position. He simply lifts the cover onto his helmet or shoulders. He may also make a mesh made of thin sticks with foliage woven in. (See figure 19-9.) The floor of the fighting position is covered with foliage to prevent detection from the air.

In sectors with undergrowth, care is taken not to strip away too much vegetation for the fields of fire. The vegetation should just be thinned out a bit. Otherwise, the fields of fire become apparent to the enemy, forming bowling alleys for enemy fire. This is a common mistake with soldiers, because they feel that every bit of foliage must be cleared for them to fire. It is actually better to leave as much foliage as possible to make it more difficult for their own position to be spotted.[16] Once the camouflage is complete, soldiers occupy the position for guard duty or combat only and live at the squad bunker. To lessen the erosion of terrain at the position itself, soldiers enter and exit the position from one path only.

The Deception Plan. The main intent of the deception plan is to delude the enemy leader. The commander wants to present a certain picture of the defense that will lead the enemy to false conclusions. To this end, the commander wants to filter information in the same way a con artist executes a sting operation.

The defender cannot prevent the enemy from gathering intelligence on the defensive sector. Security measures will not completely seal off the sector. Enemy reconnaissance (ground, aerial) will continue until the enemy commander is satisfied with the tactical picture (defending disposition, size, and composition). But the defender can control the type of information the enemy commander receives if he uses ingenuity.

The first task is to delude the enemy as to the location of the main line of defense. The defender wants the attacker to concentrate his firepower on unoccupied ground.

FIGURE 19-9

Aperture Camouflage

For instance, if the company has established a reverse slope defense, then it should establish a deception line on the forward slope. The deception line consists of dummy positions, which fall into two categories—the obvious and the believable. The obvious positions are meant to catch the enemy's attention and delude him into believing that this is the deception plan. The dummy positions are located in the open, built carelessly, only one or two feet deep, and have little or no camouflage. Enemy reconnaissance will dismiss these immediately and continue to scan the area. The believable dummy positions are located on the wood line with good fields of fire, about three feet deep, some with prominent overhead cover, others without. The bottoms of dummy positions must be lined with leaves to provide the illusion of greater depth to aerial reconnaissance.[17] Another technique is to pour oil in the bottom and burn it; the soot residue will also give the illusion of depth. All are well camouflaged, but the camouflage is taken from the surrounding area and is allowed to deteriorate with time.[18]

These measures alone are not sufficient to complete the ruse. The enemy must see activity in and around these positions. Soldiers must occupy these positions during daylight and withdraw to the main line after early evening nautical twilight (EENT). With this type of activity, the positions become worn and more easily seen. The vegetation becomes matted along routes leading into and around positions, and insignificant amounts of soldierly debris (MRE refuse, weapons cleaning materials, equipment, and so forth) become more noticeable over time. Captured and destroyed weapons and equipment reinforce the scam. During the enemy attack, soldiers can rig these weapons and fire from the safety of their real positions. Fake positions can also be incorporated into the main line of defense. To increase survivability, the company must be clever and imaginative.

Obstacles are an important part of the deception plan. Triple-strand concertina wire, not strengthened with barbwire and stakes, can be erected quickly and appears deceptively strong. Dummy or shallow minefields add credence. These obstacles are not adequately covered by fire and thus are easily breached. A dummy strongpoint or resistance nest is very effective in attracting enemy attention and firepower. These measures lead the enemy commander to conclude that all this effort must be for the main line of resistance.

Disguising the size of the defending forces completes the deception. A company defense should appear as a platoon. Tempt the enemy to commit the smallest force with minimal fire support possible. Only one-third of the company strength works on the dummy positions in daylight, while the majority of work on primary positions is accomplished at night.

Ultimately, a successful deception plan causes the attacker to focus his supporting fires and assault in front of the main line of resistance. The goal is to have him deploy prematurely and expend his firepower on empty positions. As his attack progresses, the assault element exceeds the fires of the fire support element and emerges into the fields of fire of the main line of resistance. At this point, his disorganization, fatigue, and psychological susceptibility to the surprise effect of concentrated firepower and counterattack are at their peak.

The Obstacle Plan. The erection of obstacles is a double-edged sword. Although they help delay and channel the enemy, they also highlight the location of the main line of resistance. Additionally, obstacle plans tend to be a little overambitious, extending beyond the means and capabilities of the constructing force. A basic mistake is to attempt to erect a "Great Wall of China" and then to attempt to defend it in its entirety. Any attempt to defend the entire sector in strength results in the defense being weak everywhere. The attacker will breach where the obstacle is weakly defended or where he can apply SOSR (suppress, obscure, secure far side, and reduce the obstacle) advantageously rather than avoid it entirely. The attacker knows that the obstacles are intended to channel and delay. If confronted with a long continuous obstacle, the attackers can plan a series of deliberate breaches with small breach teams and exploit through the successful one(s). In this way, he retains his freedom of movement and protects his force.

Obstacles should complement the defense without attempting to make it impregnable. A few obstacles with limited size, placed at key locations and in depth, stand more of a chance in ensnaring the attacking force into a series of small engagement zones than one large, obvious one.

The most valuable obstacles are those that come as a surprise to the attacker during his assault. Those that the attacker can see from a distance and plan for are of lesser value. An unexpected obstacle can throw off the entire synchronization of the attack, because the attacker must stop and breach it (if he happens to have the appropriate equipment), or must make a snap judgment to bypass. This is the correct way to channel the enemy. The majority of obstacles are located within and behind the main line of resistance. The ones to the front are mostly dummy or obstacles of dissuasion (obstacles that are probably bypassed but that limit the attacker's options once battle is joined). The value of a dissuasion obstacle is that the attacker cannot easily shift the weight of the attack laterally, because these obstacles block key choke points. The obstacle plan takes into account the attacker's perspective and likely reconnaissance effort.

The type of obstacle to establish at each location depends on the type of equipment the attacker is able to bring forward at the intended point of breach. In rough terrain, where only foot infantry can reach, the typical breaching equipment consists of wire cutters, mine probes (bayonet or sharpened stick), and light demolitions. Special equipment is normally used for forward obstacles because of its weight, vulnerability, and relative scarcity. Most special equipment can only operate in open terrain and is unlikely to advance in depth once it breaches the initial obstacles. Normally, it is withdrawn and held in reserve until needed again.

TTP

For hidden-point minefields, the defenders can deceive the enemy forces by burying mines beneath the fresh tire or tread tracks.[19]

Upon the initial breach, the assaulting forces must contend with the subsequent obstacles within the main line of resistance with the tools at hand, no easy task under the withering fire of the defense. If planned correctly, the assault force hits these obstacles at points beyond the reach of the fire support element. Under these conditions, the assault element attempts a hasty breach, unsupported and exposed to defensive fires, establishes a support position and awaits the arrival of special breaching equipment, or withdraws and attempts to breach elsewhere. In any case, the attack becomes mired, giving the defending commander time to react. Worse for the attacker, obstacles in depth threaten to destroy the assault forces in detail.

For those who believe that obstacles in depth can be overcome with bangalore torpedoes, they have no idea just how heavy and burdensome these devices are. No doubt, these devices can be brought forward, but it will not be accomplished instantaneously or easily. Satchel charges are perhaps a better choice for most obstacles, but they require foresight and advance preparation. When laying out the obstacles, the defender establishes the weaker type obstacles forward in the hope that the attacker will expend most of his special equipment on these or the dummy obstacles and therefore have little left for the obstacles in depth.

The commander should also consider placing obstacles along the squad and platoon flanks leading back to the supplementary and alternate positions as well as to the front and in between these positions. Once the attacker has taken a section of the forward positions, he is committed to this sector and must press forward. These obstacles in depth severely hinder his movement and practically ensnare him.

The main problem with most obstacles is that they require significant time and labor. The best obstacle in terms of labor and time is the minefield. (See figure 19-10.)

FIGURE 19-10

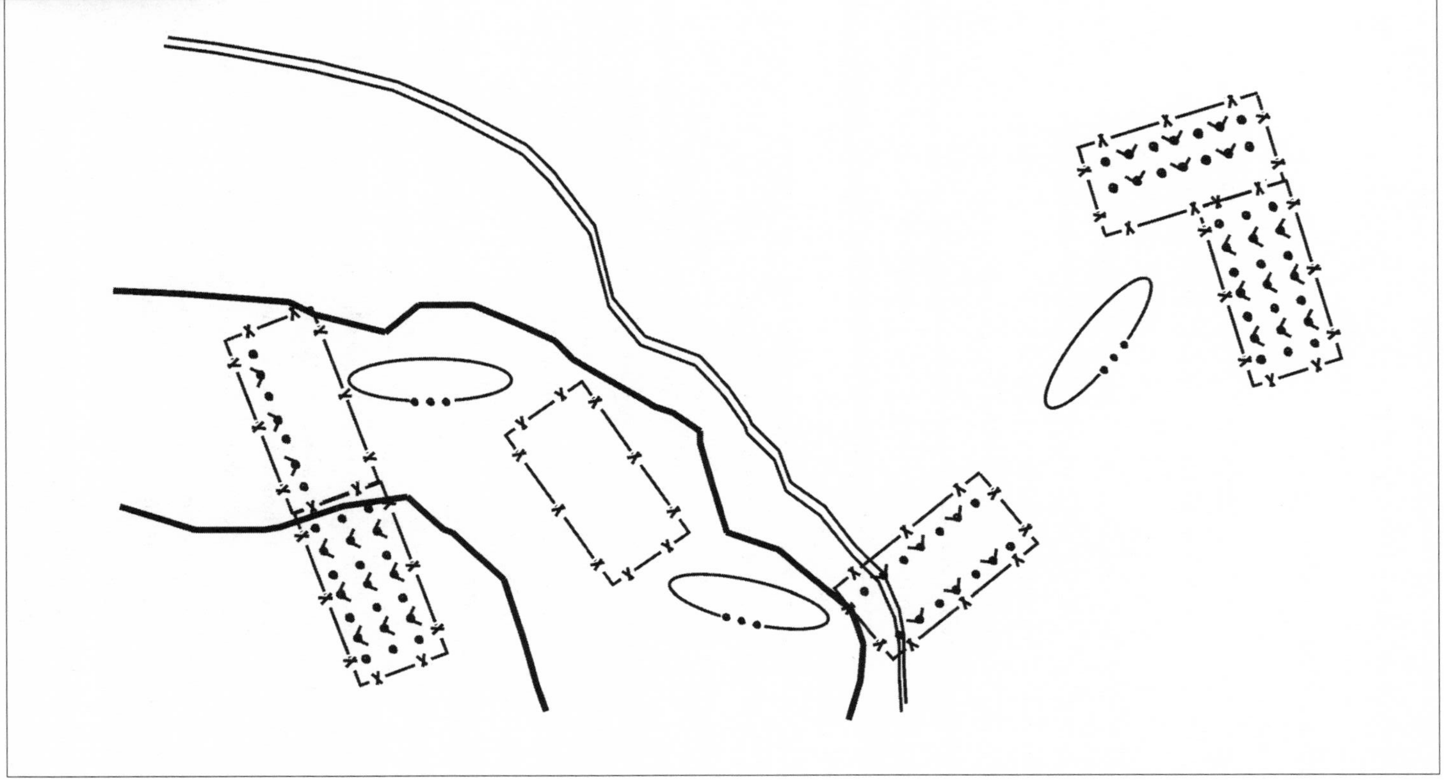

Complete, Partial, and Dummy Minefields

Minefields are most effective if the enemy recognizes it as an obstacle before actually entering it. Marked minefields may seem illogical, because the commander instinctively wants the enemy to plunge into an unmarked minefield unaware and find himself immobilized once the first vehicle or soldier detonates a mine. Unmarked minefields are normally not so effective against a seasoned enemy. Normally, reconnaissance or the advance guard discovers the minefield and alerts the main body. The commander wants to force the enemy to stop and breach the minefield while exposed to defensive fires or to channel him into engagement areas. He can do this by establishing actual and dummy minefields and clearly marking them (a strand of barb wire with a minefield marker). The breadth and width of the minefield is greater than the actual extent of the minefield proper. Even if the enemy suspects that the defender is interspersing dummy and live minefields, he cannot ignore them. The enemy commander may be quite willing to risk a thrust through a minefield in the hope that it may be a dummy, but his soldiers, those taking the risk, probably would not share his enthusiasm. The marked minefield forces the enemy to clear a lane through the obstacle whether it is real or not. In this manner, the defender can play a shell game and establish several minefields in his sector with minimum labor and time by playing on the fears of the enemy. In this manner, the marking of minefields is purely a psychological ploy.

During periods of heavy snowfall, pressure mines are not as effective, because tracked vehicles and ski troops will glide over them. Snow also dampens the explosive effect even if detonated. Use of tripwire-detonated mines is more effective; soldiers should leave sufficient slack on the wire to compensate for contractions during cold snaps. In fact, apron fences and tangle foot obstacles should remain loose, because the concussion from artillery fire snaps taut wire too easily.[20] AT mines are more effective if left unburied on hard surfaces and painted white.[21]

Soldiers should strengthen a triple strand concertina wire obstacle they are installing with barbed wire laced along its top, wrapped around each long picket, and anchored to short pickets at each end. Otherwise, the attacker can easily mash down the wire with a log or board and run across. Additionally, armored vehicles cannot overrun strengthened triple strand concertina obstacles with ease.

Claymore or command detonated mines are most effective when detonated against a massed force. Their value is as a killing weapon and not as an obstacle. They are most useful in places where enemy massing is likely (assault positions, fire support positions), and the defender is able to detonate the claymore at the moment of occupation. Another effective place is in front of the supplementary and alternate positions. During an attack, assaulting forces have a tendency to bunch up while sweeping the objective. This becomes an effective weapon against an attacker attempting to exploit a momentary advantage rapidly. Claymores also augment the firepower of the platoon, which must resort to economy of force upon withdrawal to supplementary positions because of the shifting of one or more of its squads to another point in the defense, or the refusing of the flank. Each claymore mine position should have a sandbag backstop as an added protection for the defending soldiers.

Booby traps. Referring only to those authorized under international law, booby trap devices are best employed within obstacles as long as they do not destroy the obstacle upon detonation. Within positions, these devices are not employed until withdrawing from a position. Otherwise, soldiers are likely to trip them. Their value during combat is greatly offset by the common occurrence of fratricide.

Nocturnal Detection. Battlefield illumination (artillery and mortar illumination rounds) is effective for night surveillance over clear, flat terrain, but less so in heavily wooded and rough terrain. It is also ineffective in terrain blanketed in heavy fog and smoke. Illumination rounds are normally hit or miss. They don't always catch the enemy forces moving in the open. Battlefield illumination by itself is hardly sufficient to detect the enemy. It behooves the commander to employ diverse detection systems to uncover the approach of the enemy forces at the earliest time possible.

Passive night observation devices (NODs), which enhance existing illumination, are the best comprehensive detection devices. They are light, portable, and easy to operate. They are best employed for coverage of open areas when adequate illumination but no obscurant (smoke, fog) exists. Their effectiveness deteriorates during periods or places of limited illumination. To neutralize this situation, the defender can employ chemlights in conjunction with NODs. Infrared chemlights are best, because an enemy without NOD capability is unaware that he has entered an "illuminated" place. The defense can create an invisible (to the eye) curtain of light by placing these chemlights at key areas (hanging from tree branches or on elevated stakes). The weakness of IR chemlights is their short life span, requiring replacement every few hours. Regular chemlights are the next best method but must be made directional, using opaque tape or placing them in a partially open container (can, MRE packet) in order to minimize early detection by the enemy. Chemlights can be used as a deterrent on a specific avenue of approach or a position of advantage for the enemy (fire support or assault position).

The defenders should not hang chemlights on obstacles or ring the defensive perimeter with chemlights, because it affords more advantages to the attack than the defense. However, they can be incorporated in the deception plan. Regardless of their use, soldiers should place the chemlights in front of something (tree, sandbag, fold in the ground), on the enemy side, to prevent it from blinding the viewer. A perfect position is one that produces a glowing effect, thereby silhouetting the enemy as he advances.

Thermal sight devices are effective in detecting heat signatures of vehicles and soldiers not only at night but also through fog and most smoke. However, their effectiveness is degraded by heat signatures of the surrounding environment. Vegetation and rocks (anything that absorbs heat during the day) have significant signatures at night after a hot day. Thermal sight devices are most effective in cool or cold climates or weather and during the later periods of the evening when things begin to cool off.

Pyrotechnics are limited by their short life span (flares, M-203 illumination rounds) or their one-time use (trip flares). Trip flares are particularly aggravating, because soldiers and animals seem to trip them all before the enemy arrives. Ground flares are best used in conjunction with an LP; that is, command detonated for the purpose of blinding and illuminating an intruder as the LP withdraws.

Reflecting the light from giant spotlights off a low cloud ceiling is not only effective but also difficult for the enemy to counter. Both the Allies and the Axis used this technique effectively in World War II.

Field expedient illumination devices such as bonfires fueled by wood or tires can only be used once and tend to obscure more than illuminate. Additionally, they are time-consuming to construct and not worth the effort unless the defender has nothing else available.

Even if the above devices fail to help the defender detect the enemy, they do slow the enemy's advance, and anything that delays the attack favors the defense.

The Immediate Counterattack. The real strength of the defense lies in its ability to react and counter the threat. Often units do not address the very real possibility of an enemy penetration of the main line of defense. Instead, everything relies on the main line holding and the hope that the enemy will make a frontal assault, which is not a very effective method. Once the enemy does penetrate the primary defenses, no flexible system exists to counter the penetration. Once the defense begins to crack, it is nearly impossible to disengage and fall back without the risk of suffering inordinate casualties and routing.

For such contingencies, supplementary and alternate positions are crucial. These positions give true depth to the defense and give the commander the flexibility to react to enemy assaults from every direction. Supplementary and alternate positions are planned in such a manner as to permit the company to commit unengaged squads as the reconstituted reserve or as the immediate counterattack without compromising the integrity of the defense. Fire teams form semiperimeters around their respective squad bunkers for supplementary positions, and squads do likewise around platoon bunkers for alternate positions. In this manner, a flank platoon can refuse its flank by withdrawing one or more squads to their supplementary positions. The company can refuse its flank by directing the respective platoon to withdraw its squads to the alternate positions. In a sense, the company withdraws into itself to form extra protection during the course of the defense—much like a hedgehog protects itself by curling into a ball.

The commander can reinforce a threatened sector with the reserve and reconstitute the reserve with a squad from the unengaged platoon while one or both of the remaining squads fall back to their supplementary positions, thereby refusing the flank. (See figure 19-11.) The counterattack force occupies the supplementary positions of the threatened squad sector, near the point of penetration, and counterattacks by fire, assaulting to push the enemy out if circumstances permit (that is, ineffective enemy supporting fire). (See figure 19-12.) Simultaneously, the squad leader of the threatened sector receives priority of mortar fires to suppress the enemy. Because of the proximity of friendly forces, it is prudent for the initial rounds to land to the front and walk the fires in. This serves the purpose of suppressing enemy forces making for the breach as well as the penetrating forces as the mortars close in on them. The object is to retain the ability for continued defense as events unfold without endangering the integrity of the defense.

Another point concerns those fighting positions near the penetration that the enemy has destroyed or are in danger of being overrun. The surviving occupants move laterally to friendly positions or shell holes to help contain the shoulders of the penetration, rather than rearward. The counterattack squad seals the penetration while lateral fighting positions on either side of the penetration suppress follow-on forces and any lateral movements by the penetrating force.[22]

The Deliberate Counterattack. The commander must plan for the contingency of being pushed out of his defensive position or being overrun. Unlike training scenarios, soldiers will not fight to the last man and bullet—no one wants to die for an untenable position. At a point in the fight, soldiers will see the inevitable outcome and begin pulling back. To prevent the company from routing, the commander plans for such a contingency by establishing the overrun rally point. This is located at a clearly recognizable point on the ground, far enough to the rear of the defensive sector to

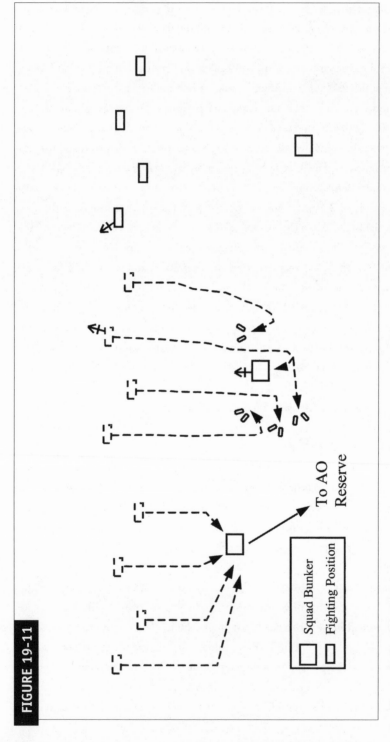

FIGURE 19-11

Refusing Flank of Platoon Sector upon Reconstitution of Company Reserve

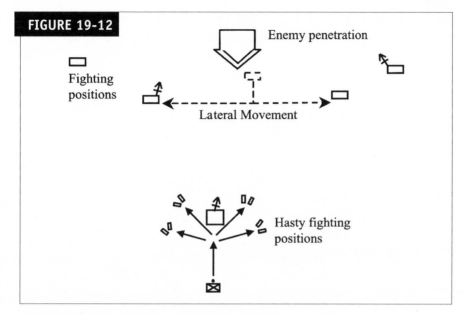

FIGURE 19-12

Immediate Counterattack Using Supplementary Positions
of Threatened Sector

preclude enemy discovery before the deliberate counterattack. Here the ranking soldier collects the company elements (and possible elements from other units) and organizes them for immediate defense, then as the company or battalion counterattack force.

During consolidation and reorganization, the enemy is most vulnerable. The counterattack plan takes time and space into account. If the enemy is given time to consolidate on the objective, the chances of a successful counterattack drastically decline. Psychologically, the defense wants the enemy to fear a counterattack every time he seizes ground. Such expectations cause him to worry more about consolidation than about exploitation of an opportunity (that is, the defense totally collapses and the entire unit routs). The defender needs time to regroup and cannot do that if the attacker continues offensive pressure. Hence, even in those situations where it is impossible to counterattack, the enemy will still expect one and will act cautiously. Of course, the main purpose of the deliberate counterattack is to repulse or limit the enemy attack. As the enemy reorganizes and consolidates, the majority of his firepower is directed forward. Even if he fears a counterattack into his rear, he will still place the majority of his firepower forward. Therefore, the counterattack envisions a rear or flank attack from terrain that makes the enemy defensive position untenable. The rear attack has the added advantage of striking the enemy's trains and CP. The commander stresses that his soldiers not attack enemy aid posts or stop to plunder enemy supplies because of hunger and thirst. Such distractions can mean the difference between success and failure.

The company overrun rally point also serves as a contact point for a battalion or higher-level counterattack. In such cases, the company commander leads the counterattack force on the best avenue of approach into the counterattack position, because his intimate knowledge of the sector provides continuity of effort and unity of command.

An intermediate solution is to launch a counterattack within five minutes of enemy seizure of the defensive positions. The commander can direct the reserve along with a few crew-served weapons to withdraw to a predesignated location and prepare to counterattack. Once the company withdraws from the defensive sector, the counterattack force strikes from key terrain on the enemy's flank. If successful, the company launches its attack and reoccupies the defensive sector. Even if not successful, the counterattack force can prevent the enemy from pursuing the company.[23]

Entrenchments. Entrenchments, as a continuation of the priority of work, are a pernicious necessity. Their value in providing greater cover is greatly offset by their conspicuousness. The longer the company remains in sector the greater the probability that the enemy will pinpoint the main line of defense. Once the enemy accomplishes this, he will bring to bear more accurate and heavier direct and indirect fire against the defense. In this environment, the defender needs stronger positions supported by lateral and rearward trenches in order to move within the defense under cover. The great paradox of World War I remains valid: In the process of protecting the defenders from enemy fire, entrenchments (as a result of spoil, obstacles, and accurate reconnaissance) draw even greater fire, thereby increasing the danger to the defender. But without entrenchments, the enemy, possessing greater knowledge of defensive dispositions, can overwhelm the defense through the firepower and assault of a deliberate attack. To solve this dilemma, the defender should use terrain to mask the line of entrenchments from the enemy point of view.[24] If the main line of defense lies along the forward edge of a terrain feature (military crest, forward trace of a town or forest), entrenchments should begin rearward (the alternate and supplementary positions) and work forward. If the main line of resistance is along a reverse slope or rear edge of a terrain feature, they can begin along the forward positions and work rearward. The bottom line is to build entrenchments where the enemy cannot spot them because they are virtually impossible to camouflage fully.

The entrenchment plan is crucial, and the commander must give it considerable thought as he develops the defensive plan. Trenches consume considerable manpower and man-hours. A long lateral trench line such as those constructed in World War I is ill advised because such a trench line must be defended throughout its length. As with any linear defense, the effort is diluted, as the defender must attempt to defend all parts of the trench, affording the enemy the opportunity to penetrate the defense at the time and place of his choosing. Once the trench line is penetrated, it becomes virtually untenable. To prevent this, the commander must modify the defensive plan to accommodate its evolution into entrenchments.

An attempt to connect all the original primary positions with trenches is ill conceived. The entrenchment plan should consist of separate squad entrenchments along the main line of resistance and platoon strongpoints at the supplementary positions, all interconnected with communications trenches, if feasible. At the very least, the squad entrenchment should have a rearward communications trench to a point that affords a good covered and concealed route to the platoon strongpoint. Because of the manpower requirements to build and occupy entrenchments, all defensive sectors should be narrowed to probably a third of their original sector. The squad entrenchment will probably need to narrow to 65–100 meters (see figure 19-13).[25]

Depending on the terrain, the platoon sector may also shrink to ensure interlocking fields of fire and to ensure the platoon does not become fragmented as a result of gaps. The commander must determine whether he can defend the original defensive sector without his platoons becoming defeated in detail or request that battalion reduce

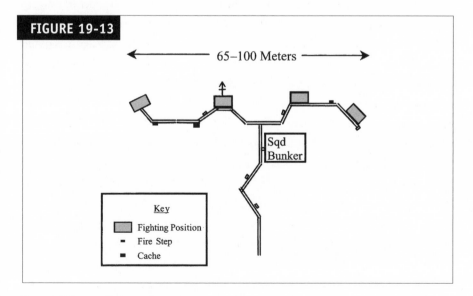

FIGURE 19-13

Squad Primary Entrenchments

his sector. He must consider the eventuality of building entrenchments as part of his plan and submit his findings and recommendations to battalion headquarters as soon as possible. If the battalion cannot change the width of the sector, the commander must speak to the battalion commander of the risks and additional manpower requirements to build and occupy the sector in proper strength.

Initially, the communications trenches are one meter deep and form a zigzag pattern (45-degree angle) every 10 meters in order to lessen the effects of enemy fire. Along the main line of defense, only internal squad positions are connected linearly by trenches. Squads build fire steps at pertinent points within the squad trench system for reinforcements to occupy. Rear berms along the rear of the trench are also crucial to prevent a halo effect of the firing position. The gaps between squad resistance nests become fire sacks, reinforced with obstacles along the perimeter.

Squad living bunkers within the trench system are dug into the rear of the trench from the communications trench to provide good protection from artillery fire and permit soldiers to occupy firing positions quickly. Soldiers must exercise caution when building bunkers. Bunkers should not be too deep or have more overhead cover than the bunker can support. The supporting beams should be dry wood, which is stronger than green wood. Cave-ins are a great danger, particularly if wet weather sets in.

Since squad entrenchments lead to the platoon strongpoint, the defense does not suffer when the primary positions become untenable. (See figure 19-14.) Of course, the company commander can elect to forego squad entrenchments altogether in favor of a platoon strongpoint serving as a redoubt. Although the geometric foundation of the strongpoint is a triangle, the overall defense must conform to the characteristics of the terrain to ensure proper coverage. The strongpoint must have multiple places to defend within in order to stem a penetration. (See figure 19-15.) The squad and platoon leaders establish embrasures (a fortified position) within the communication trenches to force the attacker above the trenches and into the fields of fire of the squad and platoon positions. Soldiers also construct and position portable barbed wire and wooden obstacles within the communications trenches. Soldiers pull these devices down into the trench and secure them (weights or long stakes) upon a withdrawal to block the attacker's advance and force him above ground. Additional obstacles within the trench system (between trenches) are located in depth to prevent the attacker from sweeping across the system quickly. Again, the decision to build entrenchments from the rear, forward, or vice versa lies in the enemy's ability to identify the frontline defensive trace.

Trenches require constant maintenance. Water and mud become the greatest enemies.[26] Damage from shelling and erosion requires repair and strengthening. Thousands

FIGURE 19-14

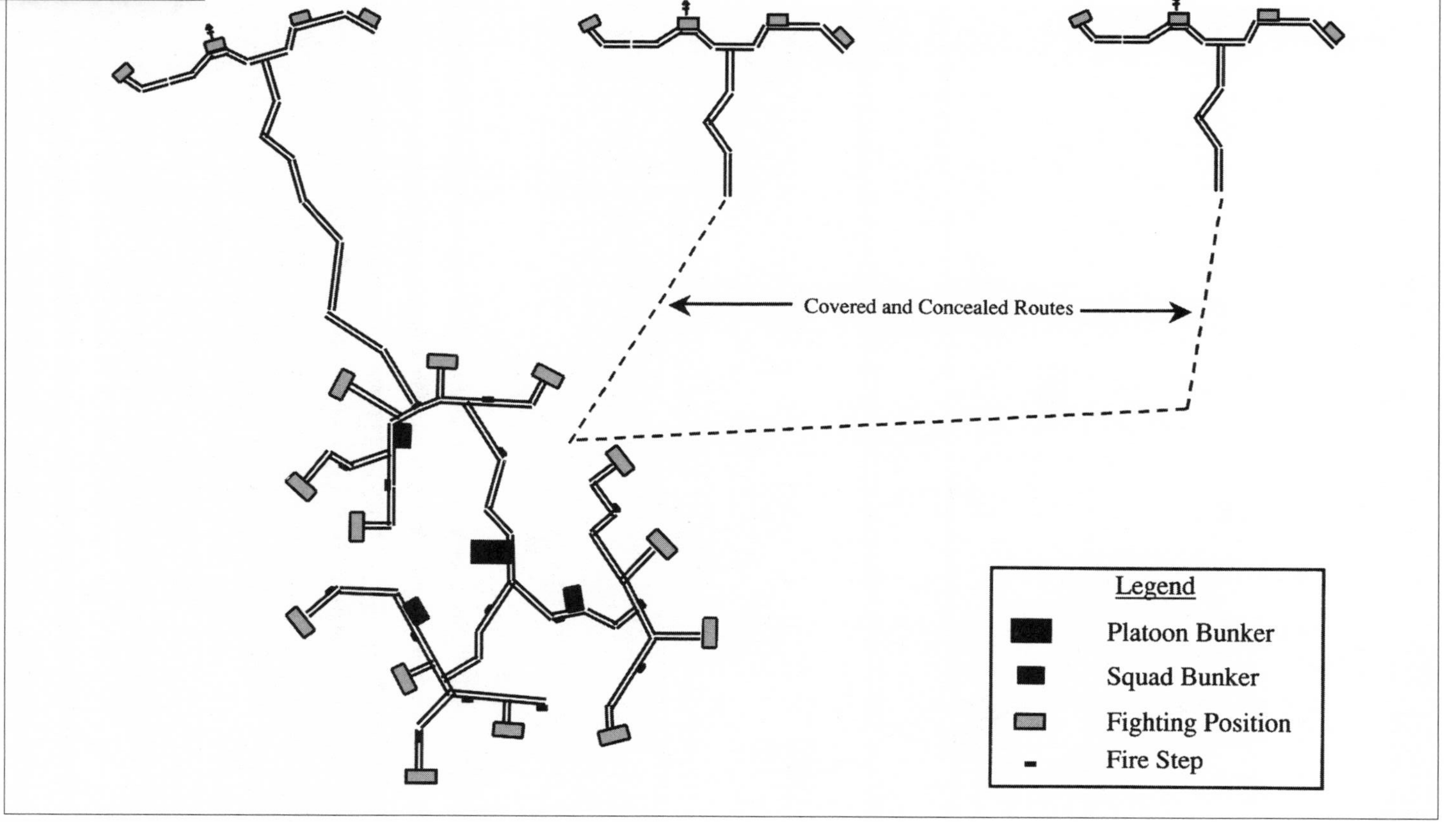

Legend

Platoon Bunker	▮
Squad Bunker	▪
Fighting Position	▯
Fire Step	-

Covered and Concealed Routes

Squad Entrenchments Leading to Platoon Strongpoint

FIGURE 19-15

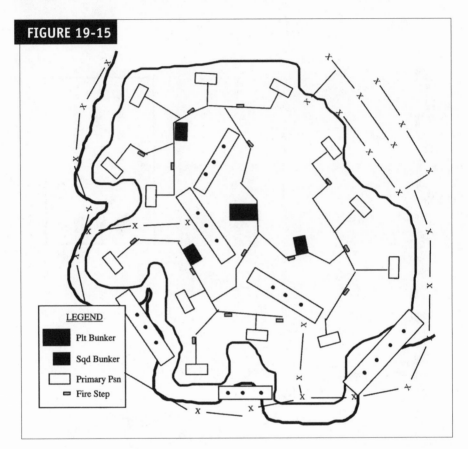

Platoon Strongpoint

of sandbags and considerable construction material are required to buttress the field works. To combat the effects of water erosion, the defenders build revetments to strengthen the trench walls. Revetments can be constructed from sandbags or wooden sticks or boards. Sandbag revetments require water portals to relieve the revetment of water pressure.[27] The defender builds wooden floors to counter the collection of water at the bottom of the trench. Wooden floors are crucial because the buildup of water and mud retards the mobility of the defenders and is the primary cause of trench foot, immersion foot, and respiratory illnesses. The need for revetments and wooden floors requires that the defender build the trench deeper and wider to make allowances for the construction and provide sufficient room for personnel passage. Work parties must pump or bail out water and shovel out mud in a constant battle against erosion. Regardless of these precautions, soldiers require rubber boots (preferably knee high) to ward off the effects of cold water and mud.[28]

Positional warfare can create appalling conditions, which cause high nonbattle casualties. Disease is a major problem if the enemy establishes battle lines nearby. Enemy fire precludes the recovery and burial of bodies, which decompose in no-man's-land and attract flies and rats, which in turn spread disease rapidly. Soldiers are unable to practice routine personal hygiene because of the danger of sudden raids and enemy fire. Typhus can break out as a result of lice infestation. Lice breed quickly among soldiers who sleep closely together—a necessity in bunkers. In such conditions, the surrounding soil can become permeated with fecal and decomposing matter, particularly if heavy shelling churns up the soil on a constant basis. This environment poses an extreme danger for wounded soldiers, because the probability of infected soil or matter entering the wound is certain, causing peritonitis and gangrene.[29] Consequently, the company must have a detailed medical evacuation and field hygiene plan. In such an environment, the company should serve one week on the front line and rotate to the rear for two weeks.

The company organizes emergency excavation parties for bunkers or trench lines that cave in from artillery fire or water damage. Deep underground bunkers are necessary during heavy enemy barrages. The excavation parties are required to rescue comrades buried alive from a near miss.

It is evident that entrenchments are highly manpower and time consuming, not only to build but also to maintain. Because of the immensity of the undertaking, company sectors will likely be narrower to accommodate the size of the modern company. As a result of the very nature of positional warfare, friendly casualties are likely to be high. It appears that this type of warfare should be avoided at all costs. Unfortunately, some situations may require that the company retain specific terrain at all costs. The commander takes great care in planning the layout, extent, and location of the trench system to strengthen the position without a concomitant exposure of the defenders to enemy fire and the elements. To this end, the reverse-slope defense favors the trench system most, because enemy observation and fire are masked. Additionally, the slope allows for greater drainage and a firmer foundation. Still, as in any protracted fight for a position, the enemy will gain accurate knowledge through patrols, limited objective attacks, and aerial reconnaissance. The commander can use this to his advantage by deploying squads into the local shell holes, bolstering them by digging bunkers in the sides and using connecting tunnels to other craters.[30] The entrenchments are still used for accommodations, but are abandoned before major attacks. Enemy artillery will be wasted on empty positions and the fight carried on from the shell holes.[31]

Position of Leaders. There is no law that requires leaders to remain at their command posts during the engagement. The CP needs to be placed rearward in a secure position. Company RTOs need a secure environment in which to conduct reporting; but the leader needs to be forward once the fight starts in order to see what is going on. Because events unfold rapidly, the commander cannot possibly get a clear picture and reach a proper decision quickly at the company CP from radio reports. An excellent alternative is for the commander to carry a small squad radio set on the company frequency. With this radio the commander maintains constant contact with his CP and higher command. During the engagement, the XO is in charge of the CP, monitoring the situation and in an excellent position to assume command if the commander becomes a casualty.

The commander positions himself to observe where he expects the enemy to assault initially (most likely enemy course of action). If this is not possible, he locates where the defense is most likely to collapse if penetrated and exploited before the company can respond (most dangerous enemy course of action). This is rarely an easy task, but the commander drills himself with this mental exercise because identification of the decisive point helps the commander maintain focus. The bottom line is that the commander is the most experienced and trained leader in the company. He can instantly recognize a threat or opportunity and use his position of authority to direct combat support in response. For the commander, a good position may be one of the platoon CPs. Likewise, a platoon leader could position himself at one of his squad CPs.

Training. In a training environment, the defense enjoys far too many advantages over the offense as a result of the inherent limitations and constraints of simulated combat. This gives soldiers a false sense of security. Fighting positions are virtually impregnable to existing training devices. Attackers cannot very well fire M203 grenades and AT training rounds at positions because of the risk of injury. Likewise, inert grenades hurled like a baseball cause injuries. Artillery suffers the worst degradation. The physical and psychological shock of artillery and mortar impacts is absent. Minefield and wire obstacles are impervious to artillery effects too. Lastly, the deafening sounds of combat and sense of chaos are absent. Granted, the intensity of combat cannot be replicated; but the commander, as chief trainer of the company, needs to put some thought into enhancing the value of training. He constantly seeks new ideas to improve realism without sacrificing safety. A few ideas are presented below.

Each fighting position and bunker is outfitted with a multiple integrated laser engagement system (MILES) vehicle harness with the yellow warning light mounted within. In this way, antitank simulators (TOW, Dragon, Viper) can be used against fighting positions and bunkers. Each soldier wears a flak vest, helmet, and a riot mask shield for protection from inert hand grenades. The commander may consider constructing barriers out of chicken wire and wooden frames to protect soldiers in fighting positions from the effects of M203 training rounds. Some of these suggestions do present some risk, so good protective measures must be used. The point is to come up with something that allows greater use of organic weapon systems during training.

The commander strictly supervises the control of artillery and mortar battlefield damage assessment (BDA). The engagement does not begin until all artillery observer/ controllers (O/C) are present and have radio contact with each other and with the exercise control center (ECC). This is a major aggravation and rarely seems to come together, but the commander must not yield even if it amounts to a 24-hour delay.

First, the commander meets with the artillery O/Cs to discuss the use of BDA. The O/C does not assess BDA until the ECC transmits the following:

- Caliber of round
- Number of rounds fired
- Fire pattern (linear, area, and so forth) and radius of effect
- Type of fuse used
- Grid of impact point
- Whether mission is observed or unobserved
- Whether corrections are observed or unobserved

If the latter two criteria are unobserved, the artillery fire is automatically scattered.

The O/C may exercise the right to have a round impact at a place other than the given grid in order to exercise the observer's fire correction procedures. Additionally, the unit receiving fire should have the opportunity to respond to the incoming fire before it becomes a fire for effect. In this manner, a defending unit does not become wiped out unrealistically from a trigger happy O/C. If the forward observer can successfully shift the fire onto the intended target, BDA is assessed; if not, BDA is assessed at the point of impact.

Next, the O/C considers the fire pattern, radius of effect, and type of fuse employed. For example, 105 mm and smaller caliber rounds with variable time (VT) fuses have little effect on obstacles and positions with overhead cover. O/Cs use their best judgment when assessing BDA. If the forward observer can bracket indirect fire onto the target, the O/C destroys it; but he avoids assessing BDA on a position where there is an outside chance of a round impacting directly on it. These things do happen, but they occur so infrequently that this should not be assessed in training. As for obstacles, the O/C should assess only partial damage no matter how devastating the supposed damage. In this manner, the O/C forces the attackers to acknowledge the limitations of artillery and that artillery alone cannot cut a path through obstacles.

NODs are an excellent means of enhancing the use of MILES at night. Because NODs allow the user to detect the impact of the laser beam, the user can direct the fire against specific targets. This training tactic is particularly effective with machine guns and tends to replicate the use of tracers. Naturally, the use of laser-directed beam devices (PAQ-4) help leaders focus fires on targets for their soldiers equipped with NODs.

CONCLUSION

Rarely will a defense be 100 percent ready before the attacker launches his attack. So, the defender must phase his preparations in such a manner that he is always ready for an attack but grows progressively stronger over time. If, in training, the defender is given 12 to 24 hours to prepare the defense, he will neglect security and readiness under the assumption that the enemy will never attack before the allotted time. If he approaches the defense as a progression of sequential and simultaneous tasks, he will be more prepared to face an enemy attack rather than becoming enslaved by an inflexible timetable. In this sense, the battalion commander and S3 should allot more time to the defense rather than trying to maximize the number of training tasks in an exercise. The defense must have the capability to respond to an enemy attack from any direction and have the flexibility to adapt to evolving events without collapsing. The defender must view the defense as a continuation of activities linked to longterm events. Rest and health are vital to sustainment. A soldier who cannot fight because of exhaustion or illness is as much a casualty as a wounded or dead soldier. Proper care of the soldier is an integral part of the plan. Lastly, the commander weighs the advantages of extra cover with the disadvantages of greater exposure to enemy observation. There exists a practical medium in every situation; the commander should know when this point is reached beforehand and should ensure that everyone knows when the priority of work should stop.

TABLE 19.1 OCCUPY DEFENSIVE SECTOR

Task	Responsibility
1. Security halt:	
a. Halt company 200–300 meters from assigned defensive sector and occupy the defense assembly area, which offers cover/concealment and covers the most dangerous avenue of approach. If terrain is favorable, form assembly area as a hasty ambush.	CO CDR
b. Organize leaders recon: Commander, CO RTO, FIST officer, platoon leaders, XO, AT section leader, mortar section leader, NBC NCO, trail platoon's NBC team (if NBC threat), attachment leaders, and two squads from lead platoon for security. Exchange Five Point Contingency Plan with 1SG: 1) The task 2) Return time 3) If the leaders' recon does not return by appointed time, attempt to establish radio contact. If no contact is made, occupy defensive sector in force. 4) If leaders' recon is attacked, it will return to defense assembly area, using the running password. 5) If the main body is attacked, defend in place. Move to overrun rally point if forced to withdraw.	CO CDR
c. Upon departure of leaders' recon, adjust perimeter and reestablish 360-degree tactical posture for main body.	1SG
c. During leaders' recon, brief soldiers on the following: 1) Five-point contingency plan. 2) Overrun rally point and code word for withdrawal. 3) Sign/countersign. 4) Number combination and running passwords. 5) Sectors of fire on avenue of approach.	1SG, PSG, SQD LDR
2. Leaders' recon establishes initial security (hasty perimeter at tentative Co CP).	SL I

(continued)

TABLE 19.1 *(continued)*

a. Clear defensive sector, using preferred clearing technique (Box, zigzag, cloverleaf, etc.).	SL II
b. Check area for NBC contamination if battalion HQ indicates a threat.	NBC NCO, NBC TM
c. Check for and remove or mark obstacles and mines.	SL II
d. Establish and occupy LP/OP position(s) with radio (one per platoon sector).	SL II
3. Conduct leaders' recon:	CDR
a. Assign platoon sectors. Brief each platoon/section leader on left and right limits of each platoon sector if time permits; if not possible, select a key location where the entire defensive sector can be viewed and brief leaders from there.	CDR
b. Select squad sectors, CP, and crew-served/key weapon positions.	PLT LDR
c. Verify Co mortar location(s).	MORTAR SECTION LEADER
d. Verify AT firing positions and designate CP.	AT SECTION LEADER
e. Verify Co CP location.	CDR
f. Select company trains location.	XO
g. At designated time, return to Co CP to conduct back-brief of defensive plan and make adjustments. Establish N + 0 time for priority of work in the defense.	LEADERS
4. Occupy defensive sector:	
a. Leaders' recon (minus security squad) returns to main body. Each leader guides his element into assigned sector CP (subelement release point).	PLT LDR/ SEC LDR
b. Emplace LP/OPs.	PSG
c. Conduct leaders' recon within respective sectors.	PLT LDR/ SL/SEC LDR
5. Begin **Priority of Work of the Defense.** (chapter 19)	ALL

TABLE 19.2 PRIORITY OF WORK IN THE DEFENSE

Task	Responsibility
1. Establish security:	1SG
a. Select permanent OPs [Day] with covered and/or concealed route back to defensive positions. (N + 30 minutes).	PSG
b. Select permanent LPs [Night] and emplacement of early warning devices (Platoon Early Warning System, trip flares, etc.) prior to EENT.	PSG
c. Emplace NBC alarms. (N + 30 mins)	CO NBC NCO/PLT NBC TM
d. Designate and brief clearing patrols for BMNT and EENT (sent out within 300 meters of positions to disrupt enemy in surveillance or assault positions) (before EENT/BMNT).	PLT LDR
2. Position crew-served/key weapons:	CDR
a. Select positions and designate sectors of fire. Rucksacks maintained at CP. (N + 30 mins)	PLT LDR/AT SEC LDR/MORT SEC LDR
b. Hasty fighting positions prepared: firing platform (to lower weapon profile), position 18 inches deep, sector stakes, fields of fire cleared, and grazing fire (MG) established. Camouflaged if immediate threat exists. (N + 45 mins)	PSG/AT TM LDR/ MORT SL
c. Verify and submit **Range Card** to Plt/section CP. (N + 1 hr)	PSG/AT TM LDR
d. Mount and zero NVD (before EENT).	PSG/SL/AT TM LDR
3. Occupy positions:	
a. Designate company reserve location. (N + 30 mins)	CDR
b. Squad emplaced: select fighting positions and designate sectors of fire. Rucksacks maintained at squad CP. (N + 30 mins)	SL
c. Supervise preparation of hasty fighting positions: 18 inches deep with sector stakes and camouflaged if immediate threat exists. (N + 45 mins)	TM LDR

(*continued*)

TABLE 19.2 *(continued)*	
d. Establish company aid post with ambulance and brief PSG/AT TM LDR/Mort SL on location. (N + 30 mins).	XO
e. Deliver pioneer tools and overhead cover LOGPAC to platoon/section LDRs. (N + 1 hr)	1SG/SUPPLY SGT
f. Secure flanks through coordination with units on left and right flanks to include: friendly flank positions, sector of fire, LP/OP positions, and establish responsibility for avenues of approach. (N + 1)	PLT LDR
3. Develop defensive plan:	CDR
a. Submit R&S plan (limited to 1 km from FEBA) to Co CP. (N + 1)	PLT LDR
b. Produce company fire support plan and distribute to platoons, AT, and mortar section. (Plan TRPs forward of, within, and to the rear of defensive line). (N + 1)	CDR/FIST/MTR SEC LDR
c. Submit one copy of squad **Sector Sketch** to platoon CP. (N + 1.5) (chapter 20)	SL
d. Submit one copy of platoon **Sector Sketch** to company CP. (N + 2.5) (chapter 20)	PLT LDR
e. Submit one copy of each AT range card to company CP. (N + 2.5)	AT SEC LDR
f. Verify overrun rally point from OPORD for immediate counterattack at platoon and company level. (N + 2.5)	CDR/PLT LDR/AT & MORT SEC LDRs
g. Conduct back-brief with soldiers on defense plan (from OPORD): (N + 3)	TM LDR
1) Location of adjacent positions, including where the sectors of fire interlock with their own sectors of fire.	
2) Co overrun rally point (using map as orientation, point out the direction and distance to the RP or physically point out the RP if within view.	
3) Challenge and password; code words from OPORD.	
4) Tentative alternate and supplementary positions, including routes. Update once this plan is complete.	

TABLE 19.2 *(continued)*	
5) Locations of Sqd, Plt, Co CPs, Co aid post, and supply post.	TM LDR
6) Lanes through obstacles and when/how lanes will be closed.	
h. Back-brief soldiers on security plan (from OPORD): (N + 3)	TM LDR
1) Percentage of security during daylight.	
2) 100 percent security during stand-to: from ____ to ____.	
3) Percentage of security during night.	
i. Back-brief soldiers on alert plan: (N + 3)	TM LDR
1) Stand-to for beginning morning nautical twilight (BMNT): 30 mins before BMNT until 30 mins past BMNT. Soldiers eat one packet of sugar and perform 20 push-ups to assist them in waking up.	
2) Stand-to for EENT: 30 mins before EENT until 30 minutes past EENT. Inspect all NOD and thermal devices. Withdraw OP to LP positions.	
3) Soldiers in full uniform with all equipment donned throughout stand-to.	
j. Produce obstacle plan on key avenues of approach and distribute to platoons and AT and mortar sections [only if directed by higher HQ]. (N + 3)	CDR/ENG LDR/ PLT LDR
k. Submit one copy of company **Sector Sketch** to battalion CP. (N + 4) (chapter 20)	CDR
4. Establish communications:	COMMO NCO
a. If no radio commo or if on radio listening silence, send one runner to Co CP until wire commo is established. (N+ 0)	PSG/AT TM LDR/ MORT SQD LDR
b. Lay commo wire to Plt/MORT/AT CP and Co Trains. (N + 1)	PLT/AT/MORT/ CO RTOs
c. Lay commo wire to LP/OPs. (N + 2)	PLT RTO

(continued)

TABLE 19.2 *(continued)*

5. Conduct sustainment (times are subject to battalion TAC SOP; otherwise sustainment to begin once the defense is established):	1SG
a. Conduct physical security checks on all sensitive items by serial number and submit report to Co CP during BMNT and EENT stand-to.	ALL
b. Submit personnel strength, casualty reports, and witness statements 0800 and 1800.	PSG/SEC LDRS
c. Submit equipment status report and maintenance requests 0800 and 1800.	PSG/SEC LDRS/ SUPPLY SGT
d. Submit supply status report and resupply requests 0800 and 1800.	PSG/SEC LDRS/ SUPPLY SGT
e. Supervise maintenance of equipment and weapons. Priority of effort to weapons and radios.	ALL
f. Establish resupply plan (N + 5):	1SG
1) Establish company logistic resupply point (LRP):	SUPPLY SGT
a) Aerial: prepare, mark, secure LZ/DZ; located to company rear.	
b) Vehicle: covered and concealed unloading and storage site to company rear.	
2) Resupply plan allows for resupply to each platoon/ section LRP, using covered and concealed routes and without compromising security.	PSG/SEC LDRS/ SUPPLY SGT
3) Proper tools for opening crates, cutting wire bands, etc. are maintained at Co trains (wire cutters, crowbars, etc.).	SUPPLY SGT
g. Ensure each rifleman has his allotment of rifle magazines and that each magazine is filled to capacity. Excess ammo is stowed for quick access without compromising the position.	TM LDR
h. Health and hygiene:	1SG
1) Monitor food and water consumption of soldiers.	NCO CHAIN
2) Establish sleep plan: 4 to 6 hours of sleep per soldier are required. Leaders establish schedule to trade off sleep among themselves.	PSG/SEC LDRS/SL

TABLE 19.2 *(continued)*	
3) Random check of soldiers' feet, hydration, and general health.	CO CDR/PLT LDR with MEDICS
4) Sick call conducted at Co aid post immediately following breakfast.	1SG/PSG/ MEDICS
5) Monitor personal hygiene of soldiers.	NCO CHAIN
6) Field sanitation team designated for each platoon.	PSG
7) Local water source checked for potability. Water purification tablets are on hand for every soldier in unit.	FLD SAN TM
8) Establish and supervise construction of one field latrine per platoon located at least 30 meters behind defensive positions, away from water sources and from feeding area. Inspect assembly area and outside perimeter for feces and urine.	FLD SAN TM
9) Inventory field sanitation kit for items to protect soldiers from insects and rodents: repellent, traps, poison.	FLD SAN TM
i. Establish tactical feed plan:	SUP SGT
1) Maintain local noise and light discipline.	NCO CHAIN
2) Enforce five-meter interval between soldiers in serving line and eating site. Ensure soldiers use available cover and concealment.	NCO CHAIN
3) Maintain security of perimeter during feed IAW security plan. Enforce consumption of rations at Squad CPs and not at fighting positions.	NCO CHAIN
6. Prepare primary positions:	1SG
a. Positions armpit deep (Dragon psns, waist deep), about two M16 lengths long, shoulder width wide; grenade sumps about one entrenching tool (folded in pick configuration) deep; berm 18" (+) thick for front, sides, rear protection; primary and alternate positions defined with sector stakes. (N + 12)	PSG/SEC LDR/SL
b. Overhead cover LOGPAC delivered to Plt/section LRP. (N + 12)	1SG/SUPPLY SGT

(continued)

TABLE 19.2 *(continued)*	
c. Positions complete with 18″ of overhead cover (Two sandbags max, beams and plywood); position camouflaged with sod and natural materials so that it blends in with the immediate surroundings. Ensure surrounding area not scarred as a result of position preparation. (N + 18)	CDR/PLT LDR/ SEC LDR
d. Company CP, supply point, and aid post have 360 degree protection with positions waist deep, overhead cover, and camouflaged.	XO
7. Plan supplementary and alternate positions:	CDR
a. Conduct leaders' recon of alternate and supplementary defensive sectors and verify withdrawal plan IAW commander's intent. (N + 12)	CDR/PL/ SEC LDR
b. Conduct recon of alt/supp defensive sectors and routes. (N + 15)	PLT LDR/SL
c. Select alt/supp fighting positions. Hasty positions prepared. (N + 24)	SL/SEC LDR
8. Obstacles emplaced:	CDR
a. Obstacle material deposited at each obstacle site. (N + 8)	1SG/SUPPLY SGT
b. Obstacles completed in order of priority:	
1) Bn and higher directed. (N + 12)	CDR
2) Co directed. (N + 24)	PLT LDR
3) Plt directed. (N + 36)	SL
9. Prepare alt/supp positions. Same criteria as primary positions if possible. (N + 72)	CDR/PL/SEC LDR
10. Conduct rehearsals: (N + 36 onward)	CDR
a. Actions on enemy contact (probe and deliberate attack)—Plt level.	CDR/PL
b. Casualty evacuation to company aid post—Plt level.	XO/PL/SL/SEC LDR/MEDIC
c. Defend obstacle/target turnover—Plt level.	PL/SL
d. Request for fire on TRP/shift from a known point/ FPF. Plt and below level.	FIST OFF/PLT LDR/SL/FO

TABLE 19.2 *(continued)*	
e. Actions in the event of overrun (overrun rally point)—Co level.	CDR
f. Withdraw to alt/supp psns under pressure—Plt level.	PLT LDR/SL/ SEC LDR
g. Execution of company plan—Co level.	CDR
h. Commit the reserve—Co level.	CDR/RES LDR
11. Establish withdrawal plan from defensive sector or relief in place (day and night):	CDR/PLT LDR
12. Continue improvement of defense (N + 72):	CDR
a. Communication trenches (1 meter deep) back to CPs.	PSG/SL
b. Bunkers completed for all CPs, supply post, and aid post.	1SG/PSG
c. Revetments for fighting positions emplaced and wooden floors built for bunkers. Adverse weather protection.	CDR/PLT LDR
d. Trench system established with strong points and resistance nests.	CDR/PLT LDR

NOTES

1. ATEP 7-10-MTP, 5-70–5-76, 5-172–5-175, 5-204–5-208, 5-209–5-212. "Defend" (7-2-1021), "Maintain Operation Security (7-2-1057), "Perform Personnel Action" (7-2-1037), and "Perform Logistical Support" (7-2-1048) form the foundation document for this portion of the SOP. FM 7-71, 5-10–5-21.
2. DA Pam 20-269, 241.
3. Wray, 80. German bunkers on the eastern front accommodated six men only because large bunkers were too structurally weak to withstand near misses.
4. DA Pam 20-269, 246. During the Russo-German War, one German commander had huge bonfires built and maintained for 36 hours in order to thaw out the frozen ground. Once excavated, he covered the bunkers with heavy logs and covered them with the excavated clay, which hardened the bunkers once frozen. To deceive Russian observers, he had several deception fires built. Wray, 74. German soldiers often used grenades and explosives to penetrate the frozen layer of earth in order to make fighting positions.
5. Lucas, 96; Wray, 76, 80.
6. Wray, 76.
7. Ellis, 13–14. The need for revetments was a frustrating necessity in World War I.
8. DA Pam 20-269, 111.
9. Wray, 16; English, 97.
10. DA Pam 20-269, 60, 112; Lucas, 54, 124.

11. English, 109. The Germans called this grouping a Pakfront *(Panzer Abwehr Kanone Front)* in order to increase the probability of hit and kill on each tank.
12. Ibid., 220, 225. Guy Sajer also mentions this training in *The Forgotten Soldier.*
13. Lucas, 23–24; Wray, 16–17; English, 112–13. The Germans frequently used hunterkiller teams and awarded individuals with tank destruction badges for the personal destruction of tanks.
14. DA Pam 20-269, 36, 56; English, 81.
15. Ellis, 13. In World War I, the rear berm was called a parados. U.S. War Department, *Handbook on German Military Forces* (Baton Rouge: Louisiana State University Press, 1990), 231. Hereafter referred to as War Department. The Germans stressed using the rear berm to preclude the silhouette effect.
16. Ibid. The Germans would keep the existing underbrush one to three yards in front of their positions.
17. Ibid., 229.
18. DA Pam 20-269, 111. The Germans were impressed by the Russian use of such deception.
19. Ibid., 244.
20. U.S. Army War College, *German and Austrian Tactical Studies: Translations of Captured German and Austrian Documents and Information Obtained from German and Austrian Prisoners —from British, French and Italian Staffs,* compiled and edited by the Army War College (Washington, D.C.: Government Printing Office, 1918), 131. Hereafter referred to as War College.
21. Wray, 82–83.
22. Lupfer, 15. Containing penetrations in this manner was very effective for the Germans in World War I.
23. English, 144. The Germans frequently launched such counterattacks with 10 to 20 men within five minutes of the loss of a position. Normally, the counterattack was by fire on the attackers' flank.
24. Lupfer, 15, 30. The Germans were quick to point out that large and elaborate fortified positions drew enormous amounts of artillery fire. Deep dugouts were not a viable solution, because soldiers remained in them too long and often were overwhelmed by the attackers before they could occupy their positions. Placing the main line of resistance along the reverse slope significantly contributed to the German successful defenses in 1917.
25. War Department, 230. The typical squad strongpoint was a linear trench with a length of 45–60 yards.
26. Ellis, 44.
27. Ibid., 13–14.
28. Ibid., 48.
29. Ibid., 52–59, 112–13. During World War I, the British suffered 3,528,486 casualties as a result of sickness and disease.
30. English, 17; War College, 122, 124, 127–28. One key lesson for the Germans in World War I was that troops were better protected in shell holes than trenches because the amount and destructive power of artillery obliterated even the heaviest fortified lines. In this sense, shell holes provided better protection because the enemy could not distinguish between fortified and empty shell holes.
31. War College, 122, 128.

Chapter 20

The Sector Sketch

Sector sketches and range cards are rarely produced and even more rarely are accurate. The reason for this lies in a lack of knowledge of the basic requirements and the difficulty in producing them. The sector sketch is not a bureaucratic requirement used for briefing one's superiors. It is the commander's tool for analyzing the completeness and strength of the defense. It helps identify weaknesses in the defense that will require additional resources or a readjustment of the defense. The commander uses platoon sector sketches as the basis for the company sector sketch. If the platoon products are incomplete or inaccurate, he cannot produce a company sector sketch. As such, the company must have a systematic and standardized method of making sector sketches for each leader's reference.

The Range Card. The antitank range card is not as detailed as the machine gun range card and requires no explanation, other than demanding that the range fan portray the weapon's maximum and minimum ranges and areas that offer the enemy cover and concealment. The machine gun range card requires more work. If a final protection line (FPL) is used, the gunner sets the machine gun on the FPL azimuth and rotates the tripod at an angle that prevents the machine gun from traversing beyond this azimuth (called metal-to-metal contact between the traversing and elevation [T&E] mechanism and the tripod). Once determined, the gunner hammers the tripod into the ground and anchors it with sand bags. Next, the assistant gunner (AG) walks along the FPL while the gunner sights in on the gunner's waist (one meter above the ground), using the T&E mechanism. The gunner records the reading on the T&E mechanism onto the range card. Using a compass, the AG maintains a pace count as he walks along the FPL. The gunner commands the AG to halt whenever his waist falls below the one meter mark and records the pace count distance at this point. The AG continues until his waist rises back to the one meter mark, at which point the gunner commands him to halt once again while he records the pace count distance again—this area represents dead space in front of the machine gun. The AG continues out to 600 meters, because this is the maximum distance for grazing fire. By doctrine, if the machine gun cannot establish grazing fire out to 600 meters, the machine gun should establish a principal direction of fire (PDF) instead of an FPL. The AG and gunner continue this procedure on the other limit of fire (left or right), other azimuths to targets, and the alternate fields of fire. The gunner uses the range card to focus fires on specific targets during periods of limited visibility. The range card and tripod remain if the unit is relieved in place. In this manner, the new gunner can maintain the same fires. The gunner makes a duplicate range card for the squad leader for his sector sketch.

The Squad Sector Sketch. The squad sector sketch portrays each position with the fields of fire for each weapon. It cannot be drawn to scale, because the sector is too large for the portrayal of detailed information. At a minimum, the sector sketch must have the grid coordinates of the far left and far right positions and sectors of fire grid azimuths for the platoon leader's sector sketch. The squad leader includes salient terrain features (with distances), obstacles, target reference points (TRPs), and supplementary positions with grid coordinates, if needed. The squad leader provides each position with the specified instructions to reduce confusion.

The Platoon and Company Sector Sketches. The platoon and company sector sketches are drawn to scale, using the following preparation guidelines:

- Draw grid lines that encompass the assigned sector.
- Using grid lines as guides, draw the major terrain features.
- Draw all man-made objects (roads, bridges, structures, streams, wood-lines, etc.).
- Draw subordinate positions (one level down) with left and right sectors of fire gird azimuths, supplementary and alternate positions, trains, mortar positions, CPs, aid posts, and listening and observation posts (LP/OPs).

- Draw TRPs to scale:

   Battalion—Alphanumeric designation.

   Company—Two-digit number designation.

   Platoon—Letter designation.
- Draw FPL with grid azimuth and distance to scale.
- Draw obstacles by type, dimensions, and scale.
- Draw passage points and obstacle lanes.
- Draw adjacent unit positions on left and right flanks.
- Draw trigger lines, phase lines, and engagement areas.
- Draw other pertinent data—nuclear-biological-chemical (NBC) alarm, platoon early warning system (PEWS) locations, and so forth.

All leaders and gunners maintain a sector sketch and range card format with an example on the back for reference.

> **TTP**
>
> Laminate blank range and sector sketch cards to make them more durable and weatherproof. Use alcohol pens to complete the cards.

CONCLUSION

Producing range cards and sector sketches takes practice, because they are time consuming and are difficult to make. The commander must demand them as a matter of routine for the priority of work in the defense and assembly areas. If done correctly, sector sketches help to increase the situational awareness of each leader. They allow the commander to fine-tune the defensive arrangements and alert higher command to problems that have come to light as a result of the submitted sector sketches. Without them, the leadership is blind. Ignorance may be bliss for some, but it is death to the combat soldier.

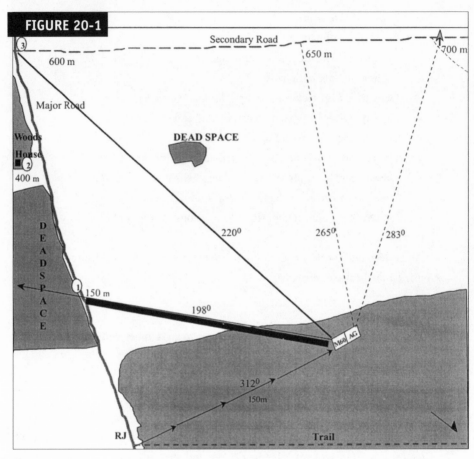

FIGURE 20-1

Sector Sketch

Range Card with Final Protection Line					
Weapon: M-60 MG	**Team: 1**	**Platoon: 1ST**		**Company: A**	**Date: 6 June 98**
No.	**Direction**	**Elevation**	**Range**	**Description**	**Remarks**
1		−25/3	150M	FPL	−4
2	L180	−50/45	400 M	House	
3	R235	−25/25	600M	Road Intersection	
Notes: Not a proper FPL, because it does not extend out to 600 m.					

Sector Sketch

Sectors of Fire Data

Unit	Location	From	To:	Remarks
SAW/M16/SL	CM59301766	201°	275°	Responsible for closing lane
ATL/M203	CM59331768	216°	265°	
SAW/BTL	CM59291770	224°	285°	
M60/AG	CM59321772	198°	283°	
M203/M16	CM59251775	224°	275°	

Specified Instructions

1. Initiate fires: MG Fire/"Hannibal"	6. Move to overrun rally point: "Bastogne"
2. Lift or shift fires: GSC/"Boston"	7. Consolidate & reorganize: "Widowmaker"
3. Execute FPL/FPF: WPF/"Guadalcanal"	8. Close obstacle lane: "Vicksburg"
4. Move to supp/alt psn: "Gumby"	9. Engagement priority: Eng, Tank, Cmd, IFV
5. Initiate counterattack: "Rommel"	10. Switch to alternate frequency: "Wolfhound"

Map Scale 1:50,000	Squad: 1	Platoon: 1	Company: A	Date: 6 June 1998

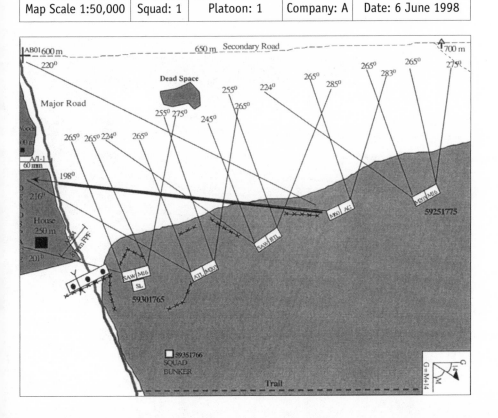

Sector Sketch

Sectors of Fire Data

Unit	Location	From	To:	Remarks
1 Squad	CM59291770	201°	275°	Responsible for closing lane
2 Squad	CM59301790	205°	273°	
3 Squad	CM59271780	210°	285°	
Dragon	CM59271780	207°	270°	

Specified Instructions

1. Initiate fires: MG Fire/"Hannibal"	6. Move to overrun rally point: "Bastogne"
2. Lift or shift fires: GSC/"Boston"	7. Consolidate & reorganize: "Widowmaker"
3. Execute FPL/FPF: WPF/"Guadalcanal"	8. Close obstacle lane: "Vicksburg"
4. Move to supp/alt psn: "Gumby"	9. Engagement priority: Eng, Tank, Cmd, IFV
5. Initiate counterattack: "Rommel"	10. Switch to alternate frequency: "Wolfhound"

Map Scale 1:50,000	Squad: N/A	Platoon: 1	Company: A	Date: 6 June 1998

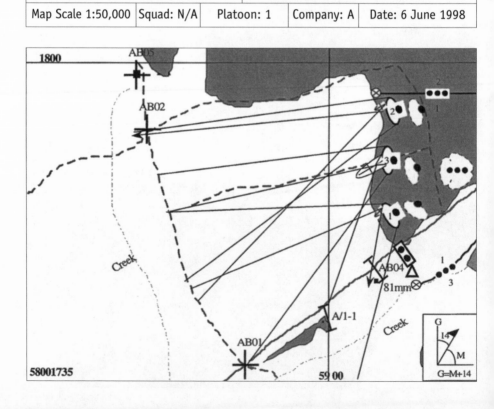

Sector Sketch

Sectors of Fire Data

Unit	Location	From	To:	Remarks
1 Platoon	CM59301780	201°	275°	Responsible for closing lane
2 Platoon	CM58801795	155°	235°	
3 Platoon	CM59551730	283°	299°	
Reserve	CM59651775			Dragon Attached

Specified Instructions

1. Initiate fires: MG Fire/"Hannibal"	6. Move to overrun rally point: "Bastogne"
2. Lift or shift fires: GSC/"Boston"	7. Consolidate & reorganize: "Widowmaker"
3. Execute FPL/FPF: WPF/"Guadalcanal"	8. Close obstacle lane: "Vicksburg"
4. Move to supp/alt psn: "Gumby"	9. Engagement priority: Eng, Tank, Cmd, IFV
5. Initiate counterattack: "Rommel"	10. Switch to alternate frequency: "Wolfhound"

Map Scale 1:50,000	Squad: N/A	Platoon: N/A	Company: A	Date: 6 June 1998

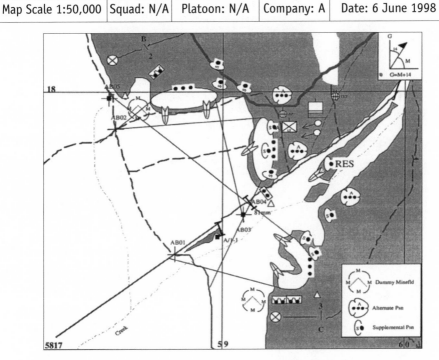

Source: U.S. Department of Army, *Light Infantry Platoon/Squad*, Field Manual No. 7-70 (Washington, D.C.: U.S. Government Printing Office, September 1986), 7-28–7-32.

Chapter 21

Relief in Place

A relief in place is a common military operation, routine but treated as a mission in itself. The enemy is rarely deceived for long that a relief has occurred. The key is to execute it before he has time to form a plan of action.

Initial Coordination. Initial coordination serves to energize the stationary (to be relieved) unit as well as to exchange information. The relieving company takes the lead during this operation and initiates the exchange of information as well as planning. The relieving company executive officer (XO) uses this time to inform the relieved company representatives of the general relief plan. He makes the initial coordination for the transfer of supplies and exchange of specific equipment. Units on the defense acquire supply stocks ranging from extra rations to obstacle material. Knowing what supplies are available for transfer lessens the logistical load of the relieving company. The exchange of equipment, such as machine gun tripods with traverse and elevation mechanisms, platoon early warning systems (PEWS), commo wire, and other non–serial number items, contributes to a smooth transition. The XO emphasizes that exchanged equipment must be operational and well maintained before the exchange takes place. Platoon sergeants inspect all equipment before the exchange is final. The transfer of sector sketches, the obstacle plan, mine data cards, and range cards, serve to familiarize the relieving company with the dispositions of the defense as well as the terrain in sector. This initial coordination for the relief-in-place meeting provides the unit to be relieved with a suspense time for meeting information requirements for the relieving company. To assist the company to be relieved, the XO provides a copy of the **Relief in Place** SOP to serve as a checklist and orientation on how the company approaches a relief in place.

Warning Order. The commander issues the warning order of the pending operation. He may choose to wait until the XO returns, in order to provide greater detail, but this is generally not necessary for relief operations. More important is getting the company working on preparations for the pending mission. The commander considers deploying the company closer to the company to be relieved in order to save time, but not so close as to compromise security. When he has returned, the XO issues a warning order covering more specific information.

Relief-in-Place Meeting. At the designated time, the commander and his key leadership move to the relieved company CP for the relief-in-place meeting. Accompanying key leaders exchange essential information and sketches and range cards while both commanders concentrate on the relief plan itself. Foremost is the time the relief is to begin (H hour). Executing the relief during night or inclement weather enhances security and surprise. The relieved commander provides information on the enemy and the defensive sector that will enhance the conduct of the relief. Enemy habits and peculiarities help determine when he is most inactive. The company sector sketch greatly assists the relieving commander with the relief plan. With it, the relieving commander can develop a plan with staging areas in a matter of minutes.

In consultation with the commander to be relieved, the relieving commander decides on a sequential or simultaneous relief. A sequential relief means that the relief is done in phases, such as from left to right, and is the slower of the two methods. This type of relief may be necessary because of the company's disposition and the terrain. Naturally, a simultaneous relief is preferred, because a fast swap-out enhances security.

The Relief Plan. Upon his return to the company assembly area, the commander issues the relief operations order (OPORD), using the **Defense OPORD Format** as reference. This method not only imparts the relief plan but also familiarizes the company with the plan of defense. The commander may not be satisfied with the existing defense but accepts it until after the relief. Once in possession, he changes it as he sees fit. The important thing is to occupy the sector quickly (about an hour) and provide an immediate and effective defense. The relieving commander is responsible for the relief plan, although the relieved commander offers input. (See figure 21-1.) The relief plan must include the company assembly area and staging areas for each platoon and squad for both the relieving and relieved company. The soldiers of the relieved unit do not use squad staging areas but move directly to their assigned platoon staging area to prevent congestion and confusion near the front line. Because of the phased nature of the relief, the relieved unit may use the platoon staging areas and company assembly area of the relieving company.

The identification of platoon, section, and squad staging areas and guides is the basis of the plan. Guides prepare the staging areas for occupation following the guidelines of **Occupy Simple Perimeter** (see chapter 15). While waiting for guides, platoon,

FIGURE 21-1

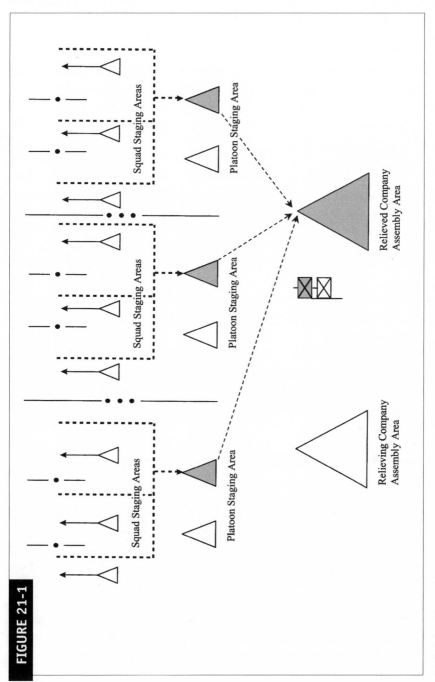

Relief-in-Place Plan

section, and squad leaders become familiar with the sector sketches and range cards of their prospective sectors. Squad leaders take this time to task organize their squads to match the composition of each position. The squad leader labels each position (1, 2, and so forth) from the sector sketch and assigns each of his designated position teams this label. When the team leader guide arrives, the squad leader familiarizes him with the organizational label to ease the occupation. Platoon leaders of the relieved and relieving companies personally emplace the machine gun teams to assist in laying the guns with the range cards and to transfer the tripods and traversing and elevation (T&E) mechanisms. Once the machine guns are ready, the relieved platoon leader moves to his platoon assembly area.

Once the front line is occupied, the platoon sergeants of the relieving and relieved unit emplace the listening and observation posts (LP/OPs). Leaving LP/OP occupation for last decreases the probability of enemy detection of the relief. Moreover, there is less danger of soldiers in LP/OPs being left behind by mistake if the platoon sergeant himself recovers them.

Establishing a distinct transfer of authority over the defense ensures that both company commanders do not clash with each other during a critical time. If the enemy happens to attack just as the transfer of authority is to occur, the commander to be relieved retains authority until the crisis passes. Transfer of authority occurs once two-thirds (two squads per platoon) have occupied positions. Platoon leaders report this action immediately.

The company to be relieved conducts clearing patrols a couple of hours before the relief takes place. This decreases the probability of enemy attack at a critical time. Of course, if the enemy detect the clearing patrol at an unusual time, he will suspect something is up. But it is unlikely that he will be able to react with an attack before the relief takes place.

Deception Plan. The deception plan is important for exposed defensive sectors. The relief should take place at night to take advantage of the darkness. In such cases where cover and concealment are lacking, weapons and munitions may be used to mask the sound of movement and keep the enemy's head down. Use of battlefield illumination blinds enemy use of night observation devices, and smoke obscures his vision. The commander should weigh the advantages of using munitions to mask and obscure the movement, because they can become indicators to the enemy over time that something is afoot.

Service Support. Both supply sergeants inspect which items are to be handed over to the relieving company. Once he has a complete inventory, the supply sergeant informs the first sergeant (1SG) what materials are on hand for improving the defense.

The relieving company taps into the existing casualty evacuation plan to ensure rapid evacuation of its own casualties. To enhance continued medical support, the company aid post does not collocate with the stationary company aid post but near the existing company casualty collection point of the stationary company.

Command and Signal. The relieving company adopts the relieved company's command and signal so as not to alert the enemy. Using the same radio net allows the transmission of instructions to all subordinates without the need to change frequencies. Each subordinate element of each company retains its call sign in order to differentiate between the two companies. The relieving company maintains radio listening silence to keep radio traffic at normal levels. Once the platoons and sections inform the company CP that the relief is complete, everyone switches to the normal signal operation instructions (SOI) frequency.

TABLE 21.1 RELIEF IN PLACE	
Task	**Responsibility**
1. Conduct initial coordination with representative of unit to be relieved:	XO
a. Exchange unit identifications.	
b. Time and location of coordination meeting between unit commanders.	
c. Exchange frequencies and call signs.	
d. Decide on method of occupation—sequential (left to right) or simultaneous.	
e. Discuss staging plan for relief—platoon, squad, and section staging areas.	
f. Determine and recon location of tentative company assembly area for relieving unit.	
g. Coordinate for relief-in-place meeting between key leadership-counterparts. Arrange time and linkup point for guide to lead relieving company leadership to meeting.	
h. Identify what supplies (CL I, III, IV, V) and sector sketches and range cards the relieved unit hands over, and what equipment (Machine gun tripods, commo wire, early warning devices) it exchanges with the relieving unit.	
i. Identify fire support available during relief.	
j. Provide representative with company **Relief in Place** SOP to serve as guide.	
2. Issue warning order upon return to the company.	XO
3. Conduct relief-in-place meeting between each company's key leadership:	CDR

(continued)

TABLE 21.1 *(continued)*

a. Ensure key leadership attendance—CDR, PLT LDRs, FIST OFF, SUPPLY SGT, COMMO NCO, CO MEDIC.	XO
b. Time (H hour) in which the relief in place is to begin—signaled by company elements departing company assembly area for subsequent staging areas.	CDR
c. Provide local enemy situation: • Enemy peculiarities and habits. • Known or suspected positions. • Known or suspected ambush sites. • Latest enemy activity. • Known or suspected obstacles.	RELIEVED CDR
d. Provide company sector sketch and essential information: • Platoon/section positions, LP/OPs, and CPs. • Fire plan. • Fire support call signs/frequencies. • Obstacle plan. • Description of terrain in sector.	RELIEVED CDR
e. Provide platoon leaders with platoon and squad sector sketches and range cards.	RELIEVED PLs
f. Finalize method of relief—sequential or simultaneously.	CDR
g. Finalize relief plan referring to sector sketch to include use of guides from company assembly area down to squad staging areas:	CDR
1) Platoon/section guides report to company assembly area at H hour and lead respective elements to their staging areas.	PLT LDR/SEC LDR
2) The relieved platoon leader guides platoon machine gun teams into position. The relieving platoon leader accompanies them into position, verifies the range card data, and supervises the exchange of tripods and T&E mechanisms.	PLT LDR
3) Relieved squad leaders lead relieving squads from platoon to squad staging areas. Referring to the squad sector sketch, the relieving squad task organizes to match each position by weapon and number.	SL
4) Relieved team leaders guide relieving personnel into each position.	TM LDR

TABLE 21.1 *(continued)*	
5) Once all fighting positions are transferred, the relieved and relieving platoon sergeant guide LP/OPs into position.	PSG
6) Relieved personnel move as a unit to the relieving squad and platoon staging areas and then to the company assembly area.	SL
h. Coordinate transfer of authority during relief:	CDR
1) Transfer occurs when 2/3 of the company (2 platoons) occupies assigned positions.	
2) Contingency plan for enemy contact during relief—controlling commander directs the defense, elements not in primary defensive positions report to assembly area. The other commander assumes control of elements in assembly area, acting as the local reserve.	
i. Coordinate transfer and exchange of company-level equipment and supplies.	SUPPLY SGT
j. Coordinate security measures for relief—clearing patrols before H hour.	RELIEVED CDR
k. Coordinate deception plan—use of weapons/ordnance to mask sound, use of battlefield illumination to blind or smoke rounds to obscure.	RELIEVED CDR
l. Coordinate casualty evacuation support.	RELIEVED 1SG
m. Coordinate command and signal procedures:	CDR
1) Exchange call signs and frequencies, challenge and password, and running password.	COMMO NCO
2) Relieving company CP collocates with relieved company CP.	CDR
3) Both companies operate on relieved company's frequency. To distinguish between both companies, the relieving company adopts the RTO call sign.	COMMO NCO
4) The relieved company maintains normal level of radio traffic. The relieving company maintains radio listening silence.	COMMO NCO

(continued)

TABLE 21.1 *(continued)*	
5) Once in position, relieving squads inform their platoon CP, using landline or messenger. Platoons/ sections notify the company of closure using the code word from the OPORD. The company then informs the battalion CP of closure.	SL/PL
6) On order, the entire company switches to its normal SOI.	CDR
n. If time permits, the relieving company leaders and their counterparts recon their sectors.	CDR
4. Issue OPORD, using **Defense OPORD Worksheet.** (chapter 18)	CDR
5. Execution. The company completes the relief at H + 1 and begins Priority of Work in the Defense.	CDR

Source: ATEP 7-10-MTP (Washington, D.C.: U.S. Government Printing Office, 1988), 5-79–5-88."Perform Relief in Place" (7-2-1067) forms the foundation document for this portion of the SOP.

Chapter 22

Health and Hygiene

Health and hygiene challenge the company leadership on various levels. It is in this arena that a bond of trust and respect is created between leaders and subordinates. For the leader to understand the capabilities of his subordinates under various conditions, he must have experienced the same hardships. Only through this enlightenment can he bridge mission tasks and actual capabilities, leading to the development of sound judgment.[1]

Leaders must be aware that nonbattle casualties will eclipse battle casualties if preventive measures are not taken. Cold, heat, and disease will quickly reduce the unit to combat ineffectiveness unless leaders proactively seek remedies. The commander must not assume that the highest levels of command will ensure that his infantryman will be properly protected from the weather, particularly in winter. Most campaigns entail risk. Desiring to maintain the momentum of an offensive to a conclusion, the theater commander will normally place fuel, ammunition, replacements, and other immediate needs ahead of winter stocks. For the theater commander, risk becomes the probability of concluding the campaign or occupying a predetermined defensive line before winter sets in. This risk is justified. If he can force the enemy forces to capitulate before the army reaches its culminating point, casualties will be lower. If he miscalculates and the weather turns harsh, the soldier will pay the price. It is incumbent on the company leadership to anticipate such events and have a plan of action. Solutions to problems often lie at the lowest level of command.

Because the leader prepares the soldier both physically and spiritually for combat, he wastes no time with bravado or theatrics. The leader is a quiet professional, who is not interested in increasing his political standing with superiors. He takes every opportunity to talk to his men quietly to reassure them and provide confidence. This does not require long speeches or pep rallies. In a coaching role, he reminds them of tactics, techniques, and procedures (TTPs), new developments concerning the enemy forces, and what to expect during contact.

During an engagement, the leader is concerned with maintaining the soldiers' confidence. He does this by speaking calmly and steadfastly. He gives orders slowly and clearly so that soldiers—who are pretty distracted by the immediate danger—understand what is required for the moment. Aware that soldiers are monitoring his behavior at all times, he is never pessimistic, or too cheerful, or lackadaisical. Soldiers will react to these characteristics in a negative manner.[2]

He does not try to inspire his men by needlessly exposing himself to fire to demonstrate his courage—a sophomoric approach to leadership. Of course, new leaders must demonstrate that they do not shirk danger, and so they must undergo a baptism of fire. Thereafter, the leader's courage is really taken for granted—his real worth to his soldiers is setting and maintaining the conditions for success. If his sole worth is always to be out front as a form of motivation, then his contribution to the unit is severely limited. His avoidable loss to the unit is actually a betrayal to the unit—leaders who constantly expose themselves cause great anxiety to the soldiers, who begin to worry more about their leader than with the task at hand. The leader must strike a balance between danger and doing his duty to his soldiers.[3]

Knowing that nonbattle casualties can represent the majority of losses to the unit if left unchecked, the leader studies the factors that can adversely affect the soldiers' health and remains vigilant to the danger signs. He attempts to minimize his soldiers' exposure to detrimental weather or incorporates a recovery period into the plan. He understands the draining effect of constant danger on the mental health of soldiers and works out strategies to allow the soldiers to recover. Finally, he understands that various forces act on soldiers that degrade their performance; a pronounced deterioration of one physical or mental element will likely lead to a cascading effect on the other elements, causing a temporary or permanent collapse of the soldiers' performance.

Although mental and physical preparation in peacetime is the best preventative medicine for combat stress, leaders must not delude themselves into believing that long and unnecessary hardship on soldiers will make them stronger. Infantrymen endure daily misery during training and combat already. Leaders make training tough and challenging, but exposing soldiers to nonsensical discomfort will drain them—train hard, but allow recovery. Like a good coach, the leader constantly looks for opportunities to help keep his men in peak form.[4]

Physical Training. Physical health begins with a good physical training (PT) program. The company conducts physical training every workday morning, with strength development on Monday, Wednesday, and Friday and aerobic development on Tuesday and Thursday. Sports are not a substitution for morning PT. The commander should schedule sports once a week because of the sense of competition and teamwork it instills, but for physical development and maintenance, it is no substitute. The commander uses master fitness trainers to help develop the program and ensure that exercises are performed correctly in order to prevent injuries. Each session is preceded by 15 minutes of warm-up exercises and followed by 15 minutes of cool-down exercises to improve flexibility and prevent injuries. Increased strength and endurance are the goals.

The primary benefit of physical conditioning is the increased resistance against fatigue, because fatigue exacerbates the effects of fear. Besides increasing the soldier's stamina, strength, and confidence, physical conditioning allows the leadership to see what is possible physically. Driving men beyond their level of physical conditioning creates conditions for insubordination and surliness. The leader, exerting his will with unrealistic expectations, destroys the soldiers' confidence in their judgment, which leads to selective compliance to his orders.[5]

Good physical conditioning also allows the body to function better in hot and cold conditions. Good circulation and body tone helps the soldier maintain a constant body temperature. The soldier becomes acclimatized more quickly and hence immediately available for missions upon entry into a conflict area. For an example of a good weekly PT program, see table 22-1.

TABLE 22.1 WEEKLY PT SCHEDULE

Monday PT			
Type Exercise	**Set 1**	**Set 2**	**Set 3**
Wide-arm push-up (4 ct)	10	9	8
Regular sit-up (4 ct)	10	9	8
Regular push-up (4 ct)	10	9	8
Rocky sit-up (4 ct)	10	9	8
Atomic push-up (4 ct)	10	9	8
Abdominal crunches (4 ct)	10	9	8
Rest 45 sec	45 sec	—	
30-minute run			
Tuesday PT			
40-minute cross-country run			
Wednesday PT[6]			
Type Exercise	**Set 1**	**Set 2**	**Set 3**
Push-up	30	25	20
Mountain climbers	25	20	20
Bottoms-up	25	20	20

(continued)

TABLE 22.1 *(continued)*

Push-up	10	10	10
Body twist	10	10	10
Flutter kick	100	100	70
Scissors kick	30	30	30
Flutter kick	30	30	30
Rest 45 sec	45 sec	—	
Sit-up	—	—	25

30-minute run

Note: all exercises are 4-count at a moderate cadence.

Thursday PT

Once per month: 6-, 8-, or 12-Mile Foot March

Interval run:
Set 1—440-meter sprint, 440-meter jog, 440-meter sprint, 440-meter jog, 880-meter sprint, 880-meter jog.
Set 2—880-meter sprint, 880-meter jog, 440-meter sprint, 440-meter jog, 440-meter sprint, 440-meter jog.

Friday PT

Type Exercise	Set 1	Set 2	Set 3
Wide-arm push-up	60 sec	45 sec	30 sec
Regular sit-up	60 sec	45 sec	30 sec
Regular push-up	60 sec	45 sec	30 sec
Mule kick	60 sec	45 sec	30 sec
Atomic push-up	60 sec	45 sec	30 sec
Bicycle kick	60 sec	45 sec	30 sec
Rest	45 sec	45 sec	—

30-minute run

Unit PT in combat is normally possible only in rest areas. However, the commander can direct that soldiers perform 50 push-ups just before stand-to and 50 sit-ups afterward. Push-ups are excellent for waking up the soldier and help fight off the morning shivers.

Hygiene. Disease and sickness can temporarily deplete a significant portion of company strength, particularly during the first months of deployment. Immunizations will help combat common diseases during the initial period, but the unit must maintain its health for sustained periods. Personal hygiene is degraded in combat, so the leadership must establish practical counters to disease and illness. The first step is to establish a latrine area in each platoon or separate entity. In training, a unit gets away with not establishing latrines, because it remains in a particular area only for a short time or has portable latrines provided. In combat, a unit may remain for weeks or months in the same place, and although it may rotate in and out of the area, it inherits the hygienic practices of the former occupants and passes them off to successors (what goes around comes around). The latrine is located within the perimeter, away from water sources and feeding areas. The field sanitation team checks each latrine trench for correct dimensions. Each leader checks the outer edge of the perimeter for feces and urine.

> **TTP**
>
> The entrenching tool can be used as a seat for defecation. Fold it at a 90-degree angle, orient the E-tool with the handle resting on and perpendicular to the ground, and rest one buttock cheek on the flat blade.

Daily washing helps prevent infections and disease. Soldiers should carry a sponge or wet napkins in a zip-lock bag to wash their hands and face each morning. Body sponges are very useful in that they are compact, make washing easier with little water, and dry quickly. Washing keeps pores open, allowing the skin to breathe properly, which is important for warding off the effects of heat and cold. Moreover, sponging helps waken soldiers during stand-to and takes minimal time to accomplish. The hygiene team NCOIC also ensures that a wash container is available at the ration distribution point and ensures that each soldier washes his hands.

Sleep. No one can function properly on little sleep. Historians have caused a great disservice by lauding Napoleon's ability to function on little sleep. The truth is that his sleeping and eating habits ruined his health, which in turn affected his judgment and forced him from the battlefield during some critical moments.[7] It is almost comical to see leaders sustain themselves with only a couple hours of sleep per night during field problems. Usually, by day three, they have become incoherent, are unable to focus and make good, rapid decisions, and need 12 or more hours of sleep before they can function again. Leaders need to understand the effects of sleep deprivation and find methods of combating it.

After 72 hours of no sleep, task performance drops to 50 percent. Moreover, the lowest period of performance for the average person is between 0300 and 0600 h—the perfect time for an attack.[8] Put these two factors together, and the commander is inviting disaster. There are too many stories of the NTC opposing forces (OPFOR) driving right through defenses weakened because of sleep deprivation. A sleep plan, in which everyone averages 6 to 8 hours of sleep per 24-hour period over a sustained period, is the ideal goal. Leaders should try to produce a schedule that establishes a sleep period for each soldier—meaning that each soldier sleeps at a prescribed time of the day.[9] Infantrymen, particularly in combat, rarely have this luxury—sleep being perhaps one of the most treasured necessities. Soldiers can maintain performance for one to two weeks with 4 hours sleep per 24-hour period (5 hours, if sleep has been interrupted) before they need a 12-hour sleep recuperation. Since shifts such as 8 hours on/8 hours off interrupt sleep cycles, it is actually more productive to work for 20 hours with 4 hours rest—especially if the rest period is between 0200 and 0600. If during continual combat sleep deprivation occurs, leaders schedule 15-minute power naps for soldiers during a lull. Note, however, that in such a condition, soldiers may take 5 to 15 minutes to become fully cognizant upon waking.[10]

Leaders especially need their sleep, because good judgment depends on proper rest. During periods of intense activity, leaders can take 15 to 30 minute power naps to keep functioning for short periods. Eventually, leaders will need a 12-hour sleep recuperation.[11] With the proper delegation of tasks and trust in junior subordinates, the leader should have no trouble getting sleep.[12]

For subordinates, lack of sleep rapidly leads to fatigue, which can trigger the fear-fatigue cycle. Moreover, tired soldiers are less security-conscious, exposing the company to a surprise attack. Realistically, soldiers must endure fatigue in combat—it has been a constant theme in recorded warfare. The leader must attack this scourge of the soldier by enforcing a sleep plan and looking for signs of undue fatigue in the soldiers.

Manpower—Quality versus Quantity. High-quality combat soldiers are required to succeed on the battlefield and endure its conditions.[13] Since the modern battlefield creates an environment of isolation as a result of a need for greater dispersion, the emphasis must be on high morale and training standards. A small force of well-trained soldiers is more effective than a large force of reluctant or demoralized soldiers.[14] A 100-man company of reliable soldiers is more effective than a 120-man company with only 100 reliable soldiers. The 20 misfits lower morale, absorb the leaders' time with their various problems, and endanger the company with their lack of discipline, sloppy work, and lackadaisical attitude. Invariably, they will fall asleep on sentry duty, avoid danger at the expense of their comrades (sick call, unnecessary chaplain visits, and chronic mental hygiene visits), and generally shirk their duties. The Army is no longer in the business of disciplining the miscreants of society. It is the commander's duty in peacetime to discharge soldiers who cannot or will not perform. Those who pass off their problem soldiers to other units perform a disservice to the Army as a whole and create a permissive environment of institutionalized mediocrity. The hitherto poor soldier who performs well and bravely in combat is the exception. The probability of such a soldier coming around is so low that it is not worth the leadership's time to put up with him. Rather than try to bring poor soldiers up to the minimum standards, it is better to devote more time and resources to raising the standards of the good soldier. This is how units become elite.[15]

Body Hydration. The average person is composed of about 11 gallons of water. He will die if he loses about 8.8 quarts of this water without replenishment.[16] Even a loss of two quarts will result in dehydration, which lowers the soldier's stamina, alertness, and strength.[17] Such a loss can also lead to internal organ damage. At moderate temperatures (68 degrees Farenheit), the average adult loses two to three quarts of water daily. Naturally, the amount of loss beyond the above minimum depends on the intensity of activity and extreme climatic conditions. Soldiers require about 8 quarts of water per day (16 to 32 in hot climates).[18] When performing work details in hot climates, they will need about two quarts per hour, because they lose one and a half to two and a half quarts of sweat per hour.[19]

Before a detail begins, the NCOIC force hydrates each soldier by having him drink one quart of water. To maintain fluid levels, soldiers must consume water in small drinks regularly, even when not thirsty. If they only drink when thirsty, they are already dehydrated and subject to a heat injury. Leaders must monitor this continuously by questioning their soldiers and monitoring the water level of canteens and company resupply sources (water buffalo, water blivets, water cans). If water is plentiful, the first sergeant (1SG) could ensure that proper intake is maintained by having soldiers drink a canteen of water while in the chow line (or directing that a canteen be consumed at each meal as a matter of habit). The problem is that water is not always available and must be rationed. One method of maximizing water usage is to instruct soldiers to take a mouthful of water, tilt the head back, and allow the water to trickle down the throat. This allows the body time to absorb water (reducing loss through urine) and slackens thirst faster than normal drinking. Water discipline requires only two mouthfuls at a time. The leadership reminds soldiers that clear urine is a good sign of proper hydration and darker urine an indication of dehydration. In between consumption breaks, soldiers can slake their thirst by inducing saliva. The most popular methods are a pebble under the tongue or chewing on a sliver of leather. As an added precaution, the commander should emphasize that medics carry extra IV solution bags during missions, allotting one IV bag per soldier.

Water purification is another matter of concern, because nonpotable water can take an entire unit out in less than 24 hours and keep it out for at least 48 hours. Leaders must check each soldier for water purification pills and order more through the medics when the bottles are less than half full. The commander must emphasize this requirement to the company medic to ensure this item is always available. In emergencies, the company should boil water rather than hoping that a water source is potable.

In hot weather, soldiers should expose as little skin as possible to hinder perspiration. Keeping clothing loose, sleeves rolled down, a bandanna around the neck, and a helmet or cap help keep the body cool and limit the amount of perspiration evaporated. The webbing inside the helmet should allow sufficient air space between the head and top of the helmet to create a cooling pocket of air. To create a similar pocket of air in the BDU cap, the soldier can insert a balled-up handkerchief. When possible, all work should be performed at night.[20]

Leaders must be vigilant for heat injuries and take immediate measures when a soldier shows signs of heat cramps, exhaustion, or stroke. A soldier who has suffered a heat injury is forever susceptible and has less stamina.

To increase water rationing, soldiers should maintain a small sponge as part of their kit for sopping up water from leaves, grass, or shallow water sources during dry spells. In all cases, water must be purified with tablets or chlorination. The field sanitation team manages water purification.

To maximize the productivity of soldiers without exhausting them, the commander should follow the following work-rest guidelines:

Heat Condition Black—20 minutes work, 40 minutes rest—no more than 5 hours of work per day.

Heat Condition Red—30 minutes work, 30 minutes rest—no more than 6 hours of work per day.

Heat Condition Yellow—40 minutes work, 20 minutes rest—no more than 7 hours of work per day.

Heat Condition Green—50 minutes work, 10 minutes rest—no more than 8 hours of work per day.[21]

Obviously, these are ideal work conditions, which the leadership will probably need to violate, depending on the urgency of the tactical situation. This is all a part of the risk assessment that the leader considers during a mission. It is not criminal to violate the guidelines, but it should be a conscious decision. The commander should figure this into his priority of work in the defense and share it with the battalion commander so he is aware of the revised time schedule.

Cold. Freezing temperatures incapacitate soldiers quickly, and frostbite casualties will decimate a unit in a matter of hours.[22] The effect is not so bad for soldiers on the move, but if they are stationary, protective measures are crucial. Uppermost in the commander's mind must be the availability of shelter for his soldiers. The survival of the unit could be at stake, and the commander should never avoid built-up areas, when available, because they offer more advantages than not. They not only offer immediate warmth, but they also allow the unit to heat food and thaw ice or snow for drinking water. Shelter also protects wounded personnel, who otherwise would quickly die from exposure or gangrene. More important, soldiers can take care of personal hygiene, which thwarts the spread of lice and hence typhus.[23] Prolonged exposure to cold makes soldiers listless and unable to think clearly. After 48 hours of combat in extreme cold, soldiers will lose about 50 percent of their efficiency.[24] The degree of cold temperatures determines the length of sentry duty on the defense. In extreme cases, the length of duty may be only 15 to 30 minutes. In general, sentry duty in cold weather should not extend past 60 minutes.[25] Thereafter, the soldier's focus is on maintaining warmth and not on vigilance.

The key to thwarting cold weather injuries is to cover exposed skin, to layer clothing in order to create pockets of warm air between the body and the outer clothing, and to wear garments that allow body moisture to escape. The simple use of a scarf to cover the face and ears plus the wearing of socks will reduce the chances of frostbite.[26] Underclothing such as polypropylene draws moisture away from the body, thereby helping to keep the body dry. If polypropylene is not available, a field jacket liner underneath the BDUs will suffice. Gore-tex is excellent for an outer garment, because it allows body moisture to escape and protects the body from wind. Newspaper makes a very good windbreaker. A field expedient measure for cutting wind and gaining insulation is to wrap newspaper around the legs, arms, and feet and padding the outer garment with crumpled newspaper or straw.[27] Note that newspaper is only effective for stationary work; if it becomes wet with sweat it loses its insulating quantities. Soldiers must also be vigilant to keep their outer garments dry.[28] Wool socks or a commercial product like Thinsulate and Gore-tex keep feet dry. Leaders must be vigilant against the wearing of cotton socks, because they retain moisture. Soldiers need to change socks daily and should wear Gore-tex boots or overshoes over leather boots to prevent shoes from getting soaked. If overshoes are not available, boots wrapped in burlap sacks will deter trench foot and frostbite.[29] The soldiers should wear headgear even while sleeping, because 80 percent of his body heat escapes through the head. A few drops of liquor in a canteen will prevent water from freezing.[30] Leaders must guard against soldiers overdressing during physical activity (foot marches, digging, and work details). Soldiers will be less miserable if they conduct physical activity with few garments on and then layer more on afterward.

TTP

During the Civil War, the World Wars, and Korea, socks were a constant problem for infantry soldiers; because of supply priorities, the infantry couldn't get them with any regularity, if at all. To ensure a good supply of wool socks, soldiers should have their families send a replacement pair on demand. Normally, the mail system operates better than the supply system in such cases.

A proper mattress pad for the sleeping area is also essential for proper rest. The body must have insulation from ground cold—the thicker the warmer. If the issued pad is not available, the soldier can make a tick mattress with his ponchos stuffed with leaves, hay, grass, and so forth. Belly bands (polypropylene or wool) protect the kidneys from the cold at night and add another layer of protection for the soldier.[31]

In defensive positions, each squad and platoon bunker should have a stove for heat. The danger of attracting enemy fire does not offset the immediate and assured danger of a cold weather injury. The leadership supervises measures to reduce light and smoke signatures rather than resorting to idiotic directives.

Keeping Dry. Moisture reduces body heat, causes injuries (trench foot, immersion foot, chaffing, and rashes), and acts as a vehicle for diseases. As mentioned above, soldiers should wear any materials (polypropylene, Thinsulate, Gore-tex) that pull moisture away from the body and "breathe" to reduce moisture buildup. After long movements or periods of strenuous activity, soldiers should towel off and change T-shirts. During an attack, drying off is a prudent measure while in the objective rally point (ORP), because soldiers will be stationary for a few hours awaiting H-hour.

During rainy weather, soldiers should wear Gore-tex garments and change socks daily, placing the wet socks next to the body to allow their body heat to dry them off.

To prevent trench foot, soldiers need to remove their boots and socks at least once a day and massage the feet vigorously for five minutes. The use of buddy teams in the past has proven effective.[32] The commander needs to rotate squads daily to warm shelter (tent with heater, bunker, or building) for them to dry off and consume hot food and drinks.[33] Use of ponchos as shelters (low to the ground) with an air or tick mattress as well as a ground sheet will help soldiers keep dry.

In tropical climates, soldiers can wear an OD green towel around the neck and shoulders for padding and wiping of the brow as a counter to perspiration. Sharing heavier loads (machine guns) during movements will also reduce the amount of perspiration per man.

Talcum powder is particularly effective in reducing body moisture and should be applied at least daily. The soldier should take care to wash away old powder before applying more, because a buildup of old powder can clog pores and cake clothing.

Insect Bite Prevention. Mosquito, horsefly, chigger, and tick bites are the prevalent miseries for the soldier. Basic preventive measures are to avoid scented soap or shampoo, deodorant, and aftershave or cologne. After three days in the field without thorough washing, natural body oils and odor help dissuade insect attraction. Using Permethrin to treat a uniform is an effective insect repellent. Reapplication after ten washings is recommended for continued protection. The U.S. Army Center for Health Promotion and Preventive Medicine (http://chppm-www.apgea.army.mil/) and the U.S. Army Medical Zoology Branch (http://www.cs.amedd.army.mil/dphs/MedZoo/Homepage.htm) websites are excellent resources for researching health issues and vectors.[34]

Mosquitoes are attracted by the breath of its host and follow the carbon dioxide stream to the mouth. The soldier can deter being swarmed by placing a bandanna over his mouth and covering exposed areas of the skin with loosefitting clothing or a mosquito net. For more protection, insect repellent with DEET is excellent.[35] Mosquito head nets are also good for protecting the face and neck without obscuring one's vision. If those are not available, the soldier can hang numerous thin strips of cloth (or string with corks hanging from the end) from the helmet camouflage band or the brim of a hat as an excellent field expedient protection for the face and neck from most flying insects. It may look silly, but it is better to have the soldiers focusing on the surrounding area rather than constantly swatting at buzzing insects. Covering exposed skin with mud, oil, and fat provides limited protection.[36]

Chiggers and ticks are attracted to the moist and hot body regions, where clothing and boots are close fitting. Insect repellent applied around the ankles, waist, and wrists is effective temporarily. A sulfur mixture (one-third sulfur powder and two-thirds body powder) kept in a sock and applied to the body is also effective. Lastly, multi-B vitamin complex tablets (50 mg tablets twice a day) contain riboflavin, which emits a body odor that repels most insects. To ease itching, sprinkling monosodium glutamate (MSG) on the bite breaks down the irritant compounds.[37]

Morale. So many factors affect the morale of a unit that volumes can and have been written on the subject. The leader focuses his study and thoughts on the practical aspects of morale—which factors will cause the unit to fall apart and which will maintain cohesion. The defining moment of any unit is when disaster strikes. Morale is always high when riding the wave of success. It is only when confronted with failure that the true fiber of the unit is tested. This is the moment the leader prepares himself and the unit for.

Difficult, stressful, and challenging training is the foundation for building cohesion. It is the best way to tell the soldier that the leadership cares for his welfare.

The soldier will complain on the outside. He will grouse to his friends at the apparent nonsensical activities; why should a unit do a 25-mile foot march in the age of mechanization? But inside, he will feel a sense of pride in his accomplishments. He will become inured to hardship and fatigue, as well as seeing the connection between standards and duty. Finally, such training identifies soldiers weak in character. If he refuses or cannot mature, he is not worth keeping—chapter him.

Soldiers need a clear moral compass in the leader. Upon taking charge, the leader must clearly and directly tell all his subordinates what he believes are unethical and illegal actions—he cannot omit even the most obvious repugnant activities. Moreover, he must tell them what actions he will take against violators; there must be no ambiguity. Certain soldiers will test the leader to see what the parameters are—what they can get away with. The leader should quell any talk that rationalizes atrocities or hypothetical unethical behavior. The U.S. soldier is not incapable of committing atrocities.[38] Too much evidence from World War II and Vietnam reveals that both officers and enlisted soldiers committed atrocities and other immoral acts, including execution of prisoners, looting, rape, and wanton destruction of civilian property. It is remarkable how easy it is for a unit to commit such acts, and once begun, such an attitude can become commonplace, even mundane. As is often the case, most atrocities originate out of a sense of mounting frustration, dejection, fear, hatred, and outrage. The enemy forces may have "broken the rules," resorting to perceived dirty tricks, or mutilating and executing our soldiers or innocent civilians. Fear, dejection, and hatred can cause a group of soldiers to suddenly snap and cut down prisoners as a release. It takes a firm leader not to allow the line to be crossed. Each leader reads the pulse of his subordinates and reminds them that the acts of the enemy do not justify following suit. The leader must tell them unequivocally that he will arrest everyone involved in an atrocity even if it affects the mission. It is better to address this problem preemptively rather than suffer the loss of a soldier when manpower concerns will be acute.[39]

The leader keeps his soldiers informed of everything. Danger coexists with the infantryman in every operation. The command must not withhold the true nature of the situation, no matter how dire, simply because such knowledge might lower morale. If the leadership is strong, it should not have to worry about outside effects on morale. A unit with strong cohesion is inoculated from failure and danger. Fear resulting from the unknown is more dangerous than knowledge.

The operational or fragmentary order (OPORD/FRAGO) is an excellent vehicle for disseminating information, but it is not enough. The commander must seek out the company of the soldiers as much as possible. He must take an interest in the situation at hand. The commander should try to hold "the rumor of the day" meetings with his platoons and sections. Rumors always surround a unit and are held in such veracity that they become a reality.[40] Bringing rumors out in an open forum is beneficial because it allows the commander to address and quash (or verify) them quickly. He also strongly requests that higher commanders visit the soldiers to help allay their anxieties, disseminate news, quell rumors, and allow soldiers to voice their concerns. These visits also apprise higher command of frontline conditions and provide it with an appreciation of the tactical difficulties and opportunities. By his position, he can help the front line succeed. When soldiers realize that higher command is taking an interest in solving a problem at the front, they will endure inadequate rations and shelter more readily.[41]

The commander always welcomes new soldiers personally to assuage their fears. The commander tells them that they are fortunate to be a member of the best company in the U.S. Army, and that the skills they will hone in training will keep them alive.

The commander makes a moral contract with his soldiers regarding the recovery of dead and wounded. The company policy is to recover and evacuate all soldiers from the battlefield regardless of their status. Soldiers need to be confident that the unit will not abandon them, alive or dead. This may seem a small matter, particularly for the dead. But it is the small things that count in combat. Leaders are not to become embroiled in hypothetical situations in which abandonment becomes an issue. The policy is to recover everyone, period.

In a related matter, commanders should allow the 1SG to deal with casualty evacuation. Commanders who are focused too much on casualty evacuation and not on winning the battle first will likely prolong the battle, cause valuable resources to be diverted, and contribute to even higher casualties.[42]

Leaders must understand and exploit those factors that keep soldiers on the battlefield. Cohesion results from shared hardships, established friendships, and military discipline. Some soldiers actually enjoy battle—both the danger and the cruelty. Although these types are effective, leaders must watch them to ensure they remain under control. Younger soldiers (under 30 years old) are resilient, and endure discomfort, loss of sleep, hunger, thirst, and physical exertion more than older soldiers. They also seem to handle terror, anxiety, separations, and bereavements better. War also brings out the sense of adventure. Comradeship, the thrill of the hunt, the excitement of a successful operation, and the fun of irresponsibility compensate for the less-appealing aspects of war. Some soldiers remain out of a sense of spirituality, both religious and addictive. Mental preparations before a battle (military rituals, proclamations, and moments of recollections) strengthen the soldier's resolve.[43] Above all else, the leader cultivates the relationship of soldiers with their closest associates, because camaraderie animates and sustains the soldier.[44]

Lastly, the commander should encourage humor. It is an enduring feature of the American soldier. Attempts to quash humor definitely will lower morale. Even if sarcastic, it helps to soothe the soldier's misery.

Battle Effects. Leaders discuss the effects battle has on the soldiers in order to help them understand and mentally prepare for them. All leaders must accept and impart to their soldiers that friction and fear are a natural part of battle. To help them prepare for battlefield stress, leaders remind soldiers that the unexpected is normal, that war is an emotional event, and that it brings unrelenting mental strain. If soldiers can foresee what the environment is like, they more likely will withstand the initial shock.[45] They can develop contingencies during the planning process and address them during the rehearsal. Additionally, leaders must explain that chaos, clutter, and apparent inaction are normal on the battlefield and are not a sign of defeat. With this knowledge, the soldier can look for signs of organized progress and seek opportunities to exploit success.[46]

Leaders must understand that fatigue and fear are closely linked. They feed off each other until the soldier is consumed with panic and exhaustion.[47] Soldiers forced to remain immobile under heavy shelling become exhausted to the point of enfeeblement, because adrenaline courses through the body with no opportunity of release through activity.[48]

Soldiers must understand the effect that the "empty battlefield" has on their psyches.[49] Once a soldier perceives that he is alone during an engagement, he is susceptible to the irrational terror of the unknown. Fear and incoming fire compound this terror. The one factor that keeps a soldier on the battlefield is the knowledge that his friends are nearby.[50]

After a battle, leaders and soldiers need the opportunity to vent their emotions through talking sessions, particularly if hit hard. The commander does not take this time for critical analysis. Talking is the best way for a shaken unit to recover its morale.[51] Leaders must remind soldiers that morale fluctuates in combat as a result of various stimuli. During these discussions, leaders must also be attuned to the soldiers who perform during hardship and danger. Courage within an individual soldier may remain a cipher until a crisis arises.[52] Some rise to the occasion, while others do not. The performers must be selected for more responsibility.

Leaders need to note the signs of combat exhaustion and take action. Some distinctive symptoms are dejection, filth, and weariness (of course, for the infantry, this is "business as usual"). The soldier may be depressed to the point of tearfulness, have trembling hands, or stare mutely into space. A veteran may suffer from "old sergeant syndrome," in which he is constantly irritable, apathetic, and lackadaisical about mission accomplishment. Treatment requires expedient evacuation to the battalion aid station for a few hours rest, some hot food, and a chance to clean up. A delay in immediate treatment will result in either a longer absence for recovery or the complete loss of the soldier to the unit. In continuous combat, rotating units out of the line every two weeks probably is the best means of maintaining peak performance. Rotation after 30 days is critical to simply maintaining combat effectiveness. Thereafter, the decline is so precipitous that mission accomplishment becomes problematic.[53] This guideline is more applicable in positional warfare and can be suspended with little risk during a victorious advance, because combat is infrequent and the unit has more occasions to rest, albeit after long movements.

Leaders must understand that a soldier continuously exposed to intense combat will probably lose his willpower and courage. Exposure to intense danger inevitably drains the reserves of courage. Once a man reaches his limit, he will no longer be able to function properly. This is not cowardice; it is a fact of combat.[54] Leaders need to recognize when a man is used up, as opposed to exhausted from combat, and have him reassigned to different duties.

On a related matter, leaders need to have a session with soldiers who have witnessed the aftermath of atrocities. Gruesome and incomprehensible acts against people can become indelibly etched into the mind of a soldier and his reaction may not manifest itself until later—usually detrimentally. Leaders need to have soldiers vent this experience rather than hold it in. Such a session will also alert the leadership of soldiers who may require more extensive and professional help in coming to terms with the experience.

It is the commander's duty to keep the battalion commander apprised of the combat effectiveness of his soldiers. Since a few days of continuous combat cause a decline in performance, the commander must not be afraid to advise the battalion that a short break will help the unit regain its peak.[55] In positional warfare (entrenchments), higher command will likely adopt a rotation policy for units on the front line. If not, the commander needs to alert higher command of this necessity.[56]

In conclusion, the commander must protect the soldier physically and spiritually. The company must be prepared to sustain itself in a long conflict. If such precautions are not taken, it will suffer such a turnover rate from attrition that its combat effectiveness will be compromised. Once the company loses its trained veterans, the unit will never recover; henceforth it will take longer to accomplish its missions, find success more fleeting, and suffer higher casualties.

NOTES

1. Marshall, 175–78.
2. Doubler, 238.
3. Ibid., 23, 238. During World War II, the commander needed to select a point on the battlefield from which to best observe the terrain and enemy. "The best frontline commanders were aggressive and took chances but conscientiously avoided unneeded risks to themselves and their soldiers. 'Dead heroes are of little further use to their units' and 'bravery must be supplemented by brainwork' were constant themes in leadership training."
4. Ibid., 26. During World War II, leaders learned that "subjecting soldiers to avoidable misery sapped the emotional and physical strength needed for combat." "Most soldiers felt that field exercises with tough, realistic battle conditions, regular and rigorous physical fitness training, running obstacle courses, and long field marches of 25 miles with full field equipment best prepared them for battle," 249.
5. Marshall, *Men against Fire,* 173–75.
6. Colonel Linwood Burney introduced this set of exercises to 2d Brigade, 7th ID (L) in June 1988. These exercises are tough but effective, increasing the strength and stamina of the soldier in preparation for combat.
7. Frederick F. Cartwright, *Disease and History* (New York: Dorset Press, 1991), 103–10. Napoleon's poor eating habits also contributed to his incapacitating hemorrhoid problems.
8. U.S. Department of the Army, *Soldier Performance in Continuous Operations,* Field Manual No. 22-9 (Washington, D.C.: Fort Benjamin Harrison, December 1991), 2-7. See also, U.S. Department of the Army, *Combat Stress,* Field Manual No. FM 6-22.5, MCRP 6-11C, NTTP 1-15M (Washington DC: U.S. Government Printing Office, 23 June 2000).
9. Ibid., 3-1–3-3. The sleep routine follows the "circadian rhythm," which describes the work-rest cycle in a 24-hour period.
10. Ibid., 3-11, 3-17, 4-18.
11. Ibid., 3-21.

12. Lucas, 79. The need for sleep competes with the constant vigil of security. German soldiers could only snatch a few hours of sleep between patrols and sentry duty. The lack of adequate manpower on the front line certainly contributed to this problem.

13. Griffith, *Forward into Battle,* 176.

14. Ibid., 101, 103. In an observation given by Griffith of World War I, the creation of the "empty battlefield" requires a greater reliance on the duty and discipline of the individual, because command and control from above is attenuated.

15. Lord Moran, *The Anatomy of Courage* (Garden City Park, N.Y.: Avery Publishing Group, Inc., 1987), 151–61. Lord Moran, who served as a battalion medical officer in World War I, contends that the malcontents were a burden that dragged the rest of the unit down.

16. John Wiseman, *The SAS Survival Handbook* (London: Harvill, 1991), 227.

17. U.S. Department of the Army, *Desert Operations,* Field Manual No. 90-3 (Washington, D.C.: U.S. Government Printing Office, 24 August 1993), G-3.

18. U.S. Department of the Army, *Survival,* Field Manual No. 21-76 (Washington, D.C.: U.S. Government Printing Office, March 1986), 3-1–3-2.

19. FM 90-3, G-3.

20. Ibid., G-7.

21. Ibid., G-3.

22. Doubler, 241. From November 1944 to March 1945, the European Theater of Operation First and Third Armies suffered 32,163 losses to cold weather casualties compared to 115,516 battle casualties, representing about 22 percent of the total. Albert Seaton, *The Russo-German War: 1941–45* (Novato, Calif.: Presidio Press, 1990), 228. During the first winter in Russia (1 November 1941–1 April 1942), the Germans suffered 228,000 frostbite cases out of a total of 900,000 losses, a representation of about 25 percent. Wray, 195; English, 118. Some divisions suffered 40 percent casualties as a result of frostbite.

23. Wray, 74–75. Major Wray contends that the severe winter forced the Germans into establishing strong points at villages in contravention of established defensive doctrine. The immediacy of the emergency warranted such measures, but the establishment of a defensive line based on village strongpoints was not operationally sound. The solution lies in establishing strongpoints with heated bunkers on tactically sound terrain.

24. Doubler, 200, 222. Lessons learned from winter combat during the German Ardennes offensive.

25. Lucas, 80. This became a common German procedure on the eastern front during frigid conditions. Wray, 195. The Germans placed sentries in pairs or within sight of each other as an added precaution against cold weather injuries.

26. Lucas, 80. Many frostbite cases on the eastern front resulted when soldiers were forced to wear boots without socks. Layering was crucial to maintaining warmth. The cold-weather cap was particularly prized. Goggles were discarded, because they fogged up and froze to the skin.

27. Wray, 194.

28. Lucas, 81. On the eastern front, German soldiers often lined their boots with straw or paper.

29. Doubler, 222.

30. Ibid., 223.

31. Alfred Toppe, *Desert Warfare: German Experiences in World War II,* Combat Studies Institute, Research, Reprint, 1952 (Fort Leavenworth, Kans.: Command and General Staff

College, August 1991), 82. While on tour of the Bundeswehr Mountain Warfare museum, I learned that the German mountain troops still wear bellybands as protection against cold in the mountains.

32. Eisenhower, 316. General Eisenhower noted that trench foot injuries caused most of the nonbattle casualties during the fall of 1944 as a result of the worsening weather. The solution of massaging the feet eliminated this problem. Ellis, 48–49. During World War I, soldiers who wore wet boots and socks without changing over a 24-hour period contracted trench foot. The feet would grow numb, turn red or blue, and would likely turn to gangrene if not treated. Having suffered 74,711 cases of trench foot and frostbite during the war, the British found it necessary to have officers conduct regular inspections to ensure that soldiers changed their socks and dried their feet, because the pace of activity and resultant apathy from the frontline conditions made solders forget. Soldiers were required to carry up to three pairs of socks and required to change them one or two times daily. Use of a whale oil-based grease or a mixture of borated talcum powder and camphor helped combat these ailments. Patton, 322.

33. Doubler, 242.

34. My special thanks to LTC William B. Miller, Ph.D. Branch Chief, Medical Zoology, AMEDD C&S, for pointing out incorrect information in the first edition of *Command Legacy*. Notably, the use of flea collars is hazardous to one's health. Because sulfur tablets, garlic, and vitamin B12 are not proven to work, it is inadvisable to ingest them to ward off blood-sucking insects. The following manual is an excellent resource as well: U.S. Department of the Army, *Field Hygiene and Sanitation*, Field Manual No. 21-10, MCRP 4-11.1D (Washington D.C.: U.S. Government Printing Office, 21 June 2000.

35. Boga, 95.

36. Wiseman, 53.

37. Andrew Rubman, MD, "Fighting Bug Bites . . . Naturally," *Bottom Line* 21 (13 July 2000): 14.

38. John Keegan, *The Face of Battle* (New York: The Viking Press, 1976), 322. Keegan surmises that the soldier's small sense of value on the vast and dominating battlefield in World Wars I and II created conditions for cruel behavior. Since the soldier's own life was perceived as unimportant, atrocities were more likely to occur.

39. Paul Fussell, *Wartime: Understanding and Behavior in the Second World War* (New York: Oxford University Press, 1989), 267–97. Gerald Lindermann, *The World within War: America's Combat Experience in World War II* (Cambridge, Mass.: Harvard University Press, 1997). The entire book is filled with anecdotes and the reasons behind atrocities, as well as other factors that formed the soldier's psyche.

40. Fussell, 35–51. World War II seems to be the war where the art of the rumor reached its peak.

41. Marshall, *Men against Fire,* 103–105.

42. Griffith, 154–55. A tactical conclusion of the Vietnam War was the undue attention leaders paid to casualty evacuation, which diverted uninjured manpower from the fight and gave higher commanders monitoring the situation the impression that the unit was in dire straits when this was not the case.

43. Keegan, 324–27; Ellis, 97–98. World War I British soldiers were motivated by the presence of their friends. The dominating factor was the sense of composure—the fear of contempt from their friends overrode the fear of death.

44. Marshall, *Men against Fire,* 39–43.

45. Adolf von Schell, *Battle Leadership* (Ft. Benning, Ga.: *Benning Herald,* 1933), 10. Marshall, *Men against Fire,* 36–37. Marshall echoes the conclusions of von Schell, adding that uncontrolled fear must be mastered if an operation is to be successful.

46. Marshall, *Men against Fire,* 116, 181–82. It is very important for junior leaders to understand that things will go wrong even with excellent plans. As part of their development, jun-ior leaders should attempt to anticipate likely places for things to go wrong (critical events) and plan branches (modifications). In this manner they become resilient, disciplined thinkers.

47. English, 223.

48. Griffith, 172; Ellis, 64–65. World War I soldiers would often become comatose as a result of artillery bombardments.

49. Keegan, 306. The increasing lethality of small units has created the "empty battlefield," as soldiers rely on entrenchments and night for protection.

50. Griffith, 87, and Marshall, *Men against Fire,* 65.

51. Marshall, *Men against Fire,* 118–19.

52. Ibid., 182–84.

53. FM 22-9, 5-7–5-11. Treating a soldier with combat fatigue immediately may save the permanent loss of a good soldier. Keegan, 328–29, 335. Keegan observes that "The first hours of combat disable 10 percent of the fighting force." Combat exhaustion results from continuous combat exposure. Psychiatric casualties could account for as many as 30 percent of the total casualties of a battle. If handled immediately, 90 percent will return after a brief rest. Most combat line soldiers are ineffective after 180 days of combat. Their peak of effectiveness is 90 days. After 200–240 days, they contribute little. Doubler, 242–45. From 1944 to 1945, psychiatric casualties accounted for one quarter of combat casualties. A First Army report estimated that a soldier could endure only 200 days of combat before breaking down, 243. Lindermann, 356. According to some combat psychologists, the effectiveness of a soldier deteriorated after 30 days in continuous combat, and after 45 days the soldier became vegetative. Total breakdown occurred between 200 and 245 days in continuous combat.

54. Moran, xvi.

55. Doubler, 25–26. "Even the best units were capable of only limited, continuous enemy contact and . . . troops passed their peak efficiency after three days of non-stop combat. . . . Troops who had been in the line for extended periods suddenly became apathetic or emotionally unstable." During World War II, 2,000 cases of combat exhaustion were reported in Italy during the 1942–43 winter, while during the Normandy campaign, upward of 30,000 soldiers suffered from combat exhaustion. 26, 61.

56. Ellis, 28–31. During World War I, the Germans rotated battalions, "four days in line, two in support and four at rest." The British had no standardized procedure; front-line duty ranged from 8 days to 58 days. French units usually spent four days in the trenches and four days at rest. After three or four such rotations, they would spend a few days in reserve.

Chapter 23

Resupply Planning

The one constant in war and training is that the U.S. soldiers at the front line receive the least logistical support of all the branches. They learn to fend for themselves and often do without, because that's just how it is. The infantry is constantly on the move and normally in hard-to-reach places. Infantry soldiers learn not to depend too heavily on higher command to affect their state of affairs, because such matters are not worth its attention and, and it is likely to pawn the problem off as a chain of command problem. So, the infantry takes care of itself.

Logistical support requires proactive planning and coordination. Much frustration can be avoided if the company supply sergeant visits the S4 and coordinates logistical support directly. The biggest issue is the status of the company vehicle—a light infantry company needs a dedicated vehicle for impromptu and scheduled needs. It is impractical and too time-consuming for the S4 to manage the use of company vehicles, even if it violates doctrine.

After the warning order, the supply sergeant coordinates with the first sergeant (1SG) for the resupply work details. The supply sergeant recommends the number of personnel needed per detail and estimates how long the work will take. The 1SG in turn directs through the platoon sergeant or section leader (PSG/SEC LDR) how many and what time personnel from each platoon or section will report to the company trains. When possible, the 1SG uses soldiers with limited profiles, light wounds, and so forth to conserve the fighting power of the unit.

Configuring logistical packages (LOGPAC) greatly reduces the time required to resupply the unit, particularly during consolidation and reorganization. LOGPACs also limit waste by ensuring that supplies are evenly distributed, which is especially important for overhead cover, minefields, and wire obstacles. The LOGPAC is the responsibility of the S4, but the company should not wait until the S4 gets around to it. To get the process moving, the supply sergeant establishes the initial configuration of each LOGPAC with the S4 section, modifying it as a result of more experience and thought. The final package is completed and incorporated into the company SOP before the unit deploys to the field.

The supply sergeant may need to resort to shuttling supplies, including rucksacks and duffel bags, to the company front. Consequently, the 1SG gives him the resupply priorities, such as water, ammunition, rucksacks, and duffle bags. The supply sergeant limits his time spent at the battalion field and combat trains to just preparing and transporting supplies and to issuing reports and supply requests. He spends the majority of time at the company trains. Supply sergeants tend to become ensconced in the field trains and never are available when the company needs them. He lives with the company!

The bottom line is to keep the frontline soldiers as well supplied as possible and in a timely manner because they suffer the greatest deprivations in both training and war. The chain of command above the battalion level needs to focus its efforts on the well-being of combat soldiers rather than on the soldier farther to the rear, because the company is at the end of the food chain. Soldiers in the rear normally find the means to take care of themselves at the expense of the frontline soldier. The staggering amount of supplies that are diverted along the way for special reserve stockpiles is the norm in all military operations, because it is human nature for all levels of support to horde. And if anyone challenges this statement, he is more than welcome to spend time with the infantry.

TABLE 23.1 RESUPPLY PLANNING

Task
1. Coordinate type of resupply planned by battalion S4/support platoon:
a. Push forward: Support platoon automatically forwards a single class of supply (water, ammo, medical, etc.) directly to the company based on past usage and planning estimates.
b. Call forward: Support platoon attaches a vehicle to the company for specific resupply items. Operates between the battalion field trains and the company trains. Supply consists of preconfigured supply packages (LOGPAC and Preconfigured Unit Loads) tailored to the needs/mission of the company.
c. Logistical release point (LRP):
1) Transfer supply package from battalion LRP:
a) Service Line: Company vehicle transfers supply items from each Support Platoon vehicle which is loaded with a single line item.
b) One Stop Shopping: Company vehicle parks along designated vehicle and transfers pre-configured package, or company trades its vehicle for the loaded Support Platoon vehicle.

TABLE 23.1 *(continued)*

2) Distribute supplies to company units by means of company or platoon LRP or Push Forward.

3) After resupply, return to battalion LRP to return empty containers and refuse.

d. Aerial resupply:

1) Type:
- Parachute or sling load: Prepare/secure small DZ.
- Air land: Prepare and secure landing zone.
- Rope lapse: Mark upper tree canopy, using ropes to slide supplies down.

2) Recovery detail moves supplies quickly, sanitizes area and departs.

3) Security detail protects area and helps recovery detail carry supplies to company.

e. Cache: Prepositioned and concealed (preferably buried) supply points. Ideal for stay behind and unconventional movement to contact missions. Diversify and disperse caches no closer than planned patrol bases.

Logistical Package (LOGPAC)

LOGPAC	Type	Size Unit	Package Contents	Remarks
1A	Subsistence	Platoon	7 5-gallon water cans 3 cases MRE	1 DOS
1B	Subsistence	Platoon	36 separate MRE packets 28 5-quart water bags	1 DOS
1C	Water	Platoon	28 5-quart water bags	1 DOS
1D	Subsistence	Company	28 5-gallon water cans 12 cases MRE Sundries	1 DOS
20	Obstacle	N/A	61 72-inch pickets 34 24-inch pickets 2 rolls barbed wire 20 rolls concertina 20 feet banding material 1 48 x 40-inch pallet	100 meters of triple-strand concertina wire
2C	Overhead cover	Platoon	150 72-inch pickets 800 sandbags (8 bundles)	15 positions: 10 pickets and

(continued)

			20 feet banding material 1 48 x 40-inch pallet	53 sandbags per position
5A	Ammunition	Platoon	400 7.62 mm linked 1200 5.56 mm linked 1200 5.56 mm ball 60 40 mm grenades	N/A
5M	Mortar	Mort section	30 60 mm HE 5 60 mm Illumination	N/A
5MF	Mines	N/A	70 M21 antitank mines 28 M16 antipersonnel mines 1 cardboard box 40 x 40 x 24 inch 20 feet banding material 1 48 x 40-inch pallet	100 meter four–row minefield
5D	Demolition	Squad	50 1/4lb C-4 300 feet detonation cord 200 feet time fuse 40 fuse igniters 40 nonelectric blast caps	N/A
5B	Breaching	Squad	10 1/4lb C-4 100 feet detonation cord 100 feet time fuse 40 fuse igniters 40 nonelectric blast caps 4 bangalore torpedoes	N/A
5C	Crater	Squad	10 1/4lb C-4 200 feet detonation cord 100 feet time fuse 40 fuse igniters 40 nonelectric blast caps 6 40lb shape charges 6 40lb crater charges	N/A
5AT	Antitank	Company	12 AT-4 12 LAW 6 Dragon rounds	N/A
5AP	Claymores	Platoon	15 claymore mines	N/A
8	Medical	Company medics	10 IV start pack 10 IV catheter	N/A

TABLE 23.1 *(continued)*

			10 Dressing, first aid, field, 4 x 7 ins 10 Bandage, muslin, comp 10 4 x 4 gauze pads 4 Dressing, first aid, field, 7 1/2 by 8 ins 2 First aid kit, eye dressing 5 Bottles, iodine purification tablets, 50 ea 1/2 oz tube bacitracin 5 Packages band aids 30 Alcohol pads 1 Roll surgical tape 1 Ace wrap, medium 1 Bottle non-aspirin pain reliever (100 ea)	N/A

TABLE 23.1 (continued)

Planning guidance

Food: 2 MRE packets per day per soldier. 1 tray-pack meal.
Water: 4 quarts per soldier per day / 8 quarts per soldier per day for personal hygiene.
For work details, refer to heat index for water requirements per soldier:[1]

Index	Hot part of the day	Cool part of the day	Work time period
Black	2 qts/soldier/hr	1 qt/soldier/hr	5 hrs
Red	2 qts/soldier/hr	1 qt/soldier/hr	6 hrs
Yellow	2 qts/soldier/hr	1 qt/soldier/hr	7 hrs
Green	1 qt/soldier/hr		8 hrs

Source: 2d Brigade, 7th soldier Infantry Division (Light), *TAC SOP* (Mimeograph: January 1990) Annex D (Combat Service Support), Appendix 4 (Resupply), Tab B (LOGPACs), D-4, B-1–B-3. The LOGPACs came directly from the 2d Brigade SOP to illustrate how resupply planning must be organized in packages for standardization and simplification for transportation (vehicles or pallets). FM 7-71, 7-12–7-17.

NOTE

1. FM 90-3, G-3–G-5.

Chapter 24

Military Operations in Urban Terrain

Ground forces detest conducting military operations in urban terrain (MOUT), or urban combat for short.[1] Whether attacking or defending, MOUT is fraught with peril. Fire, explosions from gas leaks, collapsing buildings, and shattering glass are some chief dangers to soldiers even when not in direct enemy contact. Generally, urban combat is the infantry soldiers' fight. They can expect little combat support, meaning they must accomplish their tasks with the weapons at hand. Conceptually, MOUT operations are the same as for other operations, but the unique environment requires meticulous execution, tactical patience, and greater control of fires.

THE OFFENSE

As expected, attacking in a built-up area is more difficult than defending. The particular source of anxiety is the extreme difficulty in pinpointing enemy positions. Unlike a defensive sector in other terrain, where entrenching spoil and obstacles are difficult to conceal, a defensive sector in a built-up area is already riddled with debris, smoke, and rubble. As revealed at Stalingrad and Monte Cassino, defending soldiers can rapidly turn rubbled buildings into miniature fortresses. Further expenditure of munitions on these points appears to strengthen not weaken them. Because of the compartmentalized nature of the urban grid system and the degradation of radio communications, advancing units can quickly become fragmented. Navigation becomes tenuous, because units rushing from building to building or underground often become disoriented, and it is difficult to find recognizable terrain features in order to pinpoint a unit's position. Even with a Global Positioning System (GPS), buildings can block the signal. Another significant problem is that weapon effectiveness and observation are drastically reduced so that a unit clearing a building is deprived of supporting fires.[2]

Fighting through urban terrain is a very decentralized and time-consuming process. It is actually a set-piece battle, requiring deliberation of effort and tactical patience. Units attempting to blitz through an urban area can be easily dispersed, isolated, and destroyed in detail by relatively small enemy forces.[3] For planning purposes, the commander can expect to clear a city block with a platoon in 12 hours.[4]

Depending on the threshold of conflict, the use of firepower is subject to more constraints and limitations for fear of collateral damage. Civilian casualties and destruction of historical and religious objects, and perceived wanton damage receive immediate political scrutiny. As unfair and ridiculous as this seems to the soldier, leaders fearing a public-relations backlash are apt to prefer higher casualty rates than overwhelming firepower to neutralize a suspected enemy position. Toleration of collateral damage is linked to this threshold. If national survival is at stake, the toleration is quite high. If it is a low-intensity conflict, toleration will be nonexistent. Therefore, the company compensates by conducting the advance in a deliberate and coordinated manner, accepting decentralized control. Although the forward company may greet the prospect of plunging into a potential firestorm with dread, the commander can minimize casualties and smash enemy resistance in its path by instilling systematized procedures and drills.

The degree to which the company clears its zone depends on the urban density, the material strength of the structures, and the degree of enemy resistance. In less dense urban areas (where structures are not as sturdy), such as villages and city environs, the spearhead company cannot systematically clear every structure in its zone—unless the zone is extremely narrow and time is not a factor. It clears those structures the enemy forces are likely to incorporate in their defense (located on commanding terrain or a strongly built structure) and react to enemy fire during the advance. The defender is unlikely to strongpoint a village unless it is incorporated into the overall defense. Environs are less likely to be defended in strength, because observation and fields of fire are limited, the structures are constructed of light materials, and the attacker can bypass and infiltrate positions with ease. In dense areas with sturdy structures, such as the inner city and industrial areas, the company may be required to systematically clear every building in its zone.

Movement to Contact. Regardless of the threshold of conflict, the infantry must operate in urban terrain. In a low-intensity conflict, this may amount to peacekeeping operations with the infantry maintaining a presence in support of stability, as the British have done in Northern Ireland. Here the chief danger stems from the single shot of a sniper, a bomb or booby trap, or an ambush. In a mid- to high-intensity conflict, an urban area may or may not be defended in force. Resistance may be as minor as a lone sniper or partisan. Or it may just as well be offered by a platoon upward. Regardless of the intelligence, the lead company will have to verify the degree of resistance.

The company conducts a movement to contact as it approaches the urban area and occupies an intermediate objective or objective rally points (ORP) regardless of the type of enemy resistance. At the intermediate objective, the unit conducts limited reconnaissance and task organizes for the deliberate attack if enemy resistance warrants it. Otherwise it maintains its normal task organization.

Even in MOUT, movement to contact remains the most flexible attack formation. If some enemy resistance is expected, the commander determines those buildings that likely will serve as defensive positions and assigns these as immediate and final objectives. It is not necessary to clear every building along the route. Enemy elements likely will reveal their position by engaging the unit rather than forcing it to clean them out systematically. Follow-on forces will finish securing the area. In any case, the company advances through urban areas with a heightened sense of security but without undue slackening of its pace. The company must close with the main body of the enemy defense in order to prevent the battle lines from hardening.

Movement Formation. If there is no contact with the enemy forces, or contact is unlikely, a squad-sized advanced guard precedes the lead platoon. Regardless of their position in the order of march, all platoons, squads, and teams move with 360-degree, three-dimensional vigilance. Movement by fire teams is more secure, because soldiers moving in small teams are less likely to assume that other soldiers in the company are providing sufficient security—as a fire team, they are on their own. Because of the increased danger, each soldier maintains a five-meter interval.

The fire team moves in a staggered formation along the street or alley, maintaining an arms-length from the walls of buildings, with each man assigned a sector of responsibility. (See figure 24-1.) The point man's sector is to the front at eye level and below vertically and 180 degrees laterally; this soldier surveys the ground floors of buildings for sandbagged windows, basement openings, manhole and sewer drainage points, piles of rubble, and entrenchment spoil in open areas. The second man's sector is to the front at eye level and above vertically and 180 degrees laterally; the point man looks for buildings with sandbagged windows, windows with wire mesh, loop holes in roof tops and walls, and possible sniper positions in prominent structures such as church steeples. The last two men perform the same function to the rear. Soldiers in the center of the team orient laterally. A squad can provide excellent mutually supporting coverage by moving its teams opposite one another along each side of the street.[5]

If the unit encounters a deliberate defense of a building, the lead fire teams suppress all the windows of the building while firing two to three antitank (AT) rounds at the corner of the building to create a breach (or to use demolition charges). Use of a tank or self-propelled artillery in the direct lay mode is the preferred weapon for creating a breach.[6] In this case, a guide must go back and lead the tank forward to a good firing position and pinpoint the enemy position for the crew. Before entering the breach, the initial fire team tosses in a grenade to clear the room. The concussion not only incapacitates surviving enemy soldiers but also sets off booby traps. Even if the breach room is not occupied, it is always a good idea at least to use a concussion grenade to counter booby traps.

If at any point a full fight ensues into the depths of the urban terrain, it is better for the unit to advance through buildings, taking advantage of the cover and concealment of adjoining buildings. This entails blasting holes through the adjoining walls with demolitions. If buildings are close together, as they are in many European towns, the unit is afforded even better cover and concealment, as well as advancing on the enemy from an unanticipated avenue. This manner of clearing a town avoids the streets and open areas, which the enemy forces are likely to cover with fire.[7]

If the commander determines that the enemy forces are defending the urban area in some strength, he attempts to maintain the momentum through a hasty attack.

FIGURE 24-1

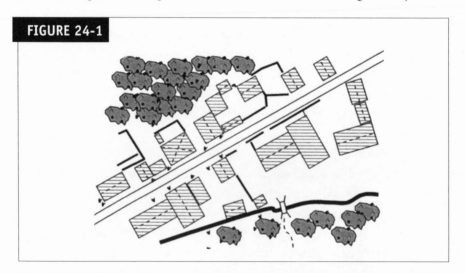

Street Movement

Once the assault element clears the first building, the assault platoon leader plans for the assault on the next objective building. He identifies fire support positions in the secured building that complement or succeed the current fire support element positions and informs the commander of his needs. The commander displaces the fire support element wholly or partially to the initial building rally point (IBRP)—the room where the assault element gained a foothold. The assault element platoon leader places the fire support element into its positions.

The company performs a hasty reorganization and consolidation for every building cleared. The first sergeant (1SG) organizes the rooms on the bottom floor and closest to the friendly side into the aid post, company trains, and EPW point. All EPWs, casualties, and logistics are funneled to the rear through the IBRP—this ensures the most secure route available. The commander selects a command post (CP) in preparation for the next assault. The assault platoon leader ensures that the building has 360-degree coverage and that all subterranean passages and other access points are covered. He also ensures that the friendly side of the building is marked (engineer tape or VS-17 panel hanging from the IBRP entrance) to indicate that it is cleared. The second assault element moves into its staging position for the assault on the next building. On order from the commander, the attack begins.

Limited Visibility TTP

Subterranean Passages. The lead soldiers don night vision goggles (NVGs) and attach an (IR) chemlight to the barrel of the lead weapon barrel or secure a penlight to the barrel with either an IR lens or a regular lens covered completely with opaque tape, which has a pinhole to allow a portion of light to escape. To increase the amount of light, the lead squad throws additional IR chemlights forward along the passage.

Clearing Rooms. Toss in an IR chemlight and a regular chemlight, or a flashlight with a cone cover so that it rolls in a circular pattern in the center of the room. Enter with NVGs.

ATTACK ON A DELIBERATE URBAN DEFENSE

An attack against a deliberate defense in an urban area will result in a moderate-to-high amount of collateral damage. In such a situation, the degree of collateral damage is secondary to protecting the force during the assault. As already mentioned, the company conducts a movement to contact to an intermediate objective or objective rally point (ORP) and conducts final preparations before the deliberate attack.

Task Organization. The company task organization is essentially the same as in any mission—security, assault, and fire support. If attacking a village, the security element occupies external positions that isolate it. If the urban terrain is larger, this becomes the task of higher headquarters. Security then devolves to the assault platoons to protect themselves from local enemy counterattacks from any direction. Depending on the degree of support from battalion headquarters, the company forms two to three assault squads, each with its own breaching capability. If an engineer platoon is attached, each platoon receives a squad. If a squad is attached, it goes to the main effort platoon. To provide maximum flexibility at the expense of centralized fire support, the commander will devolve fire support responsibility to each platoon once the company has secured a foothold in the urban area. The commander will augment the platoon's organic fire support with organic and attached fire support weapons (AT, mortars, tanks) as well as engineer assets.

The tank-infantry team is highly effective in MOUT and should not be ignored. The best way to destroy the enemy forces without suffering mutual attrition is for the infantry to identify enemy positions, guide tanks into protected support-by-fire positions, and destroy them with direct fire from relative safety.[8] Use of a dozer tank is

FIGURE 24-2

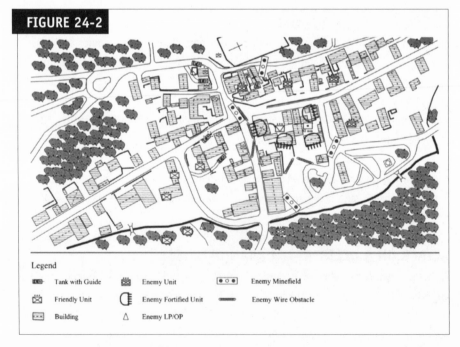

Legend

▭▬	Tank with Guide	🏛	Enemy Unit	[●○●]	Enemy Minefield
⊠	Friendly Unit	◖	Enemy Fortified Unit	▬▬	Enemy Wire Obstacle
⌷	Building	△	Enemy LP/OP		

Tank-Infantry Team in MOUT

highly effective in clearing rubble and barricades as well as creating berms out of rubble for tanks and infantry to use as support by fire positions. A useful tactic is for a dozer to create a ramp for a tank to drive up to increase its elevation of fire against enemy positions in multiple-story buildings. The commander may want to retain a fire support element to serve as a reserve as well as a security element for the CP.

In urban combat, the company commander assumes the position of providing resources for the platoons to succeed. He places himself where he can observe the progress visually. The best way for him to assure success is to keep the battalion commander apprised of progress and request additional resources to include the linkup point with the guide. In no other way can a platoon receive additional support quickly as the commander-to-commander link.

Each assault platoon organizes into a self-contained team with it own security, assault, and fire support element. A basic platoon task organization is as follows:

Platoon HQ: platoon, radio-telephone operator (RTO), two ammo bearers (extra machine gun [MG], squad automatic weapon [SAW], 40 mm, and hand grenades), two demolition men (satchel charges, mines, mine markers), two medics with canvas stretchers, and one forward observer.
Fire support element: PSG, two M-60 machine gun teams, two SAW gunners (optional), two M203 grenadiers (optional), two AT gunners (light antitank weapon—LAW/AT-4), two snipers, and 60 mm mortar (direct fire).
Two to three assault squads organized into three teams:[9]
Team 1: squad leader (smoke grenades), SAW gunner (axe), and SAW gunner (grappling hook).
Team 2: team leader (multipurpose tool), grenadier (bangalore torpedo section), and rifleman (extendable ladder).
Team 3: team leader (bolt cutters), grenadier (demolition charge), and rifleman (grappling hook ladder).

The platoon HQ follows directly behind the first assault squad. Because of the likely pandemonium and obscuration, the platoon leader must resort to visual and verbal communication. Simple hand signals to direct personnel forward and to indicate enemy positions or obstacles are effective. Use of markers (sheets, VS-17 panels) on the friendly side of the building to indicate cleared rooms or buildings provide instant recognition of progress. The ammo bearers carry a heavy load of high-expenditure ammunition. Since they are not expected to fight, this load may be up to 120 pounds.

For this reason, ammo bearers do not assume this load until they reach the ORP or the assault position, if tactically practical. The demolition team is used for breaching obstacles and buildings. The medics should carry extra supplies for burns and injuries caused by booby traps. Because urban combat results in higher-than-normal casualties, stretcher bearers are a necessity. To assure that the platoon has the necessary strength to sustain an assault, the company commander may augment it with two four-person stretcher bearer teams. If this is not possible, the platoon leader can task organize into two assault squads and use the personnel for medical evacuation and augmentation of the fire support element. The platoon leader may attach the forward observer to the snipers because of the better observation position.[10]

Since it is tactically prudent for platoons to maintain their own security, the second and third assault squads should position a fire team to the flanks and rear of the secured building. Too often platoons make the mistake of crowding everyone forward into the front rooms to maximize firepower for the next assault. This disposition is not only unnecessary, it can also lead to the destruction of the platoon if these crowded rooms are hit with enemy fire. Only the forward assault squad and fire support element occupy the front portion of the building. In securing the depths of the building, the squad leader places one or two soldiers per room and away from the windows. The soldiers do not need to have panoramic views; the overlapping observation of other rooms will ensure complete coverage. In this way, the platoon maintains proper dispersal and sufficient covering fire for the assault.

The size of the assault element depends on the size of each building. A good planning figure is to allocate a squad for each floor; hence clearing of a three-story building would require three assault squads. This planning figure gives the commander a good idea of the number of squads needed. For initial entry, the initial assault squad is augmented with an engineer squad. Once inside, each assault squad forms into its clearing and security team with team 1 performing security and teams 2 and 3 conducting clearing. Use of grappling hooks is excellent for opening closed doors or suspect areas that may be booby-trapped. Although cumbersome to carry, an extendable ladder is excellent for entering upper floors quickly. The ladder carrier comes forward, seats the ladder, rotates beneath the ladder and holds it in place for the assault squad to scale. Of course, the size of the ladder is limited to reaching the second floor of a building, but in villages most buildings are only one- or two-story structures anyway. In cities, the company will likely need to clear a building from bottom to top.

The fire support element is used to isolate each objective building, suppress local enemy positions, or provide overwatch of the assault element. If it cannot provide these tasks from its position, the platoon sergeant must have the authority to move the element into a position that provides such support. The fire support element also provides the guides for tanks moving forward to support the advance. The commander attaches the two snipers and the two AT gunners. The snipers operate independently in order to find good positions for the engagement of enemy snipers, leaders, and key weapons. The AT gunners may be brought forward to destroy enemy bunkers, strong points, and armored vehicles.

Assault plan. The initial assault aims to isolate the initial objective building with firepower, secure a foothold, and then rapidly clear the building with small teams. For the tentative plan, a matrix is helpful in assisting the commander formulate a scheme of maneuver and supporting fires. Using a sketch or sand table, he labels each house in his zone with a number or letter. Using both the matrix and the sketch or sand table, the commander is able to explain the timing and synchronization of the fight in a step-by-step manner. This ensures that he includes everyone involved in the fight without confusion. It also helps everyone visualize all the moving parts of the assault, and how the conditions are to be set. Referring to the task organization of the operation order (OPORD), the commander specifies in detail the assignment of targets pertaining to the assault on the initial building. This amount of detail deteriorates for subsequent buildings; certainly, the commander does not have the time to detail the assault on each building, particularly for large urban areas. Hereafter, the plan identifies the order in which subsequent buildings are attacked, while the commander emphasizes that the elements establish the same conditions of fire support, breaching, and assaulting as with the initial building. (See table 24-1 for a matrix example.)

TABLE 24.1 BUILDING ASSAULT MATRIX					
Element	**Initiate Fires**	**Breach**	**Lift/Shift Fires**	**Assault**	**Secure**
Fire Spt 1					
Fire Spt 2					
MORT Spt					
Arty Spt					
Assault 1					
Assault 2					
Assault 3					
Security					

The subdivision of each matrix cell is as follows: Type Signal/Building Label/Floor Number.

Seizing a Foothold. The company fire support element occupies positions that afford it the opportunity to isolate the objective building. Because buildings along the outer edge of a built-up area are likely to contain booby traps, targeting them with a few rounds of tank or mortar fire (direct mode) is a good technique for setting the booby traps off or burying them with rubble. If more than one fire position is used, the fire support element leader controls his element by use of field phones. Another line to the company CP assures full coordination. The company is forced to rely on wire, because radio transmissions are severely degraded in urban areas. As the fire support element advances to a new position, the wire is recovered to assure that commo wire is available for subsequent operations.

The initial assault squad deploys to a covered and concealed assault position as close to the point of penetration as feasible. The squad performs the final preparations on breaching equipment. The platoon fire support element within the assault platoon occupies positions that suppress enemy positions within the objective building. The commander maintains dispersion of company elements within the confines of the built-up area to preclude potential catastrophic casualties. The platoon establishes wire communications with the company CP to enhance the coordination of the initial assault.

The preparatory fires are designed to isolate the objective building, as well as to suppress and destroy enemy overwatch positions. Because artillery tends to cause significant rubble, which hinders the advance, employing time-fused warheads along enemy open access routes to the building is more effective. Because of the close nature of urban terrain, use of artillery may be too dangerous as the company moves into position. The commander may elect to employ the destructive fires and smoke against suspected strongpointed buildings and open areas (parks), which favor entrenchments, and to use the fire support element to isolate the objective building. He must select targets for smoke with care, because smoke may mask the supporting fires of the fire support element.

On the commander's order, the fire support element initiates its fire to continue the isolation of the objective building and suppression of enemy positions, particularly those in the objective building. Referring to the **U.S. Weapons Capabilities** chart (chapter 11), fire support personnel identify which weapons are appropriate for identified enemy positions. Hunter-killer teams, each consisting of an AT weapon, M203, and a machine gun with AP rounds, can identify special targets for elimination. AT gunners aim six inches below or to the side of a bunker aperture to cause spaulding effects within the bunker. A round through the aperture is likely to pass through the bunker without harm to the bunker crew. The rest of the fire support weapons concentrate on suppression. Logistically, the company cannot expend vast amounts of munitions as it advances from one building to the next. The amount of ammunition expended is controlled by a time limit, perhaps 5 seconds as preparatory fire and an additional 10 seconds of alternating fire to cover the assault squad. Thereafter, firepower is focused on those areas where active enemy fire has been identified. Machine guns can achieve sufficient suppression by alternating six to nine bursts every five seconds, or for a rapid rate of fire, six to nine rounds every two to three seconds.

As the fire support element initiates fire, the initial assault squad throws several smoke grenades along its access route to the breach point. After the smoke builds sufficiently, the squad moves to the point of penetration. The squad does not select a door or window as its entry point, because these are likely covered by enemy fire, booby-trapped, or heavily barricaded. During this time, the fire support element employs full suppressive fire to ensure the safety of the squad. The bangalore torpedo man reduces any wire obstacle in front of the breach point. The demolition man places the demolition charge in place, takes cover, and detonates it. The entry point must be large enough for swift access. If the opening is not sufficient, the squad must be ready with another charge. Solid structures employ reinforcing rods, which the assault squad cuts with the bolt cutters. Clearing the initial room with a fragmentary or concussion grenade (two-second cook off), the squad secures the foothold and signals to the platoon leader to bring the next squad forward. Grenades should always be used in order to set off booby traps. If necessary, each successive squad masks its movement with smoke grenades before deploying into the foothold. Once the assault element has closed on the building, the support element lifts fire (on order) and continues to survey the area for identified enemy objects.

Clearing the Building. The initial room in the initial building becomes the initial building rally point (IBRP). This is where casualties and prisoners of war (EPW) are processed until the entire building is captured. The lead assault squad labels the entry point (chalk or paint) with the letters IBRP and the company designator as a guide for the rest of the company and perhaps other companies. The entry points for subsequent buildings are labeled BRP.

Once the assault element has consolidated in the IBRP, it begins the process of clearing the building. The initial assault squad continues to secure the IBRP, while the rest of the assault element passes through into the interior. The assault element clears a route to the top of the building, because clearing a building from the top to the bottom favors the attackers.[11] The defenders' orientation is invariably downward and outward toward their front. By attacking downward onto them, the assault element can flank and unhinge their position without high casualties. Of course, clearing a route to the top is not easily accomplished. If at all possible, the assault squads want to avoid taking the obvious route through the stairwell. The defenders will likely blockade parts of it and have it covered by fire. An excellent technique is for the assault squad to cut its way vertically from one room through the top level (preferably not from the IBRP). This requires the squad to use a demolition charge to blast a hole in the ceiling and through the floor of the room above. The soldiers of the assault squad then climb through, using a ladder, stacked furniture, or each other as an aid. The squad then secures this room for the follow-on squads to continue the ascent to the top. Securing the room means establishing a position in the hallway to cover the most likely enemy avenue of approach.

Once the assault element reaches the top floor, the "top" squad begins clearing the floor. The assault element platoon leader controls the flow of squads to the top floor. As mentioned earlier, the platoon leader should have enough squads to allocate a squad per floor. For most buildings, the company has enough squads for clearing operations. If the building is too large, the commander requests augmentation from the battalion commander. The platoon leader maintains contact with his element, using visual, verbal, or wire communications. The "top" squad leader requests additional squads if he deems it necessary.

TTP

If the company enters the building at a point above the ground floor, it clears a vertical path to the bottom floor in order to establish an IBRP before clearing a path to the top floor.

Cutting a vertical path from room to room may not be feasible. This is a constraint in training and hence may become the approved method for the majority of building clearance operations. The lead assault squad passes through the IBRP and clears a path to the nearest stairwell. While the IBRP security squad (from the initial assault squad) overwatches, the lead assault squad moves along the hallway and clears each room along the path to the stairwell IAW the established method for clearing floors and rooms. As the squad advances, it drops off a security team (two or three soldiers) at a decision point (hallway intersection, stairwell access, tunnel or mouse hole for soldiers to crawl through) and the access point on each floor. Once the squad is depleted as a result of attrition or occupying decision points, the next squad passes through and assumes the lead. This continues until the top floor is reached and the floor-clearing operation begins. It is quite possible that the number of decision points could deplete the company manpower before it can reach the top floor. At this point, the commander either accepts risk and abandons specific decision points, or he accepts the loss of time and requests additional manpower from battalion HQ.

Floor Clearing. The initial room on each floor is the floor rally point (FRP) designed to convey reinforcements, casualties, and supplies into a secure area. (See figure 24-3.) The access point is labeled "FRP" in chalk to orient squads passing through. This also serves as the squad or platoon CP. The security team covers the room clearing teams from a central location in the hallway. The clearing teams clear rooms on both sides of the hallway simultaneously. As the clearing teams advance from one room to the next, the security team displaces forward to preclude the clearing team masking its fires. The platoon leader deploys temporary security teams to a floor to cover newly discovered hallways (decision points) until the clearing teams are ready to clear them, or he deploys

FIGURE 24-3

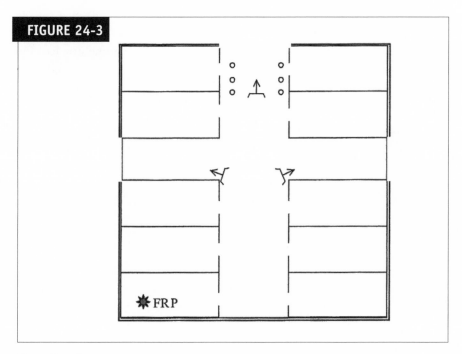

Floor Clearing

more squads to clear these hallways with dispatch. Each security team engages any enemy personnel the clearing team flushes out or just happens to appear. Security teams at decision points also engage enemy forces flushed from the above floors. Clearing teams vary the pattern of clearing rooms, breaching through doors, hallway walls, or through adjacent rooms. If the clearing team does clear the next room from another room (for example, not from the hallway), it must have a signal for the squad security team that it is reentering the hallway from another room (visual or oral). Once the top floor is cleared, the squad moves to the FRP of the next lower floor and updates the next squad responsible for clearing the next floor. The squad remains at the FRP as a local reserve while conducting resupply, EPW evacuation, and casualty evacuation.

Room Clearing. Room clearing is the most hazardous part of clearing a building. Even if the building is defended, not all rooms are occupied, and the unit does not have the munitions to simply expend as a safety measure. If the room door is open, one lead member of the clearing team takes a quick look into the room from the floor level. He next extends his fixed bayonet across the doorway to draw fire. If the room appears clear, the two-man room clearing team clears the room IAW the squad's room clearing technique. If the room door is closed, the team leader and lead member step across to the other side of the door, while the last member places his ear to the wall or door to ascertain occupation (firing, voices). While the team leader provides local security down the hall, the member closest to the door handle opens the door and steps back. As before, the other member then takes a quick look, displays the bayonet, and both clear the room. If the door is locked, the team breaches the room with a demolition charge or shoots off the lock and proceeds to clear as already mentioned. If the room is occupied, the team leader cooks off a fragmentation or concussion grenade and throws it into the room. Once the grenade detonates, the two-man clearing team clears the room.

TTP

Use of a metal mirror attached to the bayonet helps verify that a room is clear before entering.

The squad continues clearing until attrition or the discovery of too many decision points brings the operation to a halt. Once this occurs, the squad leader informs the platoon leader that reinforcements are needed. As the clearing teams advance, they mark cleared rooms and their path as a guide for follow-on squads. For rooms located on the friendly side of the building, opening the windows and toppling out sandbags or hanging sheets out of the window mark the cleared rooms.

TTP

If walls are made of concrete, the team maintains an interval of one foot from the wall in order to avoid "rabbit rounds"—rounds that ricochet and travel along the sides of hard surfaces.

Movement and Room-Clearing Techniques in a Building. The clearing team moves through hallways, hugging the wall, with two men abreast in front and the team leader in the rear providing rear security. Each has his weapon oriented along his assigned line of site, a diagonal sector of responsibility that optimizes his observation and reaction to enemy contact. If a concern, soldiers can tie sandbags over their boots to muffle their footsteps.[12]

TTP

To clear a room with a hand grenade, the grenade thrower places his weapon against the wall, faces the wall, and pulls the pin, using both hands. He cooks off the grenade for two seconds, ensuring that the spoon does not fall to the ground, alerting the enemy of the imminent assault. Throwing with the hand nearest the door, he aims deliberately to avoid obstructions. Upon release of the grenade, the entire team lies prone, with helmets oriented toward the wall (during this time, the squad security covers the hallway). The thrower anticipates recovering the hand grenade and rethrowing it if it bounces back into the hallway.

Two effective techniques for clearing a room are the criss-cross and the buttonhook. The criss-cross technique has the number one man (normally the one to the left of the door) enter the room first, firing a short burst into the room while moving to the opposite rear corner. The number two man enters immediately after the first but does not fire unless a clear target appears on his side of the room. He too moves to his opposite rear corner. While the number one man provides security, the number two man searches for enemy, booby traps, intelligence, and mouse holes. In the buttonhook technique, the number one man enters first, firing a burst while rapidly moving along the wall on the right side of the room. The number two man enters immediately after the number one man and moves along the left side of the room without firing. As in the first technique, the number one performs security while the number two conducts the search. Once a room is clear, the room-clearing team informs the team leader (who alerts the squad security team) that it is coming out and marks the room as cleared with its symbol along with an arrow indicating its direction of advance. This is essential, because the clearing team may take a route that goes from the hallway, into a room and directly into subsequent rooms before emerging into the hallway again.

THE DEFENSE

As with other defensive operations, the company performs most of the same tasks as it prepares to defend in an urban environment. The following highlights those considerations specific to urban defensive operations. Upon receipt of an urban defense mission, the commander dispatches a recon team (preferably a personal leaders' recon) to the assigned sector. Because the density of the urban area requires a close-knit defense, meaning few gaps between units, the company has little leeway regarding the disposition of the main line of defense. It must make the most out of the assigned sector even with its tactical deficiencies. The recon identifies good covered and concealed routes to the assigned sector and marks the route using the unit designator and arrow with paint or chalk. The recon also seeks city blueprints and the underground system. Such documents are located at the local city hall, utilities building, fire departments, police stations, and department of statistics centers. These centers can be located with the aid of a telephone directory. Recovered documents that pertain to areas outside of the company sector are forwarded to battalion HQ. These documents form the foundation of the defensive plan.

Planning and Preparation Considerations. Unlike with other terrain, the commander cannot develop a comprehensive estimate of the situation and tentative plan from a map. Blueprints assist the commander in determining which buildings will compose the main line of defense. But the crucial fine-tuning of the defense occurs during occupation of the sector. Furthermore, company elements regard all other buildings between the main line of defense and the forward edge of the sector (the outpost sector) as an area to disrupt the enemy's attack and to deceive him. Regardless of the depth of the defensive sector, the commander avoids placing his main line of defense along the outer row of buildings (or first block if possible). The enemy is likely to concentrate the majority of his firepower on these even if he discovers no sign of our activity here. It is better to accept degraded observation and fields of fire than to suffer heavy attrition. Once the enemy has closed on the defense, his observation and fields of fire are as reduced as the company's are; moreover, the proximity of the defensive positions makes it difficult for him to use artillery. Whenever planning the defense, all leaders must address both the fight in the streets and the fight inside of the occupied buildings. Failure to see the interrelationship of both fights will lead to ruin.

Converting a building or grouping of buildings into a strongpoint is highly effective. (See figure 24-4.) The commander chooses a location, which protects key terrain (bridge, airfield, and so forth) in sector. Because the strongpoint receives the majority of resources, the company conducts a defense in sector forward of it, meaning that the company elements conduct a fluid defense in the forward buildings to erode the enemy strength, disrupt it, and force it to expend munitions and resources before the main defense.

The strongpoint must have all-around defense and must have some separation from other buildings in order to deny the enemy a covered and concealed approach. This may require razing nearby or adjoining structures. To reduce its signature as a strongpoint, no obstacles should surround it. Since outside obstacles are so hard to conceal, enemy aerial reconnaissance can easily identify them. Obstacles may be placed against the friendly side of nearby buildings though. Underground passages are preferable for use as lines of communication within the strongpoint. Dozers with infantry work teams can create covered and concealed lines of communication from existing rubble and buildings. This is particularly important for withdraw routes.

FIGURE 24-4

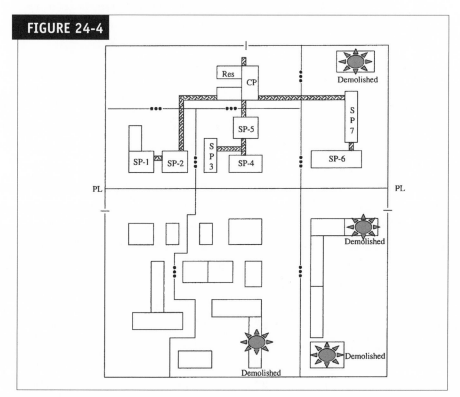

Company Strongpoint Defense

Priority of Work Considerations. During the occupation of defensive positions, each team and squad leader has all subordinates check for gas leaks and other hazardous items. The squad locates and shuts off the main gas line into each occupied building. All windowpanes are broken out. The platoon leader identifies enemy mounted and dismounted approaches into sector (underground passages, access from adjacent buildings, streets and alleys), emplaces obstacles, and covers these with appropriate weapons. As already mentioned, selection of an ideal position is limited given the attackers' ability to exploit existing cover and concealment and flank a position from an undefended building. The platoon leader anticipates that the enemy will exploit a covered and concealed route to close on the position. This may require the squad or platoon to fight within the building or withdraw along a prescribed route once the enemy has hit a trigger point for withdrawal. Having a withdrawal plan is very useful for squads deployed forward of the main line to disrupt the attack. Strict adherence to trigger points for withdrawal (attackers penetrating the building, the top floor, and so forth) are crucial in order for the squad to occupy the main line of defense in strength and before the enemy attacks. With this in mind, the leader tasks squads to identify covered routes (sewers, tunnels, adjacent buildings) between positions. He identifies buildings with basements for excavation tunnels to other positions if resources and time are available.

Key weapon positions are very effective outside of buildings and usually are not anticipated by the attackers. Open areas, such as parks, grassy areas, or areas with covering debris are excellent for machine gun positions because they are conducive to grazing fire, afford good protection against fire, and are more difficult to target. Machine gun positions in buildings are subject to degraded observation and fields of fire. These factors are further degraded by any collateral damage to the building. It is more difficult to establish positions for grazing fire in buildings. The average positions will result in plunging fire. If buildings must be used, machine guns are best employed at ground level (ground floor or basement). Machine gun positions that use basement windows, loopholes on the ground floor (those at the corner of a building particularly provide excellent positions for grazing fire), or basement passageways reinforced by sandbags and overhead cover are quite formidable.[13]

AT weapon positions require large rooms in order to avoid fratricide from the back-blast. An M-47 Dragon requires a 15 X 12-foot room with a ceiling height of 7 feet. The position itself requires a vent area of 20 square feet and an aperture height that allows a six-inch muzzle clearance. A LAW or AT-4 requires a room with 4 feet from the back wall, with a ceiling height of 7 feet, and a vent area of 20 square feet. Fields of fire are also greatly restricted the farther the firing platform is away from the window, door, or loophole. Subsequent AT fires from the same position may not be possible, because back-blast fumes and dust are likely to engulf the room. On the plus side, the enemy will find it difficult to pinpoint an AT position by its fire and back-blast unless the platform is near the window. Disposable AT weapons (LAW, AT-4) in buildings are preferred over wire-guided AT weapons. From elevated positions, LAW/AT-4s are more effective against the topsides of enemy vehicles. A wire-guided AT weapon needs more standoff and is best sighted down the axis of a long street.

Fighting positions are oriented both externally to cover avenues of approach and internally (within buildings) to fight the close battle against enemy penetrations. The external fields of fire focus on interlocking fields of fire along streets and open areas. Fields of fire also extend to buildings located to the front of the defense. These façade buildings are prepared in a similar manner as clearing fields of fire. Basically, the defense exposes the friendly side of the building with shape charges, tank rounds, and so forth in order to engage the enemy approaching or entering it. Some windows and doors are left open on the enemy side of the building to engage the enemy through the building. Other windows and doors remain shut to target enemy soldiers who open them. The amount of damage on the enemy side of the building must be limited, because the defense does not want to alert enemy soldiers that the building they are about to enter is "Swiss-cheesed" with fields of fire. Leaving the enemy side of the façade building undamaged lessens the enemy's sense of danger. The surprise effect of taking casualties without determining where the fire is coming from (it is extremely difficult to ascertain the firing point from fields of fire through a house) can actually paralyze a unit. If time permits, the defense may turn the façade building into a dummy strongpoint, inducing the attacker to expend his resources.

The preparation for the interior fight pertains to those buildings the company wants to retain, such as a strongpoint. (See figure 24-5.) The leader of the strongpoint anticipates that the enemy will penetrate into the building and plans how to counter them. The main doors are barricaded or have a prepared concertina wire obstacle to be executed on order. Anchoring the concertina wire coil to the top of the entrances with a holding retainer allows use of the doors until the last moment. Once the retaining rod is removed, the wire falls into place and is anchored into the side of the building with metal spikes. The door may be reinforced by a wooden frame filled with sandbags, which is placed on hinges behind the door. This shields the defenders against AT rounds or demolition charges placed against the external door. The unit fortifies all room positions for 360-degree defense. This requires any and all materials available—sandbags, ammo cans, crates, and ration boxes filled with sand, tightly bundled newspapers or magazines, plywood, furniture, and wooden beams (2 X 4s)—to build fighting positions. Numerous wooden beams buttressing the ceiling strengthen the structural integrity of the building. The tightly bundled newspapers, magazines, and books placed in furniture (cabinets, chests, and so forth) provide adequate protection from small arms fire. Lining the floors with sandbags provides additional protection to floors below and serves to prevent penetrated rounds from ricocheting. Of course, the additional weight makes multiple beam supports crucial for the rooms beneath. Most of the fighting positions should allow the soldier to move to any side in order to fire both externally and internally. Windows and loopholes are reinforced with sandbags to protect against small arms and AT fire. All windows are covered by wire mesh and gauze to hinder outside observation. The bottom of the wire mesh should remain free so that the defenders can drop hand grenades to the outside of the building. For AT positions, the gunner installs a pulley system to lift the wire mesh for firing. Holes with coffee can inserts in the building walls and floors allow the defenders to drop grenades below without exposing themselves.

To hinder enemy freedom of maneuver along specific hallways and stairwells, the platoon emplaces obstacles (wire, furniture, wooden barricades) covered by fire from internal positions. Absolutely no booby traps are installed in occupied houses, because the danger of accidental detonation is too high. Specific rooms are dedicated to sleeping, supplies (munitions, water, rations), and the command post. To maintain contact with the outside world, placement of a commercial radio in the CP is recommended.

The leader decides how he wants to conduct the internal fight. Each main position has a secure route via mouse holes, rooms, and hallways to subsequent positions. Each alternate position also has an escape route to an adjoining room or floor. The defenders must use portable ladders for mouse holes and withdraw them to prevent their use by the attackers. Ideally, the leader envisions a defense in sector with the defenders making a progressive withdrawal to the room with the escape tunnel or route to the rear. In this manner, the soldiers have the flexibility to withdraw without sacrificing themselves once a position is untenable. The escape tunnel is concealed in the room with a hinged door to prevent the attackers from identifying it quickly. To conceal the outside exit, a board with a support pole is used to hold the excavated dirt. (See figure 24-6.) The exit is filled with earth until it is even with the ground and camouflaged. When the escape tunnel is used, the support pole is removed with an attached rope, allowing the earth to fall into the pit and allowing egress.[14]

Although a building affords protection for the defenders, each main defensive position requires fortification. In general, the fighting position is no closer than a meter

FIGURE 24-5

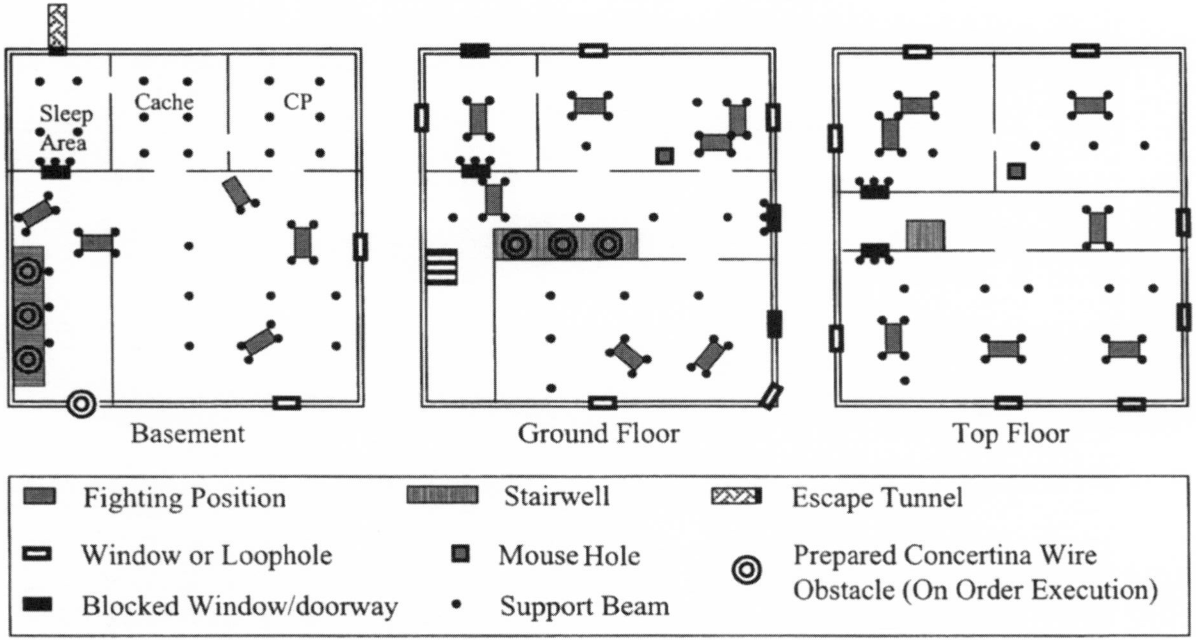

Legend:

- Fighting Position
- Window or Loophole
- Blocked Window/doorway

- Stairwell
- Mouse Hole
- Support Beam

- Escape Tunnel
- Prepared Concertina Wire
- Obstacle (On Order Execution)

Top Floor

Ground Floor

Basement

CP

Cache

Sleep Area

Strongpointed Building

FIGURE 24-6

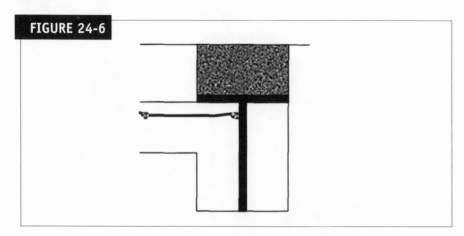

Escape Tunnel

(one M-16 rifle length) from the window or loophole so as to conceal the muzzle flash. Sandbag positions strengthened by a wooden frame transform each fighting position into a miniature bunker. Frontal protection is afforded by a sandbag wall two to three sandbags thick, three to five sandbags in length, and level with the window. For sturdiness, each sandbag layer is laid at right angles to the layer below and above it. Overhead cover with support beams, a sheet of plywood, and a layer of sandbags provides good protection. (See figure 24-7.) Additional sandbag walls to the sides of the firing position provide flank protection. Fighting positions may be made for standing, kneeling, or prone firing. Beds and tables reinforced with sandbags make excellent platforms for prone firing positions. A fighting position made out of ammo cans filled with dirt and strengthened with a wooden frame can withstand most types of fire. (See figure 24-8.) Such a position allows the soldier to orient his fires in any direction quickly without exposing him to fire. Windows that are not used are completely sandbagged and covered by a drop cloth for concealment.[15]

Obviously, the intensive preparations are reserved for the main positions. Because so many sandbags are required per position, the sandbag filling detail begins as early as possible with the 1SG supervising their transportation to each defensive building and each platoon sergeant managing the allotment of sandbags per position. Since earth and sand are not readily available in MOUT, the 1SG manages the filling and transportation of sandbags to the sector very early. The supply sergeant focuses on delivering ammunition, demolition charges, bolt cutters, axes, concertina wire, and mines. As mentioned, the soldiers use everything possible to strengthen their positions. This requires the scavenging of materials from other buildings that are not part of the defense.

FIGURE 24-7

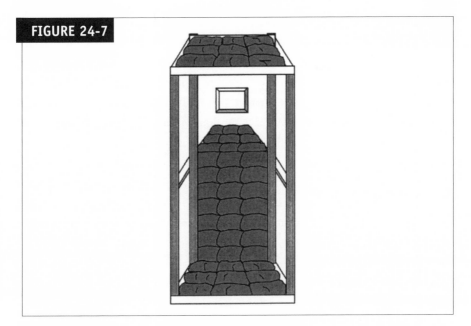

Fighting Position

With the plethora of material existing in an urban area, soldiers can fortify any building with minimum external resources.

Generally, one or two soldiers per room provide sufficient overlapping fires. To prevent soldiers from feeling isolated during the fight, the platoon leader, platoon sergeant, squad leaders, and team leaders constantly move through their buildings and rooms to keep the soldiers apprised of the situation. All internal doors that have not

FIGURE 24-8

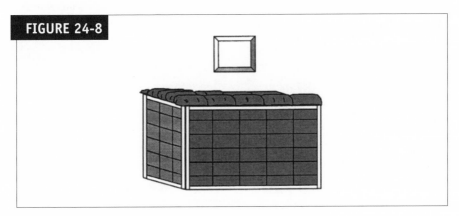

Ammo Box Fighting Position

been barricaded are removed in order to facilitate communication. The distribution of the soldiers throughout the building ensures full coverage and overlapping fires. In this manner, a squad can easily defend a good-sized building. Fighting positions over and above the complement of defenders in a room allow the leader to reinforce a sector quickly. Allotting a portion of the platoon or squad to the internal fight is prudent, because it represents the greatest danger to the unit. In addition to the 360-degree coverage for each room position, fighting positions within the building are oriented solely for the internal fight. Soldiers occupying the internal positions also act as a reserve to reinforce room positions if required. It follows that both the external and internal fight are equally important and deserve equal attention.[16]

Because fire is always a danger, soldiers stockpile sand and water in containers within each room to extinguish the flames immediately. Spreading sand on the floor also helps deter fire from spreading rapidly.

To counter attackers with night vision devices, the defense can install lights at key approaches to the building that switch on at will. Similar lights within subterranean passages and within key points in the building are also effective in thwarting night vision devices during the interior fight. Lights (vehicle headlights, small searchlights, outdoor lights) wired to vehicle batteries are plentiful in urban areas. Use of reflective devices (metal mirrors) makes it difficult for attackers to eliminate the light quickly. Forming "gator hunter teams," using mounted flashlights on rifle barrels is also effective in countering targets of opportunity. While one member spots, the gator hunter flicks on the flashlight, blinding enemy targets, eliminating them, and then switching off the flashlight quickly. With a little ingenuity, a defender can deprive the attacker of the night-fighting edge.

Counterattack forces are essential in MOUT because they break the momentum of the attack and cause its early culmination. The counterattack should occur when the enemy is least prepared. Hence, knowing the enemy's sleeping and eating habits is key to a successful counterattack. The ideal size of such a force is a squad or even a fire team. It can be used as a raiding force, building recovery force, or reinforcing force. As a raiding force, it carries only grenades, pistols or carbines, LAWs, and demolition charges. The best point of entrance is from an adjoining building using a breaching charge. If the approach must take place over open ground, it should be under darkness or smoke and with covering fire. Once in the building, the raiding force immediately advances, rapidly throwing hand grenades into rooms indiscriminately, and then withdraws before the enemy can recover. A raiding force is an excellent means of eliminating an enemy tank or a key support-by-fire position. In the recovery of a building, the force performs the same function as the raid but consolidates in a predetermined room. The main body follows closely behind and occupies the remainder of the building for immediate defense. As reinforcements, the force occupies predesignated positions in the designated building. To prevent congestion, each frontline unit reserves rooms or even an entire floor for the reinforcing force. The counterattack force discusses each of its tasks with the company commander and coordinates the plan with the frontline unit. Lastly, it rehearses each of its tasks with each frontline unit.[17]

The mortar section is best deployed in a courtyard or a roofless building. The ground base is important, because the mortar base plate must be firmly seated to ensure accurate fire. The commander ensures that the mortar section has engineer support to dig in the mortars. The fire support team (FIST) officer places target reference points (TRPs) on the roofs of defended buildings to deny the attackers access. Timed fuses are most effective in this case. For flat roofs, lining them with sandbags allows the use of light mortars in an emergency.

The company CP is located to the rear of the main effort building. This allows the commander to maintain the pulse of the fight without being intrusive. The commander may choose to maintain a squad in reserve with him to react to an enemy penetration immediately.

The company aid post and trains are one building behind the main line of defense, using courtyards when available. The company establishes caches of ammunition, water, rations, and medical supplies. Conversely, it deprives the enemy of all utilities by shutting off water, cutting telephone lines and electricity, and removing all nonperishable food.[18]

In a prolonged urban environment, health and hygiene require particular attention. Soldiers are not to assume that tap water is potable. Purification tablets are mandatory. Field sanitation teams create water reservoirs by filling bathtubs with purified water and covering them. Other containers, such as plastic jugs, are handy water reserves for individual positions. The field sanitation team inspects all bathrooms for functioning toilets. If water has been cut off, the team ensures that each bathroom has a bucket with water to flush waste. If this becomes a problem, the team blocks off or removes the problem toilets and has the platoon establish a latrine at ground level.

The most effective and reliable means of communication is the field telephone. Use of radios is too unreliable because of urban obstruction. Until the lines are laid, each platoon and section automatically dispatches messengers to the company CP. The company may be able to maintain contact with the battalion tactical operations center (TOC), using its AN-292 antenna. But if communications are sporadic, dispatching two messengers to the TOC assures contact. The commander should take advantage of a functioning telephone system whenever possible. No one should assume that the line is secure, so the battalion S3 establishes a code sheet to use in passing recurring information.

As with all defenses, obstacles are covered by fire. The abundance of debris and objects, such as civilian vehicles, construction material, and equipment, provides excellent obstacles. Reinforcing these with mines or booby traps and concertina wire make them quite formidable. The intention is to separate the enemy infantry from its armor and direct fire gun support. These obstacles are placed between and on the friendly side of forward buildings, as well as between the defensive building positions. Covering these obstacles with light AT weapons from positions above ground level may not destroy an enemy vehicle, but they will make the enemy apprehensive. Buildings with basements also make excellent tank traps should the tank crew panic and crash into such a building for cover. Access routes from rooftops also require obstacles but only if they can be covered by fire.

The deception plan strives to delude the enemy into focusing his initial efforts against the deception buildings. Placing obstacles to the front of these buildings and placing dummy positions with sandbags along the enemy's side promotes the delusion. Employing listening or observation posts (LP/OPs) or delaying teams within these also adds to the deception. Once the enemy receives fire from delay positions within, and later from fire passing through them from our main positions, the enemy will likely conclude that these buildings are heavily defended. Placing dummy positions within the main defense and its surroundings also diverts enemy fire once he enters the main defensive sector. If the deception is successful, the enemy will expend the majority of his firepower on empty ground, his combined arms coordination will be disrupted, and the defense will be able to destroy the attack in detail.

No one should make the mistake of concluding that the initial battle will be enough to repulse the enemy. Once he has a foothold, he will use this as a base of assembly, staging, and continuance of the offensive. Urban terrain requires many more units than the norm, and it is unlikely that higher command will devote enough resources to defend against all enemy thrusts. As in mountain warfare, there are simply too many points where the enemy can infiltrate and bypass resistance. The intent of the defense is to defend the main approaches, force the enemy to expend time and effort finding these alternate routes, and erode his strength to such a degree that he is ripe for a counterattack or even a counteroffensive.

CONCLUSIONS

Whatever the threshold of conflict, infantry companies will operate in urban terrain on numerous occasions. With focused training and careful preparation, the commander can keep casualties low despite the hazardous environment. Moreover, the company can inflict serious casualties on significant enemy forces if they are not well trained in MOUT. Despite anticipated advances in technology such as improved communications capabilities, improved munitions, and the ability to divulge enemy positions within buildings, technology alone will never be a substitute for good ground tactics in urban terrain. The commander must conduct MOUT training at least once a quarter and fight the bureaucratic tendencies of the division administrators to keep the MOUT site as a showpiece. All commanders at all levels must protest vehemently against regulations requiring that units rake gravel, place smoke grenades only in approved predesignated containers, keep the building walls clean, and not damage furniture. The division commander quashes such ridiculous tendencies and allocates a MOUT maintenance team for the purpose of repairing and replacing damaged buildings and furniture after training. Each MOUT building should have prefabricated mouse holes, loopholes, and breach holes at numerous places. These may be filled with molded hardened foam, which is removed once proper munitions have been used to create the hole. In this manner, the leader can reinforce good tactics rather than accepting poor habits caused by unimaginative MOUT facilities. If soldiers are forced to conform to the artificial constraints during training, they will make the same mistakes in combat.

NOTES

1. FM 90-10-1. This FM forms the foundation for this portion of the SOP.
2. Doubler, 89–90, 92.
3. Even the Russians, with their vast experience in urban combat, suffered a stinging setback in Grozny, Chechnya, because they precipitously punched mechanized battalions into the city center. Even though armed with small arms, mortars, and RPGs, the Chechnians easily isolated and wiped out the professionals.
4. Patton, 326.
5. Ibid., 326.
6. Ibid., 326.
7. Doubler, 91. This proved to be the solution for infantry units clearing Germans out of towns and cities. The Germans could not counter this tactic.
8. English, 105. Use of the tank-infantry team in MOUT became standard practice in World War II.
9. Doubler, 93. U.S. units were often task organized into assault and close support elements during urban combat. The assault element comprised two Browning Automatic Rifle (BAR) teams and a bazooka team. Attachments of demolition and flamethrower teams

were the norm. Some units organized the assault team into searching and covering parties. The searching party comprised four riflemen further divided into two "buddy" teams. The covering party had one BAR and three soldiers under the control of the assistant team leader. In this manner, the clearing of buildings was deliberate and systematic. I derived the foundation of the platoon and squad task organization from numerous observations of the Bundeswehr 353d Light Battalion, which provides demonstrations of MOUT operations at BONNLAND, the German Infantry School's MOUT site, February 1999 to July 2000. Hereafter cited as BONNLAND.

10. BONNLAND.
11. Doubler, 91. This differs from how U.S. infantrymen cleared buildings in World War II. The order of clearance was ground, top floors, and basement.
12. English, 105.
13. Doubler, 91. Infanterieschule, *The FIBUA-MOUT Training Guide* (Hammelburg, Germany: Infanterieschule, 1994), 15–18. This training pamphlet from the German Infantry School in Hammelburg provides TTPs for the infantry platoon and squad. Many of the lessons of World War II are apparent in this document.
14. Observations of the BONNLAND strongpointed building demonstration model, July 2000. Hereafter cited as Strongpoint.
15. Strongpoint.
16. Ibid.
17. English, 137, 106–107. The Russian use of counterattack forces in MOUT during World War II provides an excellent model.
18. Doubler, 91. This was a part of German urban combat doctrine during World War II.

Chapter 25

Counterinsurgency

As Bard E. O'Neill recounts, insurgencies vary by their ends, ways, and means. Despite their prevalence globally, the United States will rarely become involved *directly* in insurgencies and most likely only as a result of conventional wars ending with regime change.

Unlike conventional warfare, in which states resolve political issues with established armed forces, following international laws, rules, and customs of warfare, insurgencies are domestic conflicts and are not governed by conventions. Since insurgents are non-state actors, they are under no obligation to adhere to the accepted ways and means of warfare.

In the vast majority of insurgencies involving the United States, the population is the center of the conflict. So, insurgents will use terrorist acts, assassinations, threats and the like to intimidate the population into acquiescence. Under these circumstances, the population will remain passive and noncommittal, waiting to see which side will prevail. Unfortunately, the population is caught in the middle, as insurgents operate in the population centers and government security forces clash with them, resulting in collateral damage.

The proper way to conduct counterinsurgency is for the coalition to provide security at the local level, that is, in every urban area. Here, the term "urban" connotes hamlets, villages, towns, and cities and not the geographic differential between urban and rural areas. The development of national military and police forces are necessary institutions for the state's long term interests, but it cannot be the sole focus in counterinsurgency. The insurgents will be well ensconced by the time these institutions mature. Hence, establishing local security is the most effective way to prevent the insurgency from growing and can be thought of as an urban denial strategy.

Government security and coalition forces cannot cover completely every urban area because it will result in overextension. The recourse is for the people to provide their own local security forces (e.g., auxiliary police and militia). Unfortunately, these forces can easily become as bad as the insurgents if they lack values and discipline. The exigencies of the emergency may dictate this solution initially, but only to the risk of long-term stability. Hence, in order to separate the insurgent from the population, the coalition should adopt some form of an urban denial strategy.

Missions. Under an urban denial strategy, the coalition assigns geographic areas of responsibility for units to establish security. As such, a company may either be assigned clearing missions (formerly called search and destroy) or hold and build missions. Clearing operations rely on typical conventional missions, such as patrolling, ambushes, raids, movements to contact, etc., so this chapter need not explore them in greater detail. Hold and build missions are unique to counterinsurgency, and hence serve as the focus here.

In hold and build missions, the company is assigned a district or part of a district (depending on its geographic size) to secure. In high threat areas, that is, in areas where insurgents are active, larger maneuver units should clear the area of apparent insurgent forces before the company assumes its hold and build responsibilities. The commander can expect a brief interval before insurgents attempt to reassert themselves in the assigned area. Hence, the commander must formulate his plan according to insurgent responsiveness to the coalition's activities.

Within his assigned district, the commander can assume a permanent presence in every urban area or rotate training teams to the urban areas from a company central base. A permanent presence in each urban area has the most enduring and effective impact on denying insurgents access though. Under a permanent presence approach, the commander assigns a four-man cadre to each urban area for the purpose of raising a local security force. In essence, this approach is the reverse of insurgent tactics.

Cadre Framework. The cadre meets with the local urban authorities (e.g., mayor, council, chieftain, etc.) for the purpose of explaining its presence and gaining their acquiescence. The first task is to raise a local police force with input from the local authorities. The cadre recruits, organizes, trains, equips, and pays the salaries of the police force. Conceptually, a twelve-man police force is adequate for most villages and towns. The idea is to provide the force with basic skills so it can respond to insurgent activities in the immediate area. Finally, the cadre instills discipline and values into the force so it does not morph into a death squad, an oppressor, or a criminal conspiracy. This danger is the best reason against having local authorities establish local security forces on their own. After a period of evaluation, the cadre selects promising leaders for formal police training. Upon their return, the trained leaders take over cadre responsibilities, permitting the cadre to focus on the next phase—construction and development.

Information Operations. The cadre becomes a major source of news for the inhabitants. Its task is to inform, reassure, and persuade the people as well as to rebut insurgent propaganda. Providing daily news creates a hungry audience for information. In the meantime, the cadre needs to reassure the inhabitants that a return to normalcy is the cadre's primary objective. The cadre must also persuade the locals that the coalition has a superior strategy and commitment, but ultimately, the people must seize control of their own destiny. The cadre must respond immediately to insurgent propaganda and rumors, otherwise, the locals will believe them, no matter how ludicrous. Ultimately, information operations seeks to instill hope and confidence among the people so they will start sending their children to school, start seeking work, and start investing in the community.

Intelligence. The main source of intelligence in counterinsurgency is human intelligence (HUMINT). The population is the greatest resource for HUMINT, but it must have the confidence, ways, and means to provide that intelligence without suffering insurgent retribution. With the assistance of the local authorities, the cadre establishes a neighborhood watch network by appointing a chief administrator. In abstract, the system is a pyramid network, with the chief administrator dividing the urban area into two sectors and appointing two sector administrators. This system of division continues down to a basic group of five families under the responsibility of a block administrator. Each family reports suspicious activity or the arrival of new people into the neighborhood to the block administrator, who in turn alerts the police and submits a report up the network. Notably, the administrators should be prominent citizens (e.g., businessmen, councilmen, and the like) who have a connection with the community. The cadre meets with the chief administrator frequently for intelligence collection. In this manner, the company provides intelligence to the battalion TOC, local provincial reconstruction team (PRT), and the local Special Forces detachment.

Cadres should be wary of initial information and not leap to conclusions. Some people may falsely accuse others of insurgent activities out of personal vendettas, ethnic or tribal hatred, or basic rivalries. The local police should investigate accusations since they have intimate knowledge of the people. Admittedly, justice under this approach will not always be served, but the cadre cannot assume the role of policeman for the plethora of disputes in the community. It can only establish the conditions for law and order to take root over time.

The commander determines whether the intelligence indicates organized insurgent activity in the district. If he suspects such activity, the commander requests higher headquarters conduct a clearing operation for the purpose of pushing the insurgents away from the urban areas. Eliminating insurgents is not nearly as important as denying them access to the population.

Security. Establishing good rapport with the local authorities and population provides a fair degree of security. In some cultures, local authorities feel obligated to safeguard guests so insurgents will not be able to attack the cadre with impunity. At a minimum, the cadre needs the people to provide warnings of possible attacks. The cadre needs a safe house with security guards for immediate security, or it might prove expedient to locate the safe house in the police compound for additional security.

The cadre develops an escape and evasion plan to a link-up point with the company quick reaction force (QRF). An escape tunnel from the safe house to the evasion route is an effective technique. The QRF should avoid moving straight into the urban area because insurgents may set up an ambush or place explosive devices on the likely approach routes. From the link-up point, the cadre can lead the QRF into the urban area with the least amount of confusion.

Company Central Base. The commander may opt to keep the company consolidated in a central base and have training teams visit the urban areas frequently in order to increase security as well as command and control. The commander must realize that any degree of separation from the population hampers progress towards establishing local security. Moreover, the frequency of visits is dependent on available transportation. Actionable intelligence is likewise inhibited or retarded due to this separation. If insurgents are able to gain a foothold, sooner or later they will begin targeting the company directly. Hence, the decision to consolidate the company out of concern for greater force protection may create the conditions of greater vulnerability in the long run.

The number of urban areas in the assigned area of responsibility may prevent the allocation of sufficient cadres to every village or hamlet. In this case, the commander may opt to assign a number of villages or hamlets to each cadre. While this arrangement is not optimal, it is preferable to the company central base framework.

Construction and Development. While the cadres are establishing local security, the commander and executive officer visit the local PRT, NGOs, IOs, UN Assistance Mission, Special Forces detachment, civil affairs unit, and USAID. Through these discussions, the commander formulates a construction and development plan for the purpose of providing immediate impact projects in his district. Priority of projects goes to communities that have achieved a functioning police force and security. He also speaks with local authorities to determine their greatest needs (e.g., schools, clinics, wells, roads, etc.).

Construction projects should utilize local labor as much as possible. Training provides useful occupational skills, salaries bolster the local economy, and participation instills pride of ownership. All three factors encourage the locals to remain loyal to the government rather than joining the insurgency.

The commander should coordinate for capacity building training (e.g., teaching, technical, maintenance, agricultural, etc.) with appropriate organizations and agencies. Creating capacity has an essential long term impact, and the commander must be the chief advocate.

The cadres do not administer justice. For observed criminal and morally repugnant activities, the cadre should submit reports to the company commander so he may bring the issue up with the appropriate UN agency and other organizations. With this approach, the cadre does not become involved in local sensitive issues which may compromise its position. Besides, UN agencies have intimate knowledge of such issues and can deal with them appropriately. The cadre's unique position as a provider of information should not be compromised.

Development begins with creating a commitment to the community. As a consequence of insurgents' use of terrorist acts, intimidation, and subversion, the people withdraw into themselves out of a sense of paranoia. Insurgents instill this type of behavior because it undercuts organized resistance and encourages passivity. The cadre needs to reverse this behavior so people can renew trust and form bonds with one another. Organized sports (e.g., soccer and volleyball), social clubs (e.g., sewing, arts and crafts, and cooking), and school activities are superb ways to create a sense of community and fraternity. The police should also participate in these activities because their commitment to the community is a critical factor in their resolve to defend their friends and families. This commitment distinguishes local police from outside forces, such as auxiliary and national police and should not be downplayed.

The 1SG can arrange for medical activities (e.g., Medical Civil Affairs Program, MEDCAP) once the security environment permits. MEDCAPs should occur in villages that have yet to experience construction projects. In this manner, the villagers do not feel left out and frustrated. MEDCAP personnel are responsible for running the event. Other than arranging for one, the company's involvement should be minimal.

Cadres must anticipate that a few police recruits are unsuitable either due to corruption, mixed loyalties, or incompetence. Cadres must either consult or inform the local authorities whenever they fire a policeman and provide an explanation. If the cadre can, it should retain the former policeman in some capacity. If his loyalties are suspect (e.g., an agent for a local power broker, crime boss, or insurgent cell), he must be watched.

The company commander must meet with the local authorities in his district frequently. His goals are to apprise them of projects as well as insurgent activities. He should encourage them to preach patience and tolerance in their communities in regards to the other communities. He should warn them that wanton acts against other communities that have received construction projects are counterproductive and will likely lead to retribution. The commander should control the meetings by asking the authorities for their top three needs; otherwise, the authorities will fill the time with a long list of demands and grievances.

Service Support. The XO acquires the weapons, ammunition, and uniforms or badges for the cadres to issue. Normally, finding weapons is not an issue and cadres may be able to find enough locally to equip the police forces. Ammunition may be more problematic and so the XO must use some ingenuity in finding it.

The XO must arrange for the safekeeping of the money earmarked for the cadres. Because the local PRT or military camp is the most secure area, money can be safeguarded there.

The supply sergeant manages Class I support from the local military camp or PRT to the cadres. The cadres can hire a local cook or have one assigned from the military camp if available. Except for special occasions, eating the local cuisine should be infrequent. The medical personnel in the area should provide guidance on food. Of course, only bottled water is safe for consumption.

Because of the relative isolation of the company from the military logistical system, the commander can opt for contracting support.[1] Deployed military units require a robust logistical program that can meet urgent and compelling requirements for various supplies and services. Occasionally, military units have support requirements that cannot be met by the military supply system (supplies) or internal/organic military capability (services). The majority of supply requirements are for a specific commercial item or class of supply. They include items such as batteries, office supplies, building materials and general supplies. The same holds true with services that cannot be supported by organic military engineer and other CS/CSS units. Good examples of services requested include ice production, water purification, fueling operations, ground and vertical construction, billeting, and power generation.

Support requirements are identified at all levels. For example, a company commander identifies a need for building materials in order to build a company command post (CP). He must break the requirement down into exact items and quantities. He uses as much engineering subject matter expertise that is available to validate his requirement. Once he has validated the requirement, the commander prepares a DA Form 3953 Purchase Request and Commitment (PR&C) through the company supply clerk and battalion S4. After the DA Form 3953 is prepared, the battalion S4 works with the brigade S4 to receive funding from the Division comptroller. After funding is approved and the DA Form 3953 has been signed by all approval officials, the unit representative will deliver the DA Form 3953 to the supporting Contingency Contracting Office located at the respective military camp (e.g., basecamp, FOB, or COB).

The commander can resolve services needs with the Logistical Civil Augmentation Program (LOGCAP) and Air Force Civil Augmentation Program (AFCAP). In Iraq and Afghanistan, for example, 95 percent of all service requirements are covered under LOGCAP and AFCAP contracts. All U.S. military personnel are provided with services under a specific contract task order for the time period the military unit is deployed. Typical services include Laundry and Bath, Food Service, Mortuary Affairs, Field Sanitization, Waste Management, Billeting, Facilities Management, Power Generation and Distribution, Bulk Water and Fuel, MWR, Theater Transportation, and Engineering and Construction. To request services, the company commander prepares a memorandum and Statement of Work (SOW) for the requested services and submits it for approval by an 06 or designee. The SOW must provide the 5 Ws—**Who** is being supported, **What** is the service requested, **When** (period of performance for the services), **Where** (location for the services to be performed), and **Why** (impact on the military unit if the services are not provided). After the documentation is completed, the company commander submits the package to the military camp Mayor for review and approval. The Mayor must first determine if the requirement can be accomplished with organic assets before requesting LOGCAP or AFCAP support. If unable to support with organic assets, the Mayor forwards the request thru the Facility Engineer Team (FET) and/or Department of Public Works (DPW) personnel to ensure the request does not interfere with the military camp facilities master plan (FMP). If the Mayor approves the request, it is forwarded to the LOGCAP or AFCAP office where a Defense Contract Management Agency (DCMA) contracting officer reviews and drafts a Letter of Technical Direction (LOTD) that will be issued to the CAP contractor for execution.

Command and Signal. Cadres must have radio contact with the company net, which must have radio communications with the battalion TOC, local PRT or local military camp. If communications are a problem, the commander must request more powerful radios for the respective cadres. Assured communications are essential to the cadre system.

The cadres should strive to create a radio network for the district police to the provincial police headquarters. Eventually, the national police communications system will expand to the provinces; the cadres can assist in closing the gap.

The company command post is most effective at the district or provincial capital. His location in the capital permits him to interact with the district and/or provincial authorities. Ideally, the commander's location permits him to interact easily with the local PRT as well as the other agencies and organizations.

Conclusions. Counterinsurgency requires a counterintuitive approach to security operations. Since insurgents draw virtually all of their strength from the populace, the counterinsurgency effort must focus on denying insurgents access. The longer they are denied access, the weaker they become. Insurgents are at their weakest at the beginning of an insurgency, so the faster the coalition can establish local security, the greater the chances of choking off the insurgency. Besides, the vulnerability of cadres to insurgent attacks is mitigated by the establishment of a local police force, the neighborhood watch network, and a cadre security plan.

Some areas will be more endangered by insurgency activities than others, but even in secure or stable areas, cadres must strive to inoculate their urban areas from insurgent incursions. Insurgents show a remarkable ability to shift activities when forced out of one area. Even in those regions where insurgency presence is nonexistent, cadres can focus the police forces on criminal activities.

The establishment of security at the local level is the foundation of hearts and minds. Thereafter, construction and development may progress as a means of creating incentives for positive behavior as well as returning hope and confidence in the people. Without security, construction and development are built on a foundation of sand.

In aggregate, the counterinsurgency pushes insurgents increasingly away from urban areas and to the borders. Ultimately, defeat for the insurgents is not gained by casualties, but by loss of hope. Success for the counterinsurgency results when the outer circle of insurgents goes home and settles down. They need not love the government to live under it peacefully. Insurgent incitement and propaganda aside, citizen satisfaction with the established government is not a significant factor in determining the outbreak and expansion of an insurgency.

Democracy is a process towards broader rights and freedoms and not a substitute for security. Because insurgents deprive the people of choice, such issues as elections, economic wellbeing, corruption, and development are largely irrelevant in the final outcome. The commander may need to remind subordinates of that fact when they become frustrated with the slow pace of progress. It takes much longer to normalize life once an insurgency has shattered society, the economy, and order. This is the reason counterinsurgencies are prolonged struggles. If the company can establish an enduring security in every village and hamlet, it is 90 percent successful.

NOTE

1. The author is grateful to Major (P) Lance B. Green, USA, Commander of the Defense Contracting Management Agency, Albuquerque, New Mexico, for providing the procedures for supplies and services support.

RECOMMENDED READING

Bulloch, Gavin. "Military Doctrine and Counterinsurgency: A British Perspective." *Parameters*. U.S. Army War College, Summer 1996.

Chiarelli, Peter W. and Michaelis, Patrick R. "Winning the Peace: The Requirement for Full-Spectrum Operations." *Military Review*. July-August 2005.

Fall, Bernard. *Street Without Joy*. Mechanicsburg, PA: Stackpole Books, 1964.

———. "The Theory and Practice of Insurgency and Counterinsurgency." *Naval War College Press*. Winter 1998. http://www.nwc.navy.mil/press/review/1998/winter/art5-w98.htm.

Hammes, Thomas X. *The Sling and the Stone: On War in the 21st Century*. St Paul, MN: Zenith Press, 2004.

Horne, Alistair. *A Savage War of Peace: Algeria 1954–1962*. Revised Edition. New York: Penguin Books, 1987.

Kilcullen, David. "Counterinsurgency Redux." August-September 2006. http://smallwarsjournal.com/documents/kilcullen1.pdf.

Kilcullen, David. "Twenty-eight Articles: Fundamentals of Company-Level Counterinsurgency." Edition 1. March 2006. http://www.d-n-i.net/fcs/pdf/kilcullen_28_articles.pdf.

Metz, Steven and Millen, Raymond. *Insurgency and Counterinsurgency in 21st Century: Reconceptualizing Threat and Response*. Carlisle, PA: Strategic Studies Institute, November 2004.

Millen, Raymond. "The Hobbesian Notion of Self-Preservation Concerning Human Behavior during an Insurgency," Volume 35, Issue 4, *Parameters*. Winter 2006.

———. "Thinking Small: Applying Hobbes to Counterinsurgency," *Military Review*. TBP, Fall 2007.

———. "The Yin and Yang of Counterinsurgency in Urban Terrain," Volume VIII, Issue 3, *Joint Center for Operational Analysis*. September 2006.

Nagl, John A. *Counterinsurgency Lessons from Malaya and Vietnam: Learning to Eat Soup with a Knife*. Westport, Connecticut: Praeger, 2002.

O'Neill, Bard E. *Insurgency & Terrorism: Inside Modern Revolutionary Warfare*. 2nd Edition, Revised. Washington, D.C.: Potomac Books, Inc., 2005.

Taber, Robert. *War of the Flea: The Classic Study of Guerrilla Warfare*. Washington, D.C.: Brassey's Inc., 2002.

Taw, Jennifer Morrison and Hoffman, Bruce. *The Urbanization of Insurgency*. Santa Monica, CA: RAND Corporation, 2004.

Trinquier, Roger. *Modern Warfare*. Translation by Daniel Lee. New York: Praeger Publisher, 1964.

Chapter 26

Conclusion: Empowering the Junior Leader

Readers should view this book as an inspiration for developing their own tactical doctrine. Establishing a company doctrine is admittedly time consuming and laborious for the initiator. But the benefits for his successors are exponential. With a single resource for reference and a place to deposit hard-learned lessons, leaders can now have a foundation for professional growth. This is far better than units filing after-action reviews (AARs), ARTEP results, and other evaluations away in desk drawers or trash cans because no one knows what to do with such information. The challenge for all is not to blindly copy information from manuals, military history sources, or even the doctrine of other units simply to fill up space. A good doctrine is tailored to the needs and mission essential task listing (METL) of the company. The leaders must put pen to paper and write down their own thoughts in order to understand the thought process and exercise their brains. The concepts discussed in this book will do no one any good if they are simply copied without anyone thinking about what has been said, what others think, and what is impractical. I have revised this book a great deal since starting it after company command. Many insights and thoughts I considered worthwhile initially have been modified or thrown out. Further readings inspired me to question earlier beliefs. Moreover, some of my ideas, such as the open-terrain defense or views on entrenchments may not be practical in modern combat. The reader must question everything, no matter how revered the source. Once a thought is written down, the company must experiment with it and adapt it to make it practical. The company doctrine will require frequent updates as a result of AARs, the introduction of new technologies, and new ideas. In this spirit, the reader must view this book as a starting point for his own company doctrine rather than the definitive source of all knowledge.

The challenge for commanders in the future is adapting new technologies and changing missions with company doctrine. The purpose of company doctrine is to focus training and harmonize the tactical activities of the company in sustained combat. The danger of short conflicts, which the United States has experienced since Vietnam, is that the deficiencies in training, technology, and doctrine often do not surface. Leaders would be wise to view this as a Cinderella period so as not to become complacent when dealing with future adversaries. Certainly, the British Army in the nineteenth century experienced this period up to World War I, even though the Boer War should have served as a wake-up call. The Germans too suffered from such false assurances in World War II, until the sustained operations on the eastern front not only revealed their own deficiencies but also served to decimate the German Army.

Commanders must assess the impact of technology and peace-making/keeping/enforcement missions on combat effectiveness. No one is going to say that technology is a substitute for human resources, but the commander should be wary of claims that certain technologies will allow the company to do more with less.

The commander must not stake the security of his company on intelligence platforms and similar systems in lieu of patrols and listening and operations posts (LP/OPs). No intelligence system is going to provide 100 percent coverage, 24 hours a day. Even if higher headquarters does acquire the enemy forces, they will probably make contact with the company before the intelligence filters down. When intelligence does work in the company's favor—great. The company will put it to good use. If not, the company will have its own resources in place and functioning without a deterioration in security or information.

When testing and receiving advanced weapons and equipment, the commander must provide feedback to higher headquarters concerning their effectiveness and reliability. The company must at least address the following questions.

- Does the weapon increase lethality?
- Is the weapon or equipment durable? If it is too delicate to withstand reasonable handling in combat, its use to the infantry soldier is minimal.
- Does it reduce the soldier's load?
- Is it simple to place into operation at night?
- Does it increase command and control and situational awareness?
- Does it reduce weather effects on the soldier?
- Does it provide sufficient protection from enemy firepower to give the soldier time to react?

All questions are closely linked, because a negative on one may nullify the advantages of the others. Progress is good only if it results in an improvement to the combat effectiveness of the company. Lastly, junior leaders must assess how new technologies change tactics, techniques, and procedures (TTPs). It is important for leaders not only to look at the intended uses but also to consider the spin-off possibilities. For example, the use of the 88 mm antiaircraft gun in an antitank role provided the Germans with a weapon that devastated tank units in World War II. Increased technology also tempts higher command to micromanage subordinate units—the proverbial helicopter in the sky. Commanders must firmly insist on being allowed to exercise their freedom of action. This position is difficult, because it can result in an early end to a career. But for the system to work, leaders must remind higher command that doctrine emphasizes mission-type orders and that loyalty and trust run both ways. I say this, because everyone needs to be reminded that junior leaders are placed in their position to ease the burden of command on higher headquarters and to learn their craft.

The explosion of the information age, particularly with the Internet, allows people to access a great deal of information effortlessly. With military posts, units, schools, and other centers on-line, the researcher can seek solutions to most tactical problems quickly. The ability to access manuals from TRADOC and lessons learned information from CALL makes the acquisition of current knowledge immediately beneficial to the unit. In many ways, the accession of the computer and the Internet have made the creation of a company doctrine practical. I would like to see the Army take the next step by creating an on-line forum for the free exchange of information. In this manner, soldiers can submit their tactical ideas and receive immediate feedback from other soldiers. Even though the feedback can be brutal, the criticism serves to cause everyone to think about the flaws and strengths of the original premise, inspire alternate ideas, and help the original author revise his premise.

Another resource for helping junior leaders learn operations and visualize the battlefield is computer simulations. Eventually, the Army will adopt cheap, userfriendly war-gaming simulations for the company that do not require a horde of technicians, computers, and weeks to run. The commercial market already has excellent company-level computer simulations (for instance, Steel Panthers II) at reasonable prices. First, with the capability to tailor the order of battle, terrain, and types of weapons, the commander and his subordinates can test new TTPs through several variations before applying them in the field. Second, these simulations can be set up and played without wasting time and resources. Besides being fun to play, they are also valuable tools for leaders to learn their craft.

The infantry, more than any other arm, must perform a wide diversity of tasks and missions. It is shameful that society regards infantrymen as simple soldiers. With the vast number of technical and tactical tasks that even the lowest private must perform, as well as the demanding physical and leadership aspects of his trade, the infantryman should be regarded with awe. In the great majority of cases, an infantryman departing the service is much more intelligent, savvy, and effective as a leader than when entering service. The infantry does not demand the best recruits; it molds the best soldier possible from the materials it is given. Consequently, it takes much time and training to produce a good infantryman. In fact the ability to maintain a well-trained force is probably the decisive element in any prolonged conflict. It certainly proved decisive in destroying the Wehrmacht and Luftwaffe in World War II. Throwing partially trained infantrymen into combat only increases casualties without increasing the combat effectiveness of the unit. It is folly to think that units can operate undermanned in peacetime and conduct combat operations with an influx of replacements at the last minute. The Army's main priority should be to maintain all infantry units at 100 percent strength. It should also lower the numerous administrative requirements on infantry brigades and battalions so they are not forced to overfill their staffs with assistants and clerks. It is not enough to direct that staffs be limited to a certain size if higher command does not have the good judgment to lift the stultifying *administrivia*. The Army should also form training battalions in each division for the purpose of training replacements and preparing them for the rigors of combat. Plenty of documentation exists of the German model to organize such a battalion rather than expecting companies to assimilate replacements while in the midst of conducting operations. To do otherwise is simply to create cannon fodder.

Except for the absence of immediate danger, infantry in training is the same as infantry in combat. The weather is no less compromising, sleep is a valued commodity, and physical exertion is no less demanding. Infantrymen do not earn the Combat Infantry Badge simply because they have participated in a battle; many other arms also go into battle. The badge honors the one arm that is always on the front line, exposed to imminent danger, ever vigilant, and expected to persevere—no matter how deplorable the situation, odds, or decisions or higher headquarters.

Perhaps the proudest moment of any soldier as he leaves a unit is looking back on his accomplishments. As a vehicle for the corporate effort, the company doctrine allows everyone to make a contribution and to formulate experiences, tactical judgment, and ingenuity in written form. For the junior leader, the contribution to the SOP is an enduring legacy, which not only helps the unit become better but also helps the junior leader to be a better one. For the commander, it is his legacy of command.

Appendix

Professional Reading

Ambrose, Stephen. *Citizen Soldiers.* New York: Touchstone, 1997.
———. *The Victors.* New York: Simon and Schuster, 1998.
- Good anecdotal accounts of small unit actions from Normandy to the end of the war.
- Some redundancy between both books, but both stand alone.
- Excellent evidence of the power of individual leadership in turning a battle favorably.

Boga, Steven. *Orienteering.* Mechanicsburg, Pa.: Stackpole Books, 1997.
- Use of terrain features for navigation.
- Direct lay of compass on map to determine magnetic-grid azimuth.
- Calculating rate of travel over elevated terrain.

Bolger, Daniel P. *The Battle for Hunger Hill.* Novato, Calif.: Presidio Press, 1997.
- Excellent insights into the Joint Readiness Training Center from a battalion commander who experienced two rotations within a seven-month period.
- Practical technique for rules of engagement (ROE), CCIR, and maneuvering against guerrilla forces.
- Anecdote on operations in Haiti and the real-world use of his practical ROE.

Bolger, Daniel P. *Death Ground: Today's American Infantry in Battle.* New York: Ballantine Books, 1999.
- Insightful examination on how modern warfare has changed the demands on the infantry.

Butler, Rupert. *The Black Angels: A History of the Waffen-SS.* New York: St. Martin's Press, 1979.
- Insight into the attitudes toward atrocities among elite soldiers.

- Dispels some of the myths regarding the pure-fighting image of the Waffen SS.

Cartwright, Frederick F. *Disease in History.* New York: Dorset Press, 1991. Chapter 4.
- Effect of typhus on Napoleon's 1812 campaign.
- Effect of stress and sleep deprivation on Napoleon's judgment.
- Effect of poor health on Napoleon's conduct of later battles.

Center For Army Lessons Learned, Fort Leavenworth, Kans.: U.S. Army Training Command (TRADOC).
- Excellent resource for topical tactical issues.
- Good source for practical TTPs.
- Its web site (http://call.army.mil/call.html) is worth viewing weekly.

Clausewitz, Carl von. *On War.* Edited and translated by Michael Howard and Peter Paret. Princeton, N.J.: Princeton University Press, 1976. Book 1, chapter 3-7.
- An examination of courage in combat decision making.
- Insights into the influence of friction and fog of war.
- These chapters, amounting to only 22 pages, are a good place to begin for junior leaders to understand their profession. There is much more in this superb classic, which is rarely read but often quoted by pseudointellectuals.

Doubler, Michael D. *Busting the Bocage: American Combined Arms Operations in France, 6 June–31 July 1944.* Combat Studies Institute. Fort Leavenworth, Kans.: U.S. Army Command and General Staff College, 1988.
- Detailed examination of the infantry-tank-engineer team in defeating the German hedgerow defenses.
- Good insights into U.S. military approach to solving battlefield problems at the lowest levels.

———. *Closing with the Enemy: How GIs Fought the War in Europe, 1944–1945.* Lawrence: University Press of Kansas, 1994.
- An expansion of his Combat Studies Institute (CSI) work, this book is perhaps the best analysis of U.S. Army tactical combat during World War II. This should be a tactical bible for commanders.
- Detailed examination of combined arms tactics against enemy prepared positions, military operations in urban terrain (MOUT), fortified positions, and forest combat.

Dunnigan, James E., and Albert A. Nofi. *Victory and Deceit: Dirty Tricks at War.* New York: Quill William Morrow, 1995.
- Use of deception in history.
- Illustrations on how deception can overextend the enemy or have him divert his resources away from the decisive point.

Ebert, James R. *A Life in a Year: The American Infantryman in Vietnam, 1965–1972.* Novato, Calif.: Presidio Press, 1993.

- Experiences from letters of U.S. infantry soldiers in Vietnam.
- A look at the influence that a nebulous war policy and poor strategy have on morale.

Ellis, John. *Eye-Deep in Hell: Trench Warfare in World War I.* Baltimore, Md.: The Johns Hopkins University Press, 1991.
- Life of an infantryman in positional warfare.
- Battle effects on the infantryman.
- Disease on the battlefield.
- Trench designs and purpose.

Ellis, John. *On the Front Lines: The Experience of War through the Eyes of Allied Soldiers in World War II.* New York: John Wiley & Sons, Inc., 1990.
- Superb insights on the effects fatigue have on the soldier's psyche.
- Good reference for other considerations on combat.

English, John A. *On Infantry.* New York: Praeger Publishers, 1984.
- Good study on the use of infantry in the twentieth century.
- Evolution of infantry tactics.
- Some historical background on the soldier's load.

Fritz, Stephen G. *Frontsoldaten: The German Soldier in World War II.* Lexington: The University Press of Kentucky, 1995.
- Experiences from diaries and letters of German soldiers on the eastern front.
- Insight into the attitude toward atrocities among the common soldier.
- Insight into the effectiveness of camaraderie in maintaining unit cohesion.

Fussell, Paul. *Wartime: Understanding and Behavior in the Second World War.* New York: Oxford University Press, 1989.
- Superb book on the brutal effect that World War II combat had on the Allied soldier, especially with an inept training and replacement system, as well as a nonexistent combat rotation system. Replete with anecdotes.
- Excellent chapter on daily rumors within a combat unit.
- Insight into the commonality of atrocities.

Ganter, Raymond. *Roll Me Over: An Infantryman's World War II.* New York: Ballantine Books, 1997.
- Experiences of a U.S. soldier during the last seven months of the war and aftermath.
- Some insight into American crimes against prisoners and civilians, mostly looting, wanton destruction of civilian property, and rape.
- Good insight into the poor assimilation of replacements into combat units.

Garland, Albert N. *Infantry in Vietnam: Small Unit Actions in the Early Days: 1965–66.* New York: Jove Books, 1985.
- Good case studies of small unit actions in Vietnam: intelligence, ambushes,

patrolling, attack, defend, fire support, combat support, special operations, pacification program, and leadership.

- Good insights on command leadership during the heat of battle.

Griffith, Paddy. *Forward into Battle: Fighting Tactics from Waterloo to the Near Future.* Novato, Calif.: Presidio Press, 1991.

- Excellent study on the relationship between firepower and closing with the enemy.
- Superb insights on the effect of battlefield stress on the soldier.

———. *Battle Tactics of the Civil War.* New Haven, Conn.: Yale University Press, 1989.

- Good companion to *Forward into Battle.* Dispels some of the myths of firepower in the Civil War.

———. *Battle Tactics of the Western Front.* New Haven, Conn.: Yale University Press, 1994.

- Good companion to *Forward into Battle.* Dispels some of the myths of firepower in World War I.
- Excellent study on British efforts to improve tactics in positional warfare.

Howard, Michael. *Clausewitz.* New York: Oxford Press, 1988.

- Good synopsis of Clausewitzian warfare.
- Provides good primer before reading the book.

Kals, W. S. *Land Navigation Handbook.* San Francisco: Sierra Club Books, 1983.

- Converting elevation into straight-line distance equivalent.
- Use of terrain rails for navigation.
- Direct lay of compass on map to determine magnetic-grid azimuth.

Keegan, John. *The Face of Battle.* New York: The Viking Press, 1976. Chapter 5.

- Good analysis on the battlefield effects on the soldier.

Liddell-Hart, B. H., ed. *The Rommel Papers.* New York: De Capo Press, Inc., 1953.

- Excellent collection of Rommel's tactical and operational thoughts in France, North Africa, and Normandy.
- Important insights in achieving moral superiority on the battlefield, the position of the commander on the battlefield, and the use of mobility to defeat superior forces. Also note the use of the baited trap in destroying enemy forces in detail.

Lindermann, Gerald. *The World within War: America's Combat Experience in World War II.* Cambridge, Mass.: Harvard University Press, 1997.

- The best book written on the debilitating effects of combat on the individual soldier and how continued combat affects the combat effectiveness of the soldier and the unit.
- Superb analysis of the causes of high casualties of replacements to the front.

- Revealing discussion on the banality of atrocities.
- Excellent discussion on the cause and effect of unit cohesion.

Lucas, James. *War on the Eastern Front: The German Soldier in Russia 1941–1945.* Novato, Calif.: Presidio Press, 1991.
- Good study on German tactics on the eastern front.
- Useful look at antitank tactics and use of terrain in the defense.

Lupfer, Timothy. *The Dynamics of Doctrine: The Changes in German Tactical Doctrine during the First World War.* Combat Studies Institute. Fort Leavenworth, Kans.: U.S. Army Command and General Staff College, 1981.
- Excellent study on the evolution of German defensive doctrine on the western front.
- Excellent analysis of the German flexible defense and infiltration tactics.

MacDonald, Charles B. *Company Commander.* New York: Bantam Books, Inc., 1978.
- Experiences of a company commander in World War II.
- Note the routing of his unit during the Ardennes offensive.
- Note his reaction to the execution of German prisoners.

Malone, Dandridge M. *Small Unit Leadership: A Commonsense Approach.* Novato, Calif.: Presidio Press, 1983.
- A study of leadership from Vietnam to the mid-1980s.
- Reminiscent of Leadership 101—and just as exciting—but still has value.

Manchester, William. *Goodbye, Darkness: A Memoir of the Pacific War.* Boston: Little, Brown and Company, 1980.
- Harrowing experiences of a U.S. enlisted Marine during World War II.
- Note the effects of danger and stress on his psyche.
- Interesting section on the Marine method of dealing with snipers.

Marshall, S. L. A. *Men against Fire: The Problem of Battle Command in the Future.* 1947. Reprint. Gloucester, Mass.: Peter Smith, 1978. Chapters 3–12.
- Excellent study on U.S. use of firepower in Normandy and Pacific islands. Note the effect of marksmanship training, fear of fratricide, and lack of confidence with the resupply of ammunition on the expenditure of ammunition.
- Excellent insight on the paralyzing effects of fatigue and feelings of isolation on the "empty battlefield."
- Note the position and role of the commander on the battlefield.

———. *The Soldier's Load and the Mobility of a Nation.* Washington, D.C.: The Combat Forces Press, 1950.
- Excellent study on the deleterious effects of overloading the soldier to include the link between fatigue and fear.
- Well-founded recommendations for optimum loads.

Mauldin, Bill. *Up Front.* New York: World Publishing Company, 1945.

- Good insight on the life of the soldier on the front.
- The value of humor in maintaining morale.

Military History and Publications Section of the Infantry School. *Infantry in Battle.* Washington, D.C: The Infantry Journal Inc., 1939.
- Excellent case studies of small unit actions in World War I. Many lessons here.
- Read this book first as a good starting point for professional development.

Moran, Lord. *The Anatomy of Courage.* Garden City Park, N.Y.: Avery Publishing Group, Inc., 1987.
- Excellent observations and conclusions of a battalion surgeon in World War I.
- Note the deleterious effects that poor soldiers have on the combat effectiveness of the unit.

Patton, George S. Jr. *War as I Knew It.* New York: Bantam Books, 1981. Part 3: Retrospect and Appendix D: Letters of Instruction, 6 March 1944.
- Excellent conclusions on tactical operations, with an offensive-minded commander.
- A mandatory read for junior leaders.

Rommel, Erwin. *Attacks.* Vienna, Va.: Athena Press, Inc., 1979.
- Experiences of Lieutenant Rommel in World War I with a fair analysis of small unit actions.
- Good examples of small unit leadership in combat.

Sajer, Guy. *The Forgotten Soldier.* New York: Brassey's (U.S.), Inc., 1990.
- Experiences of a soldier on the eastern front.
- Good description of the training of replacements for combat units.
- Note the effect that unit cohesion has on the individual.
- Note the chaos of the battlefield and brutal methods of the command to keep the Wehrmacht from disintegrating during the final two years of the war.

Samuels, Martin. *Doctrine and Dogma: German and British Tactics in the First World War.* Westport, Conn.: Greenwood Press, 1992. Introduction, part 1, and part 2.
- Good analysis of German offensive and defensive tactics albeit at battalion and higher level.

Swinton, Ernest D. *The Defense of Duffers Drift.* London: William Clowes and Sons, Limited, 1907.
- Set during the Boer War, good example of how to approach tactical problems.

Toppe, Alfred. *Desert Warfare: German Experiences in World War II.* Foreign Studies Series of the Historical Division, United States Army, Europe. 1952. Reprint. Combat Studies Institute. Fort Leavenworth, Kans.: U.S. Army Command and General Staff College, 1991.
- Good insight into fighting and surviving in the desert.

U.S. Department of the Army. *Leadership and Command on the Battlefield.* United States Army and Training Command Pamphlet 525-100-1. Fort Monroe, Va.: TRADOC, 1992.

- Good insights on the decision-making process from Just Cause and Desert Storm.

U.S. Department of the Army. *Small Unit Actions during the German Campaign in Russia.* Department of the Army Pamphlet No. 20-269. Washington, D.C.: U.S. Government Printing Office, 1953.

- Excellent case studies on the eastern front for infantry, armor, engineer operations as well as special operations (arctic, river, forest, and antipartisan warfare).
- Many TTPs on approaching various battlefield problems.

U.S. Department of the Army. *Survival.* Field Manual 21-76. Washington, D.C.: U.S. Government Printing Office, 1986.

- Good data for health and hygiene factors.

U.S. War Department. *Handbook on German Military Forces.* Baton Rouge: Louisiana State University Press, 1990. Chapter 4.

- Straightforward analysis of German tactics.
- Analysis does not always jibe with German practice, but the conclusions are sound.

Von Schell, Adolf. *Battle Leadership.* 1933. Reprint. Quantico, Va.: Marine Corps Association, 1988.

- Excellent case studies of leadership in World War I.

Williamson, Porter B. *Patton's Principles.* New York: Simon and Schuster, 1979.

- Good analysis of Patton's approach to leadership.
- Note how Patton trains his staff to anticipate movement in any direction.

Wray, Timothy A. *Standing Fast: German Defensive Doctrine on the Russian Front during World War II.* Combat Studies Institute, Research Survey No. 5. Fort Leavenworth, Kans.: U.S. Army Command and General Staff College, 1986.

- Excellent companion to *The Dynamics of Doctrine,* analyzing German defensive adaptations on the eastern front.
- Excellent insight into how the Germans defended vast areas with few troops and the long-term effects on the war effort.

Wynne, Graeme. *If Germany Attacks.* 1940. Reprint. Westport, Conn.: Greenwood Press, 1976.

- Excellent analysis of German World War I tactics.
- Insight into the development of the flexible defense doctrine.

Bibliography

Ambrose, Stephen E. *Eisenhower: Soldier and President.* New York: Touchstone, 1990.

Boga, Steve. *Orienteering.* Mechanicsburg, Pa.: Stackpole Books, 1997.

Bolger, Daniel P. *The Battle for Hunger Hill.* Novato, Calif.: Presidio Press, 1997.

Cartwright, Frederick F. *Disease and History.* New York: Dorset Press, 1991.

Chandler, David. *The Campaigns of Napoleon.* London: Weidenfeld & Nicolson Ltd., 1966.

Clausewitz, Carl von. *On War.* Edited and translated by Michael Howard and Peter Paret. Princeton, N.J.: Princeton University Press, 1976.

Doubler, Michael D. *Closing with the Enemy: How GIs Fought the War in Europe, 1944–1945.* Lawrence: University Press of Kansas, 1994.

Eisenhower, Dwight D. *Crusade in Europe.* Garden City, N.Y.: Doubleday and Company, Inc., 1948.

Ellis, John. *Eye-Deep in Hell: Trench Warfare in World War I.* Baltimore, Md.: The Johns Hopkins University Press, 1991.

English, John. *On Infantry.* New York: Praeger Publishers, 1981.

Fussell, Paul. *Wartime: Understanding and Behavior in the Second World War.* New York: Oxford University Press, 1989.

Garland, Albert N. *Infantry in Vietnam: Small Unit Actions in the Early Days: 1965–66.* New York: Jove Books, 1985.

Griffith, Paddy. *Forward into Battle.* Novato, Calif.: Presidio Press, 1991.

———. *Battle Tactics of the Western Front: The British Army's Art of Attack 1916–18.* New Haven and London: Yale University Press, 1994.

Infanterieschule. *The FIBUA-MOUT Training Guide.* Hammelburg, Germany: Infanterieschule, 1994.

Kals, W. S. *The Land Navigation Handbook.* San Francisco, Calif.: Sierra Book Clubs, 1983.

Keegan, John. *The Face of Battle.* New York: The Viking Press, 1976.

———. *Six Armies at Normandy.* New York: Penguin Books, 1984.

Lindermann, Gerald. *The World within War: America's Combat Experience in World War II.* Cambridge, Mass.: Harvard University Press, 1997.

Lucas, James. *War on the Eastern Front: The German Soldier in Russia 1941–1945.* Novato, Calif.: Presidio Press, 1991.

Lupfer, Timothy. *The Dynamics of Doctrine: The Changes in German Tactical Doctrine during the First World War.* Combat Studies Institute. Fort Leavenworth, Kans.: U.S. Army Command and General Staff College, 1981.

Luvass, Jay, translator and editor. *Frederick the Great on the Art of War.* New York: The Free Press, 1966.

Marshall, S. L. A. *Men against Fire.* Novato, Calif.: Presidio Press, 1947.

———. *The Soldier's Load and Mobility of a Nation.* Washington, D.C.: The Combat Forces Press, 1950.

Moran, Lord. *The Anatomy of Courage.* Garden City Park, N.Y.: Avery Publishing Group, Inc., 1987.

Patton, George S. Jr. *War as I Knew It.* New York: Bantam Books, August 1981.

Rauss, Erhard, Hans von Greiffenburg, and Waldemar Erfurth. *Fighting in Hell: The German Ordeal on the Eastern Front.* Edited by Peter G. Tsouras. Mechanicsburg, Pa.: Stackpole Books, 1995.

Rommel, Irwin. *The Rommel Papers.* Edited by B. H. Liddell-Hart. Translated by Paul Findlay. New York: Da Capo Press, 1953.

Rubman, Andrew MD. "Fighting Bug Bites . . . Naturally." *Bottom Line* 21 (13 July 2000).

Samuels, Martin. *Doctrine and Dogma: German and British Infantry Tactics in the First World War.* New York: Greenwood Press, 1992.

Schmidt, Heinz Werner. *With Rommel in the Desert.* New York: Ballantine Books, 1951.

Seaton, Albert. *The Russo-German War: 1941–45.* Novato, Calif.: Presidio Press, 1990.

2d Brigade, 7th Infantry Division (Light). *TAC SOP.* Mimeograph: January 1990. Annex D (Combat Service Support), Appendix 4 (Resupply), Tab B (LOGPACs).

The Infantry Soldiers Load. TA 1122-84A. Mimeograph, publisher and place and date of publication not given.

Toland, John. *The Rising Sun: The Decline and Fall of the Japanese Empire, 1936–1945.* New York: Random House, 1970.

Toppe, Alfred. *Desert Warfare: German Experiences in World War II.* Combat Studies Institute. Reprint, 1952. Fort Leavenworth, Kans.: Command and General Staff College, August 1991.

U.S. Army War College. *German and Austrian Tactical Studies: Translations of Captured*

German and Austrian Documents and Information Obtained from German and Austrian Prisoners—from British, French and Italian Staffs. Compiled and edited by the Army War College. Washington, D.C.: Government Printing Office, 1918.

U.S. Department of the Army. *Air Assault Operations.* Field Manual No. 90-4. Washington, D.C.: March 1987.

———. *Antipersonnel Mine M18A1 and M18 (Claymore).* Field Manual No. 23-23. Washington, D.C.: 6 January 1966.

———. *Browning Machine Gun, Caliber .50 HB, M2.* Field Manual No. 23-65. Washington, D.C.: U.S. Government Printing Office, 19 June 1991.

———. *Desert Operations.* Field Manual No. 90-3. Washington, D.C.: U.S. Government Printing Office, 24 August 1993.

———. *Dragon Medium Antitank/Assault Weapon System M47.* Field Manual No. 23-24. Washington, D.C.: U.S. Government Printing Office, 3 April 1990.

———. *Engineer Field Data.* Field Manual No. 5-34. Washington, D.C.: 14 September 1987.

———. *Fire Support for the Combined Arms Commander.* Field Manual No. 6-71. Washington, D.C.: U.S. Government Printing Office, 29 September 1994.

———. *Foot Marches.* Field Manual No. 21-18. Washington, D.C.: U.S. Government Printing Office, June 1990.

———. *40-MM Grenade Launcher, M203.* Field Manual No. 23-31. Washington, D.C.: U.S. Government Printing Office, 20 September 1994.

———. *Fundamentals of Nuclear and Chemical Operations,* PO34 COMPS. Fort Leavenworth, Kans.: Army Command and General Staff College, 1989.

———. *German Defense Tactics against Russian Breakthroughs.* Department of the Army Pamphlet No. 20-233. Washington, D.C.: U.S. Government Printing Office, October 1951.

———. *Grenades and Pyrotechnic Signals.* Field Manual No. 23-30. Washington, D.C.: 27 December 1988.

———. *The Infantry Battalion.* Field Manual No. 7-20. Washington, D.C.: U.S. Government Printing Office, 6 April 1992.

———. *An Infantry Guide to Combat in Built-up Areas.* Field Manual No. 90-10-1. Washington, D.C.: U.S. Government Printing Office, 12 May 1993.

———. *The Infantry Rifle Company.* Field Manual No. 7-10. Washington, D.C.: U.S. Government Printing Office, December 1990.

———. *Infantry Rifle Platoon and Squad.* Field Manual No. 7-8. Washington, D.C.: U.S. Government Printing Office, 22 April 1992.

———. *Intelligence Preparation of the Battlefield.* Field Manual No. 34-130. Washington, D.C.: U.S. Government Printing Office, 8 July 1994.

———. *Jungle Operations.* Field Manual No. 90-5. Washington, D.C.: 16 August 1982.

———. *Light Antiarmor Weapons.* Field Manual No. 23-25. Washington, D.C.: U.S. Government Printing Office, 17 August 1994.

———. *Light Infantry Company.* Field Manual No. 7-71. Washington, D.C.: U.S. Government Printing Office, August 1987.

———. *Light Infantry Platoon/Squad.* Field Manual No. 7-70. Washington, D.C.: U.S. Government Printing Office, September 1986.

———. *M16A1 and M16A2 Rifle Marksmanship.* Field Manual No. 23-9. Washington, D.C.: July 1989.

———. *M249 Light Machine Gun in Automatic Rifle Role.* Field Manual No. 23-14. Washington, D.C.: U.S. Government Printing Office, 26 January 1994.

———. *Map Reading and Land Navigation.* Field Manual No. 21-26. Washington, D.C.: U.S. Government Printing Office, 7 May 1993.

———. *Military Improvisations during the Russian Campaign,* Department of the Army Pamphlet No. 20-201. Washington, D.C.: U.S. Government Printing Office, 29 August 1951.

———. *Mission Training Plan for the Infantry Rifle Company.* Army Training and Evaluation Program No. 7-10-MTP. Washington, D.C.: U.S. Government Printing Office, 1988.

———. *Operations.* Field Manual No. 100-5. Washington, D.C.: U.S. Government Printing Office, 1993.

———. *Small Unit Actions during the German Campaign in Russia.* Department of the Army Pamphlet No. 20-269. Washington, D.C.: U.S. Government Printing Office, 1953.

———. *Soldier Performance in Continuous Operations.* Field Manual No. 22-9. Washington, D.C.: Fort Benjamin Harrison, December 1991.

———. *Survival.* Field Manual No. 21-76. Washington, D.C.: U.S. Government Printing Office, March 1986.

———. *Tactical Employment of Mortars.* Field Manual No. 7-90. Washington, D.C.: U.S. Government Printing Office, 9 October 1992.

———. *Tank Platoon.* Field Manual No. 17-15. Washington, D.C.: U.S. Government Printing Office, April 1996.

U.S. War Department. *Handbook on German Military Forces.* Baton Rouge: Louisiana State University Press, 1990.

Von Luck, Hans. *Panzer Commander.* New York: Dell Publishing, 1989.

Von Schell, Adolf. *Battle Leadership.* Fort Benning, Ga.: *Benning Herald,* 1933.

Wiseman, John. *The SAS Survival Handbook.* London: Harvill, 1991.

Wray, Timothy A. *Standing Fast: German Defensive Doctrine on the Russian Front during World War II.* Combat Studies Institute Research Survey No. 5. Fort Leavenworth, Kans.: U.S. Army Command and General Staff College, September 1986.

Wynne, Graeme. *If Germany Attacks.* Reprint 1940. Westport, Conn.: Greenwood Press, 1976.

Index

About the Author

Lt. Col. Raymond A. Millen is the director of European Security Studies for the Strategic Studies Institute, U.S. Army War College, at Carlisle Barracks. He has served as a platoon leader in the Eighth Infantry Division (Mechanized), as a company commander in the Seventh Infantry Division (Light) in Operation Just Cause in Panama, and as a battalion executive officer in the 101st Airborne Division (Air Assault). He is infantry, ranger, and airborne qualified and has attended a broad range of U.S. and foreign military education and training programs. A 1982 graduate of the U.S. Military Academy, Millen also earned a master of arts degree in national security studies from Georgetown University in 1991. He lives in Carlisle, Pennsylvania.